T0139783

Sustainable Agriculture Reviews

Volume 43

Series Editor
Eric Lichtfouse
Aix-Marseille University, CNRS, IRD, INRAE, Coll France, CEREGE
Aix-en-Provence, France

Other Publications by Dr. Eric Lichtfouse

Books
Scientific Writing for Impact Factor Journals
https://www.novapublishers.com/catalog/product_info.php?products_id=42242

Environmental Chemistry
http://www.springer.com/978-3-540-22860-8

Sustainable Agriculture
Volume 1: http://www.springer.com/978-90-481-2665-1
Volume 2: http://www.springer.com/978-94-007-0393-3

Book series
Environmental Chemistry for a Sustainable World
http://www.springer.com/series/11480

Sustainable Agriculture Reviews
http://www.springer.com/series/8380

Journal
Environmental Chemistry Letters
http://www.springer.com/10311

Sustainable agriculture is a rapidly growing field aiming at producing food and energy in a sustainable way for humans and their children. Sustainable agriculture is a discipline that addresses current issues such as climate change, increasing food and fuel prices, poor-nation starvation, rich-nation obesity, water pollution, soil erosion, fertility loss, pest control, and biodiversity depletion.

Novel, environmentally-friendly solutions are proposed based on integrated knowledge from sciences as diverse as agronomy, soil science, molecular biology, chemistry, toxicology, ecology, economy, and social sciences. Indeed, sustainable agriculture decipher mechanisms of processes that occur from the molecular level to the farming system to the global level at time scales ranging from seconds to centuries. For that, scientists use the system approach that involves studying components and interactions of a whole system to address scientific, economic and social issues. In that respect, sustainable agriculture is not a classical, narrow science. Instead of solving problems using the classical painkiller approach that treats only negative impacts, sustainable agriculture treats problem sources.

Because most actual society issues are now intertwined, global, and fast-developing, sustainable agriculture will bring solutions to build a safer world. This book series gathers review articles that analyze current agricultural issues and knowledge, then propose alternative solutions. It will therefore help all scientists, decision-makers, professors, farmers and politicians who wish to build a safe agriculture, energy and food system for future generations.

More information about this series at http://www.springer.com/series/8380

Ankit Saneja • Amulya K. Panda
Eric Lichtfouse
Editors

Sustainable Agriculture Reviews 43

Pharmaceutical Technology for Natural Products Delivery Vol. 1 Fundamentals and Applications

 Springer

Editors
Ankit Saneja
Product Development Cell
National Institute of Immunology
New Delhi, Delhi, India

Amulya K. Panda
Product Development Cell
National Institute of Immunology
New Delhi, Delhi, India

Eric Lichtfouse
Aix-Marseille University, CNRS, IRD,
INRAE, Coll France, CEREGE
Aix-en-Provence, France

ISSN 2210-4410   ISSN 2210-4429   (electronic)
Sustainable Agriculture Reviews
ISBN 978-3-030-41840-3   ISBN 978-3-030-41838-0   (eBook)
https://doi.org/10.1007/978-3-030-41838-0

This Springer imprint is published by the registered company Springer Nature Switzerland AG.
The registered company address is: Gewerbestrasse 11, 6330 Cham, Switzerland

Preface

Research is to see what everybody has seen and think what nobody has thought (Albert Szent-Györgyi).

Natural products are a major source of efficient drugs. Yet many natural compounds have major therapeutic limitations due to their low aqueous solubility, low permeability, short half-life, and low bioavailability to humans. These limitations can be overcome by advanced pharmaceutical technologies. This book presents innovative technologies and approaches for improving the bioavailability and therapeutic efficacy of natural products.

In Chap. 1, Proha et al. review the use of pharmaceutical technology for modulating the pharmacokinetics of natural products. In Chap. 2, Jain and Chella focus on particle engineering technologies, dispersions, complexation-based technologies, and nanotechnologies to enhance the solubility of natural products (Fig. 1). In Chap. 3, Ahuja et al. review the use of polymeric nanocarriers to deliver natural products for cancer chemotherapy.

In Chap. 4, Rana and Kumar discuss chemistry, pharmacology, and therapeutic delivery strategies of tea constituents. In Chap. 5, Sharma explains how galantamine therapy could improve cognition and function in Alzheimer's patients. In Chap. 6, Prabakaran et al. review paclitaxel delivery systems. In Chap. 7, Kumar et al. present components and methods of preparation of phytosomes for herbal drug delivery. Finally, in Chap. 8, Dewangan reviews the use of albumin as a natural versatile carrier for various disease treatments.

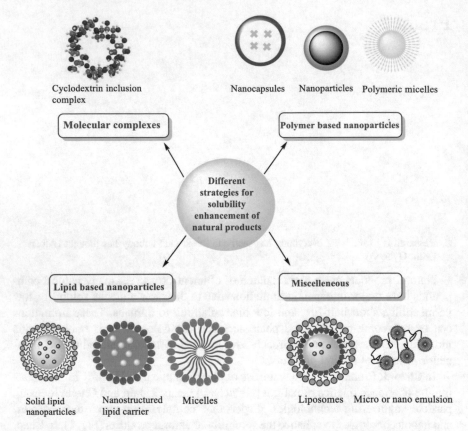

Fig. 1 Strategies to enhance the solubility of natural products. From Jain and Chella in Chap. 2

We extend our thanks to all the contributing authors and reviewers who made an excellent contribution to deliver the chapters and answer all the queries in time frame. This edited book is the result of the impactful work of all of them. We are extremely grateful to Melanie van Overbeek, Assistant Editor at Springer Nature, for her excellent support during this whole process.

New Delhi, India   Ankit Saneja
New Delhi, India   Amulya K. Panda
Aix-en-Provence, France   Eric Lichtfouse

Contents

About the Editors

Ankit Saneja currently works as Research Associate in the Product Development Cell, National Institute of Immunology, New Delhi, India. He has received his PhD from the Academy of Scientific and Innovative Research (AcSIR) at the CSIR-Indian Institute of Integrative Medicine, Jammu. He has been recipient of several international awards such as the CEFIPRA-ESONN Fellowship, Bioencapsulation Research Grant, and PSE Travel Bursary. He has to his credit several international publications. His research interest includes exploring different types of formulations for various biomedical applications.

Amulya K. Panda has been a Scientist at the National Institute of Immunology (NII), New Delhi, for the last 29 years. Currently, he is the Director of NII. He completed his master's degree in Chemical Engineering from IIT Madras and his PhD in Biochemical Engineering and Biotechnology from IIT Delhi. He has been a Visiting Scientist with Prof. Harvey Blanch at the Chemical Engineering Department, University of California, Berkeley, USA, and in the Department of Pharmaceutical Science at the University of Nebraska Medical Center, Omaha, USA. His research interest includes bioprocess engineering, recombinant fermentation process development, high-throughput protein refolding from inclusion bodies, and vaccine delivery using biodegradable polymer particles. He has published more than 130 research papers and 10 book chapters and is an inventor of more than 30 issued or pending patents. His honors include Samanta Chandra

Sekhara Award from Odisha Bigyan Academy for the year 2000; Biotech Product and Process Development and Commercialization Award 2001 from the Department of Biotechnology, Government of India, on May 11, 2001; Young Asian Biotechnologist Prize from The Society for Biotechnology, Japan, for the year 2004; Tata Innovation Fellowship 2010 from the Department of Biotechnology, Government of India; and GB Manjrekar Award of AMI at AMI 2010 Conference at BIT Mesra in December 2010. His research group was also awarded Second Prize at Intel-UC Berkeley Technology Entrepreneurship Challenge 2008 and Third Prize at World Innovation Summit, Barcelona, on June 16, 2009. He was the President of the Association of Microbiologist of India (AMI) for the year 2017.

Eric Lichtfouse, PhD is Geochemist and Professor of Scientific Writing at Aix-Marseille University, France, and Visiting Professor at Xi'an Jiaotong University, China. He has discovered temporal pools of molecular substances in soils, invented carbon-13 dating, and published the book *Scientific Writing for Impact Factor Journals*. He is Chief Editor and Founder of the journal *Environmental Chemistry Letters* and the book series Sustainable Agriculture Reviews and Environmental Chemistry for a Sustainable World. He has awards in analytical chemistry and scientific editing. He is World XTerra Vice-Champion.

Chapter 1
Pharmaceutical Technology for Improving the Bioavailability of Natural Products

Shweta Paroha, Rikeshwer P. Dewangan, and Pravat K. Sahoo

Abstract Therapeutic importance of natural products have been widely explored since ancient time for benefit of human health although the clinical uses of these natural products still a challenge for scientific community. Seventy percent of the therapeutic agents obtained from natural source are hydrophobic in nature. Natural product demonstrated several biological activities and excellent safety index therefore, used by mankind all over the world in healthcare as functional foods, medicine, pharmaceuticals and nutraceuticals. However poor aqueous solubility, instability and poor bioavailability following oral administration hindered its medical applicability. Literature demonstrated, many natural products exhibited excellent *in-vitro* pharmacological activity whereas, no or less activity *in-vivo* because of poor gastrointestinal absorption.

Pharmaceutical technologies are employed to improve poor aqueous solubility and bioavailability of natural products. Bioavailability of natural products drastically improved with the advancement in the technology. Conventional approaches for solubility and bioavailability enhancement involves particle size reduction, salt formation, solid dispersion and cyclodextrin complexation. In the past decades, novel drug delivery system paid much attention towards improvement in bioavailability and pharmacokinetics of natural products which involves phytosome technology, solid lipid nanoparticles, liposome, polymeric nanoparticles, nanocapsules, micelles, nano-suspension, nano-emulsions and self-emulsifying drug delivery systems. In this chapter, we review different approaches that have been conducted for bioavailability enhancement of the natural products.

Keywords Bioavailability · Natural products · Healthcare · Aqueous solubility · Pharmacokinetics · Phytosome · Cyclodextrin · Pharmaceutical technology

S. Paroha · P. K. Sahoo (✉)
Department of Pharmaceutics, Delhi Institute of Pharmaceutical Sciences and Research
(DIPSAR), Delhi Pharmaceutical Sciences and Research University, New Delhi, India

R. P. Dewangan
Department of Pharmaceutical Chemistry, School of Pharmaceutical Education and Research,
Jamia Hamdard (Deemed to be University), New Delhi, India

1.1 Introduction

At present, 80% of population in the world is using therapeutic agents, which are either directly or indirectly obtained from plants origin (Atul Bhattaram et al. 2002). The therapeutic application of natural products known for benefit of human health since concept of medicine started. But the medical applicability of these natural products is restricted due to hindered bioavailability. In most of the case this is because of low solubility and less membrane permeability of natural products (Md. Akhlaquer et al. 2011; Mukherjee et al. 2015). The potency of natural product depends on effective delivery to the target site to achieve desired therapeutic outcome. Many natural products have great potential but demonstrated less therapeutic action because of low solubility, eventually impaired bioavailability (Kesarwani et al. 2013). Most natural product administered by oral route which demonstrated poor bioavailability due to less gastrointestinal tract absorption eventually retarded therapeutic efficacy (Li et al. 2012). Development of suitable dosage forms of natural product is still a challenging due to several regions including limited solubility and less membrane permeability. The biological half-life plays a critical role in therapeutic efficacy of a molecule, for instance, drug with low half-life possess low bioavailability then the higher one. Similarly, the hepatic metabolism of the drug molecules undergoes more renal elimination resulting in low bioavailability (Rahman et al. 2013). Various literatures reported pharmaceutical technology plays a critical role in improving bioavailability of natural product either by conventional method or by novel drug delivery system (Kesarwani et al. 2013; Mukherjee et al. 2015).

Since last one decade novel drug delivery system emerged as successful method to improve bioavailability of natural product by encapsulating the active pharmaceutical ingredients into the nano-cargo which includes phytosome, liposomes, nanoparticles, nanoemulsion, transferosomes, ethosomes, lipid based systems, microspheres (Kesarwani et al. 2013; Mukherjee et al. 2015). Many of the natural product including, quercetin, naringenin, curcumin, hesperetin, andrographolide, ellagic acid, resveratrol, genistein, ginkgolide, bilobalide sinomenine, piperine, glycyrrhizin, showed improved bioavailability when these natural product delivered by using novel carriers (Bhattaram et al. 2002; Kesarwani et al. 2013; Mukherjee et al. 2015). In conventional method of bioavailability reducing particle size is a crucial parameter. Report suggested reducing particle size to less than 500 nm, improve absorption of the active pharmaceutical ingredients from gastrointestinal tract by passive transport mechanism (Kesarwani et al. 2013; Mukherjee et al. 2015). In this chapter we are discussing about pharmaceutical technology including conventional as well as novel approaches, which has been used for improvement in oral bioavailability of various natural products.

1.2   History of Bioavailability Enhancers

Bioavailability enhancers emerged since ancient system of medicine "Ayurveda", in which ginger, black pepper and long pepper are collectively known as "Trikatu" means three acrids (Johri and Zutshi 1992; Kesarwani et al. 2013). Bioavailability enhancers do not have drug activity its own, but it is a drug facilitator, which when used in combination they enhance the activity and/or bioavailability of a drug molecule. Yogavahi in Ayurveda describe Pippali (P. longum) and Maricha (P. nigrum), contains "piperine" (1-piperonylpiperidine) which is well known of its bioavailability enhancing effect. Piperine potentiates bioavailability of many drugs including sulfadiazine, phenytoin, rifampicin, sparteine, propranolol and vasicine (Atal et al. 1981; Bano et al. 1987, 1991).

In 1979, the term 'bioavailability enhancer' was coined by Indian researcher at Regional Research Laboratory, Jammu (RRL, now known as Indian Institute of Integrative Medicine, Jammu). The group of Indian researchers performed a series of experiment and scientifically stablished piperine as first bioavailability enhancing agent. It was reported that, in the Institute various Ayurvedic formulations has been validated for treatment of various diseases and observed that the most of Ayurvedic formulations consists of either Trikatu or at least one ingredient of Trikatu that was Piper longum (P. longum). In this study, 210 formulations were reviewed out of 370 formulations and thus the fact of Trikatu that can increase the efficacy of formulations was stablished. Based on the experiment, it was found that one of the ingredients, 'P. longum', 'Piper' improved the bioavailability of various drug which was facing poor intestinal absorption problem (Bano ct al. 1991). Bioavailability of curcumin was increased by ninefold when used with piperine (Bano et al. 1991). In subsequent research it was found that piperine was able to increase bioavailability of various drug compounds ranging from 30% to 200% (Kesarwani et al. 2013).

1.3   Need for Bioavailability Enhancement

When a drug is administered by intravenous route it showed maximum bioavailability while in oral route bioavailability problem observed because of incomplete absorption and/or first pass metabolism. Therefore, there is a need of molecule which has no activity its own but when it combines with drug molecule, the bioavailability of the drug molecule could improve. Most of the natural products have high hydrophobicity and large molecular size which is major limitation for impaired bioavailability when administered by oral route. Due to hydrophobicity the molecule not absorbed through biological membrane (Kesarwani et al. 2013). Many natural products demonstrated excellent activity when tested *in-vitro* but exert no or less activity *in-vivo*. This is mainly due to poor absorption leading to impaired bioavailability. Sometimes it was observed that the active constituents of the natural product lost their activity by gastric juice (Kesarwani et al. 2013). To overcome these challenges and to achieve optimal efficacy of the natural product the bioavailability enhancement is required.

1.4 Conventional Approaches for Bioavailability Enhancement

1.4.1 Particle Size Reduction

Particle size reduction is one of the old and widely used techniques for increase in solubility and bioavailability of drug substance. Reduction in particle size offers most promising technique to improve the bioavailability of hydrophobic drugs because it increases surface area, increase wettability of drug and enhanced saturation solubility. Therefore, this is routinely used method for bioavailability enhancement (Williams et al. 2013). 'Nanonization' is a term used for reduction of particle size of a drug to sub-micron range (Al-Kassas et al. 2017). Micronization is a technique for reduction of particle size which is commonly employed to enhance the solubility of biopharmaceutics classification system, class II drugs (Leleux and Williams 2014). By using this technique the drug particle size can be obtained between 2 and 5 mm while very few particle obtained in the range of below 1 mm size. The micronization processes involves pressure, attrition, friction, impact or shearing (Khadka et al. 2014). High-pressure homogenization, jet mills, ball mills are commonly used processing method for micronization of drugs substance (Khadka et al. 2014; Rasenack and Müller 2004). Digoxin is obtained from the leaves of a digitalis plant and commonly used to treat heart failure. The medical applicability of digoxin is limited due to its poor bioavailability. Jounela et al. performed bioavailability study of digoxin tablet in healthy human volunteers prepared with three different particle sizes *i.e.*, 7 µm, 13 µm, and 102 µm. The study was evaluated by cross-over design and alcoholic solution of digoxin was used as reference standard. The *in-vivo* result demonstrated that bioavailability of digoxin tablet for particle size 7 µm and 13 µm was 78–97%, while 39% for particle size 102 µm (Jounela et al. 1975). This, study indicated particle size is an important determinant for dissolution and bioavailability of a drug.

1.4.2 Use of Surfactant and Solubilizing Agent

Enhancement in solubility of a hydrophobic drug by use of surfactant and/or solubilizing agent is also a commonly used technique, especially in case of liquid dosage form for oral or *i.v.* administration (Kalepu and Nekkanti 2015). The of surfactant and/or solubilizing agent also used to prevent precipitation upon dilution with water (Kawakami et al. 2006). Most commonly used solubilizing agent are ethanol, polyethylene glycol 400, propylene glycol, glycerin and sorbitol. Taxol®, a marketed formulation of paclitaxel used for treatment in cancer, is an iconic example of formulation prepared by using this approach. Each mL of Taxol® solution contains 527 mg of purified cremophor® as surfactant, 49.7% (v/v) dehydrated alcohol as solubilizing agent and 6 mg paclitaxel. Later, due to hypersensitivity reaction of cremophor® other formulation of paclitaxel excluding cremophor® has been tried and get approval by food and drug administration *i.e.* Abraxane®, Genexol® (Hennenfent and Govindan 2005; Oerlemans et al. 2010). Docetaxel is an anticancer drug which is about two times more potent as paclitaxel in inhibiting microtubule depolymerization *in-vitro* system (Lavelle et al. 1995). The formulation of docetaxel (Taxotere®) is also used Tween-80 as a surfactant and ethanol as solubilizing agent.

1.4.3 pH Modification and Salt Formation

A pH-dependent solubility is observed in ionizable drugs and about 70% of drugs are shows ionization properties (Kalepu and Nekkanti 2015). Solubility of these drugs depends on ionization constant *i.e.* weakly basic drug are soluble when pH is less than its pKa while weakly acidic drug are soluble when pH is greater than its pKa (Serajuddin 2007). The pH dependent solubility is widely explored technique for formulation of hydrophobic drugs while salt formation is reported to improve stability and crystallinity of drugs (Kalepu and Nekkanti 2015; Serajuddin 2007). There are many hydrophobic marketed drugs which has been formulated by pH modification method and lactic acid used as pH modifier to improve solubility (Kalepu and Nekkanti 2015). Berberine is a naturally occurring quaternary ammonium alkaloid used to treat gastroenteritis in traditional Chinese medicine (Baird et al. 1997). In current investigation berberine has been also demonstrated as clinically effective in treatment of cardiovascular and metabolic diseases (Lan et al. 2015; Moghaddam et al. 2013; Zhang et al. 2010). The commercial available salt form of berberine is bebeerine chloride (Lu et al. 2019). Berberine chloride exists in various chemical form *i.e.* berberine anhydrate, berberine monohydrate, berberine dihydrate, and berberine tetrahydrate. The berberine anhydrous and berberine monohydrate can be transforms to berberine dihydrate at relative humidity above 12% whereas the berberine dihydrate can be transforms to the berberine tetrahydrate at relative humidity above 70% (Nakagawa et al. 1978; Yoshimatsu et al. 1981). The formation of salt cocrystal of berberine chloride with citric acid showed much stability against various humidity conditions which demonstrated a promising approach as alternate dosage form of berberine chloride tablets (Lu et al. 2019).

1.4.4   Solid Dispersion

Solid dispersions is a widely used technique to improve the oral bioavailability of hydrophobic drugs by reducing particle size, increasing wettability and eventually bioavailability (Vasconcelos et al. 2007). The product of solid dispersion is amorphous in nature which are mainly prepared by two method like melting and solvent evaporation method (Vasconcelos et al. 2007). There are several natural products which oral bioavailability has been improved by using solid dispersion technique.

Curcumin is a widely used natural product with diverse pharmacological activities but showed poor bioavailability. Wan et al. developed solid dispersion formulation of curcumin by solvent evaporation technique for improvement in its oral bioavailability. The pharmacokinetic study in rat demonstrated 7.6-fold increase in oral bioavailability at a dose of 50 mg/kg equivalent to curcumin. For instance, area under curve value was 1192.34 and 156.36 (ng/mL h) for the solid dispersion and free curcumin, respectively (Wan et al. 2012). Additionally, 5.4-fold increase in half-life 2.65 h for the solid dispersion versus 0.49 h for free curcumin) was observed (Wan et al. 2012). In another investigation, Chuah et al. reported an amorphous solid dispersion of curcumin prepared by using lecithin, hydroxypropyl methyl cellulose and isomalt through hot melt extrusion method. *In-vivo* pharmacokinetic study in rat revealed that 12.8-fold increase (area under curve value 20,685 for amorphous solid dispersion versus 1615 for free curcumin ng/mL/min) in bioavailability at oral dose of 20 mg/kg (Chuah et al. 2014).

Silymarin, obtained from the seeds of *Silybum marianum*, used as hepatoprotective agent which showed poor oral bioavailability. For improvement in its bioavailability solid dispersion formulation has been prepared by the solvent-fusion method with Gelucire 44/14 as capsules dosage form. Result demonstrated, 13-fold increase in oral bioavailability which was 7643.01, 4891.7 and 587.53 gm/mL h, for silymarin-Gelucire capsules, silymarin commercial product capsules and silymarin capsules, respectively (Hussein et al. 2012).

Paclitaxel is a natural occurring anticancer agent used against many solid tumors (Gupta et al. 2014; Nehate et al. 2014). Shanmugam et al. prepared solid dispersion tablet of paclitaxel prepared by using fluid bed technology and to evaluate its pharmacokinetics in beagle dogs. It was reported that relative percentage bioavailability of paclitaxel solid dispersion tablet increased by 132.25% as compare to Oraxol™ solution at an oral dose of 60 mg/kg (Shanmugam et al. 2015). Wang et al. prepared solid dispersions of ginkgo biloba extract prepared by hot-melt extrusion method for improvement in oral bioavailability. Pharmacokinetic study has been performed in rat and reported that area under curve value of ginkgolide, bilobalide, ginkgolide B, quercetin, ginkgolide C, isorhamnetin, isorhamnetin and kaempferol significantly enhanced following oral administration of solid dispersions of ginkgo biloba extract (Wang et al. 2015). This result suggested that the solid dispersion formulation of ginkgo biloba extract could be one of the potential options for delivery of ginkgo biloba extract in further trial.

1.4.5   Cyclodextrin Complexation

Cyclodextrins consist of cyclic oligosaccharides obtained from starch containing lipophilic cavity inside and hydrophilic surface outside having (α-1,4)-linked α-D-glucopyranose unit (Khadka et al. 2014). There are several types of cyclodextrins available depending on (α-1,4)-linked α-D-glucopyranose unit including α, β, γ, δ, and ϵ having six, seven, eight, nine and ten units of α-D-glucopyranose (Brewster and Loftsson 2007). Application of cyclodextrin in the field of drug delivery application includes enhancement of solubility of drug, improve bioavailability, increase stability, masking taste of drug and prevention of gastrointestinal irritation (Singh et al. 2011). Uekama et al. first time developed inclusion complex of digoxin with γ-cyclodextrin and demonstrated 5.4 times higher bioavailability of the complex as compare to free digoxin after 24 h in dog following oral administration (Uekama et al. 1981). Data indicated enhanced oral dissolution and absorption of the complex, which suggested decrease in dose of digoxin therapy (Uekama et al. 1981). In another investigation, β and γ-cyclodextrin complex of artemisinin, a antimalarial drug, has been prepared with 1:1 ratio each, and its bioavailability evaluated in comparison with its a normal commercially preparation (Artemisinin 250®). The study was performed in 12 healthy human volunteers aged between 22 and 44 years and body weight between 60 to 87 kg. It was reported that β and γ-cyclodextrin complex of artemisinin had higher bioavailability as compare to Artemisinin 250® (Wong and Yuen 2001).

Munjal et al. performed a study on oral bioavailability of different formulation of curcumin including, aqueous suspension, nanosuspension, micronized suspension, hydroxypropyl-β-cyclodextrin, inclusion complex, amorphous solid dispersion, spray dried curcumin milk composite and combination with piperine in rats at an oral dose of 250 mg/kg curcumin. In pharmacokinetic study, aqueous suspension observed to have area under curve and C_{max}, 26.9 ng/mL h and 28.9 ng/mL h, respectively. The study demonstrated significantly improved oral bioavailability in terms of area under curve, which was 446%, 567% and 251% and in terms of C_{max} which was 270%, 415% and 405% for the amorphous solid dispersion, hydroxypropyl-β-cyclodextrin inclusion complex and nanosuspension, respectively (Munjal et al. 2011). However, no significant difference in C_{max} and area under curve has been observed with micronized suspension and piperine while milk composite decreases oral bioavailability (Munjal et al. 2011). So, a clear impact of different formulation on bioavailability has been observed which indicates hydroxypropyl-β-cyclodextrin formulation of curcumin is superior over other formulation in terms of oral bioavailability.

Rutin is a phenolic flavonol glycoside which consisted several pharmacological effects including antioxidant, inhibition of xanthine oxidase and hyaluronidase. It is also used as antineoplastic agent (Pathak et al. 1991). Despite 10 hydroxyl groups in its structure, rutin has poor aqueous solubility. Miyake et al. prepared hydroxypropyl-β-cyclodextrin complex of rutin in a molar ratio of 1:1, with kneading method by using ethanol-water as a solvent for improvement of its oral bioavailability. The study demonstrated 2.8-fold increase in oral bioavailability by hydroxypropyl-β-cyclodextrin complex of rutin in beagle dogs after 200 mg/kg equivalent to rutin dose administration. For instance, area under curve value observed to be 642.62, 202.58 and 224.70 ng/mL h, for hydroxypropyl-β-cyclodextrin complex, β-cyclodextrin complex and free rutin, respectively (Miyake et al. 2000). This study suggested hydroxypropyl-β-cyclodextrin complex could be a suitable delivery option for rutin which provide improved solubility, fast dissolution, increased gastrointestinal absorption and thus increased oral bioavailability. In another study, bioavailability of naringenin has been improved by using hydroxypropyl-β-cyclodextrin complex. It was reported that 400-fold solubility and 11-fold transport across gut epithelium of naringenin observed by hydroxypropyl-β-cyclodextrin complexation (Shulman et al. 2011). *In-vivo* pharmacokinetic data in rat showed 7.4-fold increase in area under curve value and 14.6-fold increase in C_{max} at an oral dose of 20 mg/kg (Shulman et al. 2011). Similarly, to improve oral bioavailability of Z-ligustilide, hydroxypropyl-β-cyclodextrin complexation has been prepared by the kneading technique with stoichiometry to complex was 1:1 (Lu et al. 2014). The pharmacokinetic data demonstrated enhancement in area under curve which was 3465.72, 2877.11 and 1990.68 ng/mL h, at a dose of 100 mg/kg, 400 mg/kg and 20 mg/kg for hydroxypropyl-β-cyclodextrin complex, free Z-ligustilide (400 mg/kg) and free Z-ligustilide (20 mg/kg), respectively (Lu et al. 2014). The results clearly suggested that Z-ligustilide- hydroxypropyl-β-cyclodextrin complex is effective for delivery of Z-ligustilide in terms of stability and pharmacological efficacy.

1.5 Nanotechnology Approaches for Bioavailability Enhancement

1.5.1 Phytosome Technology

Phytosome is a novel technology came into focus in 1989. Phytosome made up of two terms, "phyto" and "some". The term "phyto" stands for plant or herb whereas "some" stands for body or cell-like structure (Kidd 2009). A phytosome is a molecularly combined form of phosphatidylcholine and polyphenol. A bond between these molecule is formed which eventually increase oral bioavailability of polyphenols because of its amphiphilic nature (Alam et al. 2013). These water soluble phosphatidylcholine molecule showed improved dispersal ability in aqueous environments and lipid soluble part directed towards epithelial cell membrane (Kidd and Head 2005; Kidd 2009; Semalty et al. 2010).

Based on the observation of polyphenols showed strong bonding ability to phospholipids in their intact plant tissues a group of Italian scientist started working on polyphenol formulation preparations which showed poor bioavailability when taken orally (Kidd 2009). That reported the bioavailability of these polyphenol formulation preparations increased while conversion into phytosome.

Siliphos® which is a silybin-phosphatidylcholine has been the first clinically available product prepared by using phytosome technology. Siliphos® showed improved bioavailability of silymarin and marked hepatoprotective activity (Kidd and Head 2005; Kidd 2009; Song et al. 2008). Silymarin which is a combination of silybin, silidianin and silicristin after forming complex with phosphatidylcholine demonstrated increased pharmacological activity and improved oral bioavailability (Alam et al. 2013). Pharmacokinetic study demonstrated after equal dose administration plasma C_{max} of silybin was <35 ng/mL, whereas, 112 ng/mL for the silybin complex after 4 h (Alam et al. 2013; Gabetta et al. 1988).

Curcumins which is well known polyphenols has various therapeutic activity including anti-inflammatory, antioxidant, anti-malignant, but its clinical applicability is hindered owing to its poor oral bioavailability (Kidd 2009). To overcome these limitations Andrea et al. reported phospholipid complexes of curcumin with enhanced bioavailability (Giori and Franceschi 2009; Kidd 2009). Mcriva®, is phytosome product of curcumin reported to have fivefold higher bioavailability as compared to native curcumin extract (Giori and Franceschi 2009; Kidd 2009). In addition to this, curcumin after incorporation into the phytosome complex showed improved hydrolytical stability (Giori and Franceschi 2009). Belcaro et al. demonstrated decreased joint pain after administration of curcumin-phosphatidylcholine phytosome complex in osteoarthritis patients and reported Meriva®, is a safe and therapeutic effective drug for complementary treatment of osteoarthritis and improvement of quality of life of the patient (Belcaro et al. 2010a, b).

Resveratrol, chemically 3,5,4-trihydroxystilbene is a polyphenol obtained from grapes reported to have cardiovascular benefit (Hung et al. 2000; King et al. 2006), estrogenic activity (Sharma et al. 2007) and anti-cancer activity (Jang et al. 1997). Resveratrol showed high absorption while taking orally but due to rapid metabolism restrict its clinical applicability due to impaired bioavailability (Mukherjee et al. 2011). Mukherjee et al. developed a complex of resveratrol with hydrogenated soy phosphatidyl choline and reported enhanced bioavailability of the drug (Mukherjee et al. 2011). There are several natural products phospholipid complexes reported to have improvement in oral bioavailability which are summarized in Table 1.1.

1.5.2 Solid Lipid Nanoparticles

Solid lipid nanoparticles are novel drug delivery system consists of a solid lipid matrix encapsulating drug molecule which stabilized in aqueous solution by emulsifiers. Solid lipid nanoparticles demonstrated several advantages over other conventional formulations, including improved bioavailability, enhanced solubility, stability, improved half-life, reduced side effect and tissue targeting effect (Kumar and Randhawa 2013; Lin et al. 2017).

Table 1.1 Effect of phospholipid complex on the pharmacokinetics of different herbal products (Mean ± SD)

Phytomolecules		Subject	Subject	C_{max} (mg/mL)	T_{max} (h)	AUC_{0-t} (mgh/mL)	$AUC_{0-\infty}$ (mgh/mL)	$t_{1/2el}$ (h)	K_{el} (h^{-1})	C_L (Lh^{-1})	(V_d) (L)	Relative Bioavailability	Reference
Quercetin	Pure	Rats	20 mg/kg	6.56 ± 0.42	3.0	27.87 ± 2.02	28.05 ± 1.95	1.40 ± 0.05	0.493	0.11	0.224	-	Liu et al. (2006)
	Complex			9.86 ± 0.52	6.0	112.87 ± 5.23	113.52 ± 6.23	3.22 ± 0.23	0.215	0.034	0.159	125.56%	
Naringenin	Pure	Rats	100 mg/kg	6.32 ± 0.41	5.0	318.07 ± 2.12	38.45 ± 2.44	2.43 ± 0.11	0.28	0.40	1.44	-	Maiti et al. (2006)
	Complex			9.35 ± 0.62	8.0	104.48 ± 5.64	107.48 ± 6.10	3.78 ± 0.15	0.18	0.17	0.95	118.55%	
Curcumin	Pure	Rats	1000 mg/kg	0.50	0.75	1.32	1.68	1.45	0.48	92.26	192.21	-	Maiti et al. (2007)
	Complex			1.20	1.50	5.90	8.73	1.96	0.35	22.33	63.82	125.80%	
Hesperetin	Pure	Rats	100 mg/kg	6.1 ± 0.30	4.0	30.47 ± 2.34	31.24 ± 2.54	1.78 ± 0.09	0.380	0.490	1.310	-	Maiti et al. (2009)
	Complex			9.2 ± 0.41	6.0	109.72 ± 6.72	151.90 ± 8.43	3.86 ± 0.13	0.140	0.130	0.970	133.31%	
Andrographolide	Pure	Rats	25 mg/kg	6.7 ± 0.54	2.50	26.24 ± 1.23	26.74 ± 1.42	1.20 ± 0.05	0.57	0.17	0.30	-	Maiti et al. (2010)
	Complex			9.6 ± 0.72	4.0	85.50 ± 2.77	87.3 ± 2.35	4.01 ± 0.12	0.21	0.055	0.26	104.24%	
Ellagic Acid (EA)	Pure	Rats	80 mg/kg	0.21	0.5	0.72	0.89	6.11	0.11	89,640.03	789,766.58	-	Murugan et al. (2009)
	Complex			0.54	0.5	2.05	2.53	8.32	0.08	31,681.15	380,301.38	2.84-fold	
Resveratrol	Pure	Rats	20 mg/kg	5.9	0.17	3.307	4.474	1.156	0.6	4.47	7.454	-	Mukherjee et al. (2011)
	Complex			7.2	0.17	4.693	10.0985	2.584	0.27	1.98	7.383	2.26-fold	
Genistein	Pure	Rats	20 mg/kg	1680	12	36,740	40,284.61	9.10	0.08	0.5	6.52	-	Mukherjee et al. (2015)
	Complex			2405	12	57,066	62,049.10	11.63	0.06	0.32	5.41	1.54-fold	
Ginkgolide A	Extract	Human	160 mg	0.0418	2	-	-	2.63 ± 0.45	-	-	-	-	Mauri et al. (2001)
	Complex			0.108	4	-	-	1.88 ± 0.13	-	-	-	-	
Ginkgolide B	Extract	Human	160 mg	0.0056	2	-	-	2.34 ± 0.38	-	-	-	-	Mauri et al. (2001)
	Complex			0.0134	3	-	-	1.69 ± 0.30	-	-	-	-	
Bilobalide	Extract	Human	160 mg	0.0376	2	-	-	2.30 ± 0.24	-	-	-	-	Mauri et al. (2001)
	Complex			0.0603	3	-	-	3.16 ± 0.35	-	-	-	-	

Reproduced with permission of Elsevier from Mukherjee et al. (2015)

In solid lipid nanoparticles the solid lipids which commonly used are highly purified glycerides, glyceride mixtures or waxes because it does not melt at body temperature. Safety is a major concern for nanoformulation when administered *via* oral route in the body (Fu et al. 2014; Hunter et al. 2012). Therefore, in solid lipid nanoparticles, the lipid matrices used are the natural or synthetic lipids including triglycerides, glyceryl monostearate, glyceryl behenate, glyceryl palmitostearate, fatty acids, waxes and steroids which can degrade in the body. The drug can be encapsulated in the solid lipid matrix which can encapsulate both hydrophilic as well as lipophilic drug depending on processing techniques, solubility and type of lipid used.

Paclitaxel which is clinically used against a variety of cancers including ovarian, lung, breast, and Kaposi's sarcoma, but bioavailability of paclitaxel is very less which was found to be less than 1% (ten Tije et al. 2003). The less bioavailability of paclitaxel is because of enzyme cytochrome P450 in liver and P-glycoprotein in the gut wall (Hendrikx et al. 2013). Solid lipid nanoparticles proved to be an effective delivery system to overcome this problem and provide better therapeutic efficacy of the drug. Baek et al. reported a surface modified paclitaxel loaded solid lipid nanoparticles in which hydroxypropyl-β-cyclodextrin has been used as solubilizing agent of the drug (Baek et al. 2012). The study demonstrated that the C_{max} and lymph node concentrations of the paclitaxel loaded solid lipid nanoparticles were (1.44 mg/mL and 11.12 ng/mg, respectively) which were higher than that of the control solution (0.73 mg/mL and 0.89 ng/mg, respectively). The drugs and solid lipid nanoparticles were given by oral route at 25 mg/kg (Baek et al. 2012; Zafar et al. 2014). In an another study, Pooja et al. demonstrated wheat germ agglutinin-coated solid lipid nanoparticles for oral delivery of paclitaxel and reported that paclitaxel showed prolonged residence time (Pooja et al. 2016). Study demonstrated that higher area under the curve of wheat germ agglutinin-coated solid lipid nanoparticles was 30 mg/mL h, as compared to conventional solid lipid nanoparticles, 16 mg/mL h, and free control, 8 mg/mL h, after oral administration at 25 mg/kg in rats (Liu et al. 2011; Pooja et al. 2016).

Curcumin which is a lipophilic polyphenol obtained from rhizome of curcuma longa reported to have anti-inflammatory, antioxidant, anti-amyloid, antimicrobial, and anticancer effects (Minassi et al. 2013). Curcumin is also used to treat parkinson's disease, malignancy and alzheimer's disease (Lee et al. 2013). Due to rapid metabolism and poor aqueous solubility the bioavailability of curcumin is less than 1% when administered *via* oral route (Prasad et al. 2014). Kakkar et al. developed curcumin-loaded solid lipid nanoparticles and demonstrated a study on mouse model having Alzheimer's disease with treatment by aluminum chloride (a neurotoxicant) *via* oral route. The solid lipid nanoparticles showed better results (97% and 73% recovery in lipid peroxidation and acetylcholinesterase) as compared to the curcumin treated group (15% recovery in lipid peroxidation and 22% recovery in acetylcholinesterase) at oral dose of 50 mg/kg (Kakkar and Kaur 2011; Kakkar et al. 2013a).

In another experiment Bhandari et al., developed isoniazid loaded solid lipid nanoparticles for improvement in pharmacokinetic profile of the drug. Developed solid lipid nanoparticles was very small in size (48.4 nm) hence bypassed reticulo-endothelial system resulted in prolonged circulation times. Pharmacokinetic studied of the prepared solid lipid nanoparticles on rat demonstrated significant improve-ment ($p < 0.001$) in bioavailability in plasma which was 6 times and in blood 4 times higher as compared to native drug after a single dose (25 mg/kg) oral administration (Bhandari and Kaur 2013). Negi et al. prepared lopinavir encapsulated solid lipid nanoparticles based on glyceryl behenate by using hot self nanoemulsification method for improvement in oral bioavailability. The study demonstrated 3.56-fold increase in oral bioavailability of the solid lipid nanoparticles over native drug after 10 mg/kg oral dose in rat (Negi et al. 2014). Efavirenz is very lipophilic compound used clinically for treatment of human immunodeficiency virus which exhibited low oral bioavailability (40–50%) (Baert et al. 2008; Chiappetta et al. 2010). Gaur et al. developed efavirenz encapsulated solid lipid nanoparticles and demonstrated improved plasma peak concentration (C_{max}) (5.32-fold) and increase in area under curve (10.98-fold) as compared to efavirenz suspension (Gaur et al. 2014).

There are several research literatures published that reported natural products encapsulating solid lipid nanoparticles which are widely explored for treatment of cancer, central nervous system related disorders and other diseases. Tables 1.2 and 1.3 are illustrating different solid lipid nanoparticles used for cancer and central nervous system related disorders, respectively, administered through oral route.

1.5.3 Liposome

Liposomes are spherical particles made up of phospholipids which can encapsulate both lipophilic as well as hydrophilic drug. Liposome consists of hydrophilic head and lipophilic tail in same molecule. The advantages of liposome delivery system for a drug includes protection from external stimuli, enhance water solubility and improve delivery efficacy (Allen and Cullis 2004; Dutta et al. 2019). In delivery of natural products, liposome delivery has been proved to be beneficial in improving ingredients bioavailability, increased intracellular uptake with altered pharmacoki-netics profiles (Ajazuddin and Saraf 2010). Some report demonstrated liposomes can improve the therapeutic activity and improve safety profile as a drug delivery system because of its ability to deliver the therapeutic agent at desired site and for prolonged periods of time (Barragan-Montero et al. 2005; Cai et al. 2013).

Silymarin is a natural origin lipotropic drug which showed impaired bioavail-ability through oral administration. El-Samaligy et al. developed silymarin encapsu-lated hybrid liposomes prepared by reverse evaporation method by using lecithin, cholesterol, stearyl amine and Tween 20 as excipients for oral improvement of sily-marin. It has been demonstrated that prepared liposome showed enhanced hepato-protective activity with decrease level of serum glutamic oxalacetate transaminase and serum glutamic pyruvate transaminase, upon buccal administration in rats. The formulation composition intended to improve oral bioavailability (El-Samaligy et al. 2006a; El-Samaligy et al. 2006b). Paclitaxel is a natural product used

Table 1.2 Formulations of orally administered solid lipid nanoparticles loaded with drugs or natural compounds against cancers and their benefits

Active ingredient	Lipid type	Surface decoration	Average size (nm)	Outcomes offered by SLNs	References
Paclitaxel	Stearic acid	HPCD	251	Improved AUC in plasma and lymph node	Baek et al. (2012)
Paclitaxel	Monoglycerides, triglycerides,and stearic acid	WGA	150–198	Increased AUC and MRT	Pooja et al. (2016)
Docetaxel	Tristearin	Tween 80 and TPGS	189–215	Improved sustained release and AUC	Cho et al. (2014)
Doxorubicin	Precirol ATO 5	Soy lecithin and poloxamer 188	217	Increased AUC and reduced cardiotoxicity	Patro et al. (2013)
Doxorubicin	Monostearin	PEG-stearic acid	153–160	Improved bioavailability and prolonged circulation time	Yuan et al. (2013)
Vorinostat	Compritol 888 ATO	None	~100	Enhanced C_{max} and AUC	Tran et al. (2014b)
Tamoxifen	Monostearin and stearic acid	Tween 80 and poloxamer 188	130–244	Improved bioavailability	Hashem et al. (2014)
Y-Tocotrienol	Compritol 888 ATO	None	105	Increased intestine permeation and AUC	Abuasal et al. (2012)
Cantharidin	Monostearin	None	121	Improved bioavailability	Dang and Zhu (2013)
Ferulic acid	Compritol 888 ATO	None	86	Increased C_{max} and half-life	Zhang et al. (2016)
Ferulic acid	Stearic acid	Chitosan	183–229	Tumor growth suppression	Thakkar et al. (2015)

Reproduced with permission of Elsevier from Lin et al. (2017)
Abbreviations: *AUC* area under curve, *HPCD* hydroxypropyl-β-cyclodextrin, *MRT* mean residence time, *PEG* polyethylene glycol, *TPGS* D-α-tocopheryl poly(ethylene glycol) succinate, *WGA* wheat germ agglutinin

clinically under the brand name Taxol®, has exhibited excellent antitumor properties against various solid tumors. The clinical use of the paclitaxel is limited due to low aqueous solubility, impaired bioavailability and associated toxicity (Gupta et al. 2014; Nehate et al. 2014). Xu et al. developed paclitaxel encapsulated liposomes prepared by thin film hydration method by using hydrogenated soy phosphatidyl choline, cholesterol, polyethylene-2000-distearoyl-a-phosphatidylethanolamine, and tocopherol as excipients for improvement in oral bioavailability of paclitaxel. It

Table 1.3 Formulations of orally administered solid lipid nanoparticles loaded with drugs or natural compounds against CNS-related disorders and their benefits

Active ingredient	Indication	Lipid type	Average size (nm)	Outcomes offered by SLNs	References
Apomorphine	Parkinson's disease	Tripalmitin	63–155	Increased bioavailability and brain distribution	Tsai et al. (2011)
Sumatriptan	Migraine	Tripalmitin	192–301	Improved AUCbrain/ AUCplasma ratio and photophobia	Hansraj et al. (2015)
Rizatriptan	Migraine	Precirol ATO 5	220	Improved brain uptake and photophobia	Girotra and Singh (2017)
Sulpiride	Psychosis	Stearic acid and Dynasan 118	256	Increased gut permeability	Ibrahim et al. (2014)
Quetiapine	Psychosis	Dynasan 118	175	Enhanced C_{max} and bioavailability	Narala and Veerabrahma (2013)
Venlafaxine	Major depressive disorderand anxiety	Monostearin	186	Increased AUC in both plasma and brain	Zhou et al. (2015)
Chrysin	Alzheimer's disease	Stearic acid	240	Improved memory loss	Vedagiri and Thangarajan (2016)
Resveratrol	Neurodegenerative disorders	Stearic acid	134	Increased bioavailability and half-life	Pandita et al. (2014)
Resveratrol	Neurodegenerative disorders	Precirol ATO 5	258	Increased C_{max} and AUC	Ramalingam and Ko (2016)
Curcumin	Alzheimer's disease	Compritol 888 ATO	135	Reduced neuroinflammation	Kakkar and Kaur (2011)
Curcumin	Alzheimer's disease	Compritol 888 ATO	135	Improved AUC and brain distribution	Kakkar et al. (2013a)
Curcumin	Alzheimer's disease	Monostearin	135	Increased jejunum permeability and bioavailability	Ji et al. (2016)
Curcumin	Alzheimer's disease	Monostearin	452	Increased C_{max} and AUC	Ramalingam et al. (2016)
Curcumin	Alzheimer's disease	Palmitic acid	412	Increased AUC and half-life	Ramalingam and Ko (2015)
Curcumin	Cerebral ischemia	Compritol 888	135	Increased AUC in brain and cognition	Kakkar et al. (2013b)

Reproduced with permission of Elsevier from Lin et al. (2017)

Abbreviations: *AUC* area under the curve, *CNS* central nervous system, *SLN* solid lipid nanoparticle

has been demonstrated that area under curve of paclitaxel liposomes and stealth liposomes increased 1.91- and 3.39-fold, respectively as compared to free paclitaxel. The stealth liposomes were long circulating which showed a half-life of 5.6 h, 13.7 h, and 34.2 h for free paclitaxel, conventional liposome and stealth liposomes, respectively (Xu and Meng 2016). In another experiment Ingle et al. developed "liposils" in which a silica coating on the surface of conventional paclitaxel containing liposome has been done to improve stability of the liposome. It was demonstrated that the paclitaxel liposils were exhibited long circulation half-life, 3.06 h, 4.27 h, and 18.55 h for free paclitaxel, paclitaxel liposome and paclitaxel liposils, respectively (Ingle et al. 2018).

Chen et al. developed curcumin encapsulated liposomes coated with N-trimethyl chitosan chloride to improve its bioavailability through thin-film dispersion method. The study in rat demonstrated that the liposome showed two-fold increases in area under curve. The N-trimethyl chitosan chloride coated liposome exhibited long circulation half-life 3.85 h, 9.79 h and 12.05 for curcumin suspension, conventional liposome and N-trimethyl chitosan chloride coated liposome, respectively (Chen et al. 2012). In another study Takahashi et al. developed curcumin loaded liposome through micro-fluidization technique for improvement in its bioavailability. Pharmacokinetic study in rat demonstrated that four-fold increase in C_{max} (liposomal curcumin 319.2 versus free curcumin 64.6 µg/L plasma). Five-fold increase in area under curve (liposomal curcumin 26502.8 versus free curcumin 5342.6 µg/min plasma) (Takahashi et al. 2009).

Quercetin is a plant flavonoid, which contains different pharmacological activities including, antioxidant, anti-inflammatory, anti-cancer, anti-aging. However, its medical applicability is restricted due to poor solubility, instability and poor bioavailability (Cai et al. 2013). Wong et al. developed liposomes co-encapsulated vincristine and quercetin to improve bioavailability of the quercetin through thin film hydration method. The study demonstrated tenfold increase in C_{max} (19.20 for liposomal quercetin versus 2.07 nmol/mL of free quercetin). The formulation exhibited long circulation half-life 1.49 h and 14.20 h for free quercetin and the liposome, respectively (Wong and Chiu 2011).

Liposomes are generally prepared with phospholipids and useful in altering pharmacokinetics profile natural products and another drug molecule. A variety of natural products encapsulated liposomal formulations has been prepared which are summarized in Table 1.4.

1.5.4 Polymeric Nanoparticles

Polymeric nanoparticles are widely used as biomaterials in drug delivery application because of their favorable properties in terms of design, biodegradability, biocompatibility, broad structures which enable to encapsulate both hydrophobic and hydrophilic drug within it (Khalid and El-Sawy 2017). There are several types of polymer employed for preparation of the nanoparticles, zincluding,

Table 1.4 Formulations of orally administered liposomes loaded with natural products for improvement in bioavailability

Natural product	Main excipients	Technique	Major outcome	References
Silymarin	Lecithin, cholesterol, stearyl amine and Tween 20	Reverse evaporation	Enhanced hepatoprotective activity (decrease SGOT and SGPT) upon buccal administration in rats. The formulation composition intended to improve oral bioavailability	El-Samaligy et al. (2006a, b)
Paclitaxel	SPC, cholesterol, PEG2000-DSPE, tocopherol	Thin film hydration	AUC_{0-t} of PTX liposomes and stealth liposomes increased 1.91- and 3.39-fold, respectively as compared to free PTX. The stealth liposomes were long circulating showing a half-life ($t_{1/2}$) 5.6 h, 13.7 h, and 34.2 h for free PTX, conventional liposome and stealth liposomes, respectively	Xu and Meng (2016)
Paclitaxel	HSL, HEL, UEL, cholesterol, stearyl amine	Thin film hydration	The PTX liposils were exhibited long circulation half-life ($t_{1/2}$) 3.06 h, 4.27 h, and 18.55 h for free PTX, PTX liposome and PTX liposils, respectively	Ingle et al. (2018)
Curcumin	SPC, TPGS, PF127, cholesterol	Thin-film dispersion	Two-fold increase in AUC. The TMC coated liposome exhibited long circulation half-life ($t_{1/2}$) 3.85 h, 9.79 h and 12.05 for curcumin suspension, conventional liposome and TMC coated liposome, respectively	Chen et al. (2012)
Curcumin	Soybean lecithins, PC, PE, PI, PA	Microfluidization	Four-fold increase in C_{max} (LEC 319.2 vs curcumin 64.6 µg/L plasma). Five-fold increase in $AUC_{0-120min}$ (LEC 26502.8 vs curcumin 5342.6 µg/min plasma)	Takahashi et al. (2009)
Quercetin	ESM, cholesterol, PEG2000-ceramide	Thin film hydration	Ten-fold increase in C_{max} (19.20 for liposomal quercetin vs 2.07 nmol/mL of free quercetin). The formulation exhibited long circulation half-life ($t_{1/2}$) 1.49 h and 14.20 h for free quercetin and the liposome, respectively	Wong and Chiu (2011)

Abbreviations: *SGOT* serum glutamic oxalacetate transaminase, *SGPT* serum glutamic pyruvate transaminase, *PEG* polyethylene glycol, *SPC* hydrogenated soy phosphatidyl choline, *DSPE* distearoylL-a-phosphatidylethanolamine, *PTX* paclitaxel, *HSL* hydrogenated soya lecithin, *USL* unsaturated soya lecithin, *HEL* hydrogenated egg lecithin, *UEL* unsaturated egg lecithin, *TPGS* D-a-tocopheryl polyethylene glycol succinate, *PC* phosphatidylcholine, *PE* phosphatidylethanolamine, *PI* phosphatidylinositol, *PA* phosphatidic acid, *LEC* liposome-encapsulated curcumin, *ESM* egg sphingomyelin

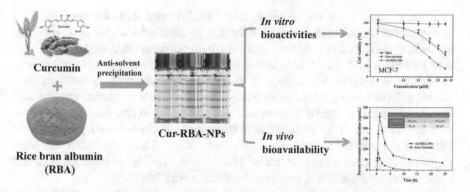

Fig. 1.1 Representation of curcumin-loaded rice bran albumin nanoparticles formulation prepared by anti-solvent precipitation method. The dose dependent *in-vitro* cytotoxicity of curcumin-loaded rice bran albumin nanoparticles versus free curcumin was evaluated in MCF-7 cells and extent of growth inhibition was measured after 48 h. The inhibition was calculated with respect to controls. *In-vivo* pharmacokinetics parameters (area under curve, C_{max}, T_{max},) of these two curcumin formulations were evaluated in rat after 20 mg/kg oral administration of the formulations. Reproduced with permission of Elsevier from (Liu et al. 2018)

poly(lactic-co-glycolic acid) (Dubey et al. 2016), poly(lactic-co-glycolic acid)-polyethylene glycol (Saneja et al. 2019), bovine serum albumin (Alam et al. 2015; Dubey et al. 2015), polycaprolactone (Youssouf et al. 2019), chitosan (Cheng et al. 2019). There are various types of nanoparticles reported to have improved bioavailability of natural product. Sun et al. developed curcumin encapsulated polybutyl-cyanoacrylate nanoparticles by emulsion polymerization method for improvement in its oral bioavailability. It was demonstrated that polybutylcyanoacrylate nanoparticles 1.7-fold increase in area under curve which was 419.62 µg/L h for curcumin nanoparticles, at dose 50 mg/kg oral versus 244.81 curcumin suspension, at a dose of 250 mg/kg oral in rat. The formulation exhibited long circulation half-life 3.85 h and 5.29 h for curcumin suspension and curcumin polybutylcyanoacrylate nanoparticles, respectively (Sun et al. 2012). In another experiment, curcumin loaded rice bran albumin nanoparticles has been developed by successive titrations method for increased *in-vitro* bioactivity and *in-vivo* bioavailability which is depicted in Fig. 1.1. It has been demonstrated that the prepared formulation exhibited 10.2-fold increase in area under curve that was 1715 ng/mL h for curcumin nanoparticles, versus 168 ng/mL h for free curcumin, at dose of 20 mg/kg, oral each in rat (Liu et al. 2018).

In the 1990s, betulinic acid, which is a pentacyclic triterpenoid was discovered as anticancer agent after screening 2500 plant extracts (Pisha et al. 1995). However, the medical applicability of betulinic acid has limited due to its poor aqueous solubility and short half-life (Saneja et al. 2017b). To overcome this limitation, Saneja et al. developed a long circulating nanoparticle loaded with betulinic acid by using polylactide-co-glycolide-monomethoxy polyethylene glycol through emulsion-solvent evaporation method for improvement in anti-tumor efficacy and bioavailability of the drug. The result demonstrated that, the nanoparticles exhibited 1.8-fold increase in area under curve which was 19232.82 ng/mL h for betulinic acid nanoparticles while 11841.31 ng/mL h for free betulinic acid, at oral dose of 10 mg/ kg each, in BALB/c mice. At the same time the formulation exhibited long circulation half-life which was 1.12 h and 8.08 h for betulinic acid and betulinic acid nanoparticles, respectively (Saneja et al. 2017a). In another study betulinic acid loaded poly(lactic-co-glycolic acid) nanoparticles has been developed and reported that the prepared nanoparticles showed 6.2-fold increase in area under curve which was 838.43 µg/mL h for betulinic acid nanoparticles while 133.81 µg/mL h free betulinic acid, at a dose of 100 mg/kg each, in rat. The formulation exhibited improved long circulation half-life of 10.73 h and 11.11 h for betulinic acid and betulinic acid nanoparticles, respectively (Kumar et al. 2018).

Agueros et al. developed paclitaxel loaded cyclodextrin-poly(anhydride) nanoparticles by using inclusion-solvent evaporation method for improvement in its bioavailability. It was reported that comparative area under curve has been observed which was 79.2 µg/mL h for Taxol® at a dose of 10 mg/kg *i.v.* while 65.9 µg/mL h for paclitaxel -cyclodextrin nanoparticles at a dose of 10 mg/kg oral in rat. Half-life reported was 4.54 h for Taxol®, and 1.48 h for paclitaxel -cyclodextrin nanoparticles, after oral administration (Agüeros et al. 2010). In another study paclitaxel loaded poly(lactic-co-glycolic acid) nanoparticles emulsified with D-α-tocopheryl poly(ethylene glycol) succinate, reported to have 9.7-fold increase in area under curve which was 8510 for paclitaxel nanoparticles while 872 ng/mL h Taxol®, at a dose of 10 mg/kg after oral administration, in rat (Zhao and Feng 2010). Similarly, bioavailability of berberine, quercetin and silymarin has been improved by delivering these phytomedicine by using polymeric nanoparticles. Yu et al. developed berberine loaded poly(lactic-co-glycolic acid)-polyethylene glycol nanoparticles by Solvent evaporation method and reported that the nanoparticles showed threefold increase in area under curve at a dose of 50 mg/kg oral each, in rat (Yu et al. 2017). In another investigation quercetin loaded poly(n-butylcyanoacrylate) nanoparticles was developed by emulsion polymerization method. The prepared quercetin nanoparticles showed threefold increase in area under curve as compared to free quercetin, at oral dose of 50 mg/kg in rat (Bagad and Khan 2015). Along with these, silymarin encapsulated nanoparticles has been prepared which showed 3.66-fold increase in area under curve as compared to the free silymarin at oral dose of 30 mg/ kg in rat (Zhao et al. 2016). Various types of natural products encapsulated polymeric nanoparticles have been prepared which are summarized in Table 1.5.

1.5.5 Miscellaneous Delivery System

There are several drug delivery systems has been developed for improvement in bioavailability of the natural products. Bapat et al. developed nanocapsules of curcumin by using lipid in which D-α-tocopheryl poly (ethylene glycol) succinate used as a stabilizer for improved oral bioavailability. The advantage of lipid nanocapsules is it can provide high drug loading and prolonged drug release. The lipid

Table 1.5 Formulations of orally administered polymeric nanoparticles loaded with natural products for improvement in bioavailability

Natural product	Polymer	Technique	Particle size (nm)	Major outcome	References
Curcumin	PBC	Emulsion polymerization	93.8	1.7-fold increase in $AUC_{0-\infty}$ (µg/L h) (419.62 for curcumin NP, at dose 50 mg/kg vs 244.81 curcumin suspension, at dose 250 mg/kg in rat).	Sun et al. (2012)
Curcumin	RBA	Successive titrations	120	10.2-fold increase in $AUC_{0-\infty}$ (ng/mL h) (1715 for curcumin NP, vs 168 free curcumin, at dose of 20 mg/kg each in rat)	Liu et al. (2018)
Betulinic acid	PLGA-mPEG	Emulsion-solvent evaporation	147	1.8-fold increase in $AUC_{0-\infty}$ (ng/mL h) (19232.82 for BA NP, vs 11841.31 free BA, at a dose of 10 mg/kg each, in BALB/c mice).	Saneja et al. (2017a)
Betulinic acid	PLGA	Emulsion-solvent evaporation	257.1	6.2-fold increase in $AUC_{0-\infty}$ (µg/mL h) (838.43 for BA NP, vs 133.81 free BA, at a dose of 100 mg/kg each, in rat).	Kumar et al. (2018)
Paclitaxel	Cyclodextrin, poly(anhydride)	Inclusion-solvent evaporation	179–310	Comparative $AUC_{0-\infty}$ (µg/mL h) (79.2 for Taxol® 10 mg/kg $i.v.$, vs 65.9 PTX-CD NP 10 mg/kg oral in rat). Half-life $(t_{1/2})$ 4.54 h for Taxol®, $i.v.$ and 1.48 h for PTX-CD NP, oral	Agüeros et al. (2010)
Paclitaxel	PLGA, TPGS	Solvent evaporation	288	9.7-fold increase in $AUC_{0-\infty}$ (ng/mL h) (8510 for PTX NP, vs 872 Taxol®, at a dose of 10 mg/kg oral each, in rat)	Zhao and Feng (2010)

(continued)

Table 1.5 (continued)

Natural product	Polymer	Technique	Particle size (nm)	Major outcome	References
Berberine	PEG–lipid–PLGA	Solvent evaporation	149.6	3.4-fold increase in $AUC_{0-\infty}$ (ng/mL h) (3499.68 for berberine NP, *vs* 1029.03 free berberine, at a dose of 100 mg/kg oral each, in rat)	Yu et al. (2017)
Quercetin	PBCA	Emulsion polymerization	166.6	3-fold increase in $AUC_{0-\infty}$ (µg/mL h) (313.97, 248, 104.22 for QT-PBCA +P-80 NP, QT-PBCA NP and free quercetin, respectively at a dose of 50 mg/kg oral each, in rat)	Bagad and Khan (2015)
Silymarin	Poloxamer 188	Emulsion solvent evaporation	107.1	3.66-fold increase in $AUC_{0-\infty}$ (µg/mL h) by silymarin NP as compared to the free silymarin at a dose of 30 mg/kg oral each, in rat	Zhao et al. (2016)

Abbreviations: *PBC* polybutylcyanoacrylate, *NP* nanoparticles, *RBA* rice bran albumin, *PLGA-mPEG* polylactide-co-glycolide- monomethoxy polyethylene glycol, *BA* Betulinic acid, *PLGA* poly(lactic-co-glycolic acid), *TPGS* D-α-tocopheryl poly(ethylene glycol) succinate, *PTX* paclitaxel, *QT* quercetin, *CD* cyclodextrine, *PBCA* poly(n-butylcyanoacrylate)

nanocapsules was prepared by antisolvent precipitation method demonstrated 12.2 fold increase in area under curve which was 1174.42 µg/mL h for the nanocapsules, while 95.64 µg/mL h for free drug in rat at a dose of 100 mg/kg (Bapat et al. 2019). In another study, W/O/W multiple emulsion of salvianolic acid extracts has been investigated for improvement in oral bioavailability. It was demonstrated that the prepared multiple emulsion showed 22.8 fold increases in area under curve which was 119.77 µg/mL min as compared to 5.25 µg/mL min for native salvianolic acid in rat at a dose of 20 mg/kg (Song et al. 2017). Berberine is an anticancer natural product that is used for treatment lymphoma but has poor oral bioavailability. Elsheikh et al. developed novel cremochylomicrons for improvement in oral bioavailability of the berberin *via* thin film hydration method and reported 2.7 fold increase in area under curve (35.09 µg/mL h while 12.75 µg/mL h for free drug) in rat at oral dose of 100 mg/kg (Elsheikh et al. 2018).

Self-emulsifying drug delivery systems played crucial role in solving low bioavailability issues of hydrophobic drugs which can solubilize the drug and enabled them for oral administration as unit dosage form (Gursoy and Benita 2004; Kohli et al. 2010). Self-emulsifying drug delivery system are isotropic mixtures of drug substance, lipids, surfactants and cosolvents (Gursoy and Benita 2004). A self-emulsifying drug delivery system droplet ranges from a few nanometers to many microns. 'Self-microemulsifying drug delivery systems' contains droplets size ranging from 100 and 250 nm while 'Self-nano-emulsifying drug delivery systems' is recently coined term in which droplet size range less than 100 nm (Singh et al. 2009). Quercetin a flavonoid having effective anticancer effects, but poor aqueous solubility and less intestinal absorption it showed poor oral bioavailability. To address this limitation, Tran et al. developed self-nano-emulsifying drug delivery systems' formulations containing quercetin by using castor oil, Tween-80, cremophor. *In-vivo* pharmacokinetic study in rat demonstrated that self-nano-emulsifying drug delivery systems formulation showed 33.5 fold increase in area under curve which was 45.35 µg/L h while 1.35 µg/L h for free drug at an oral dose of 15 mg/kg (Tran et al. 2014a). There are several types of delivery system reported for effective bioavailability enhancements of natural products are summarized in Table 1.6.

1.6 Conclusion

Natural products extensively used by human being for health benefits since concept of medicine starts. In the present scenario, many natural products have been used worldwide in the field of healthcare system that involves phytoextracts or phytochemicals as major active ingredients. However therapeutic efficacy of these active ingredients depends on bioavailability specially when administered *via* oral route. Poor aqueous solubility, poor gastrointestinal tract absorption and rapid degradation of these active ingredients, represent a major challenge for its pharmacological activity. Literature demonstrated that there are many natural products whose solubility, bioavailability and pharmacological efficacy have been significantly enhanced by involving pharmaceutical technology both conventional as well as novel approaches. In the conventional approaches, cyclodextrin complexation was found to be better choice for improvement in water solubility and bioavailability. Cyclodextrins have a unique structure with hydrophobic central cavity which can accommodate a variety of lipophilic drugs. Reports indicated many natural products including, digoxin, artemisinin, curcumin, rutin and naringenin exhibited significant improvement in bioavailability. Novel drug delivery systems are attracting much attention due to their ability to facilitate gastrointestinal absorption, protection of active ingredients from enzymatic degradation and eventually improvement in the bioavailability. Quercetin, naringenin, curcumin, artemisinin, berberine, betulinic acid, salvianolic acid, silymarin, paclitaxel, apomorphine, quetiapine, venlafaxine, chrysin and resveratrol are some of the therapeutic agent that have multiple biological activities and their bioavailability has been improved by using novel drug delivery systems which can be further investigated in human trials. Phospholipid based delivery systems like solid lipid nanoparticles, phytosomes, liposome, and noisome represents very effective delivery option for natural products as compare to the conventional approach. Therefore, novel drug delivery systems may be useful to

Table 1.6 Miscellaneous formulation loaded with natural products for improvement in bioavailability

Formulation	Natural product	Main excipient	Preparation technique	Particle size (nm)	Major outcome	References
Nanocapsules	Curcumin	TPGS, solid lipids	Antisolvent precipitation	190	12.2-fold increase in AUC (1174.42 vs 95.64 μg/mL h) in rat at a dose of 100 mg/kg	Bapat et al. (2019)
Silica nanoparticles	Curcumin	TMB, PluronicF127	Graftingmethod	100–200	Curcumin loaded within amine functionalized MSN showed improved oral bioavailability	Hartono et al. (2016)
Multiple emulsion	Salvianolic acid	IPM	Emulsification	608	22.8-fold increase in AUC (119.77 vs 5.25 μg/mL min) in rat at a dose of 20 mg/kg	Song et al. (2017)
Cremochylomicron	Berberine	Cremophor El	Thin film hydration	175.6	2.7-fold increase in AUC (35.09 vs 12.75 μg/ mL h) in rat at a dose of 100 mg/kg	Elsheikh et al. (2018)
Micelles	Berberine	SPC	W/O emulsions	330	3.1-fold increase in AUC (12.71 vs 4.10 μg/L h) in rat at a dose of 100 mg/kg	Wang et al. (2011)
Microparticle	Berberine and Betulinic acid	Chitosan	Spray drying	1–5 (μm)	Enhanced AUC (7.26, 53.86, 819.35, 5724.39 μg/L h for BA, BA-SD, BBR and BBR-SD, respectively, in rat at a dose of 100 mg/kg)	Godugu et al. (2014)
Nanosuspensions	Naringenin	PVPK-90	Antisolvent sonoprecipitation	117	1.8 fold increase in AUC (1015.7 vs 662.8 ng/ mL h) in rat at a dose of 20 mg/kg	Gera et al. (2017)
Nanoemulsion	Quercetin	Capryol 90, Labrafil M 1944 CS	Aqueous phase titration	19.3	33.5-fold increase in AUC (45.35 vs 1.35 μg/L h) in rat at a dose of 40 mg/kg	Pangeni et al. (2017)
SNEDDS	Quercetin	Castor oil, Tween-80, Cremophor RH 40	O/W nanoemulsions	208.8	33.5-fold increase in AUC (45.35 vs 1.35 μg/L h) in rat at a dose of 15 mg/kg	Tran et al. (2014a)

Abbreviations: *TMB* 1,3,5-trimethylbenzene, *IPM* isopropyl myristate, *SPC* Soybean phosphatidylcholine, *PVP* polyvinylpyrrolidone, *BBR* Berberine, *BA* Betulinic acid, *SD* spray dried, *MSN* mesoporous silica nanoparticles, *SNEDDS* self-nanoemulsifying drug delivery system

achieve desired pharmacological efficacy at lower dose which could also be accomplished with reduced side effects.

Acknowledgments The authors wish to express their gratitude to Indian Council of Medical Research (ICMR) New Delhi, Government of India, for providing financial assistance through senior research fellowship (ICMR-SRF) to Shweta Paroha.

References

Abuasal BS, Lucas C, Peyton B, Alayoubi A, Nazzal S, Sylvester PW, Kaddoumi A (2012) Enhancement of intestinal permeability utilizing solid lipid nanoparticles increases γ-Tocotrienol oral bioavailability. Lipids 47(5):461–469. https://doi.org/10.1007/s11745-012-3655-4

Agüeros M, Zabaleta V, Espuelas S, Campanero MA, Irache JM (2010) Increased oral bioavailability of paclitaxel by its encapsulation through complex formation with cyclodextrins in poly(anhydride) nanoparticles. J Control Release 145(1):2–8. https://doi.org/10.1016/j.jconrel.2010.03.012

Ajazuddin, Saraf S (2010) Applications of novel drug delivery system for herbal formulations. Fitoterapia 81(7):680–689. https://doi.org/10.1016/j.fitote.2010.05.001

Alam MA, Al-Jenoobi FI, Al-mohizea AM (2013) Commercially bioavailable proprietary technologies and their marketed products. Drug Discov Today 18(19–20):936–949. https://doi.org/10.1016/j.drudis.2013.05.007

Alam N, Dubey RD, Kumar A, Koul M, Sharma N, Sharma PR, Chandan BK, Singh SK, Singh G, Gupta PN (2015) Reduced toxicological manifestations of cisplatin following encapsulation in folate grafted albumin nanoparticles. Life Sci 142:76–85. https://doi.org/10.1016/j.lfs.2015.10.019

Al-Kassas R, Bansal M, Shaw J (2017) Nanosizing techniques for improving bioavailability of drugs. J Control Release 260:202–212. https://doi.org/10.1016/j.jconrel.2017.06.003

Allen TM, Cullis PR (2004) Drug delivery systems: entering the mainstream. Science 303(5665):1818–1822. https://doi.org/10.1126/science.1095833

Atal C, Zutshi U, Rao P (1981) Scientific evidence on the role of Ayurvedic herbals on bioavailability of drugs. J Ethnopharmacol 4(2):229–232. https://doi.org/10.1016/0378-8741(81)90037-4

Atul Bhattaram V, Graefe U, Kohlert C, Veit M, Derendorf H (2002) Pharmacokinetics and bioavailability of herbal medicinal products. Phytomedicine 9:1–33. https://doi.org/10.1078/1433-187x-00210

Baek J-S, So J-W, Shin S-C, Cho C-W (2012) Solid lipid nanoparticles of paclitaxel strengthened by hydroxypropyl-β-cyclodextrin as an oral delivery system. Int J Mol Med 30(4):953–959. https://doi.org/10.3892/ijmm.2012.1086

Baert L, Schueller L, Tardy Y, Macbride D, van't Klooster G, Borghys H, Clessens E, Van Den Mooter G, Van Gyseghem E, Van Remoortere P (2008) Development of an implantable infusion pump for sustained anti-HIV drug administration. Int J Pharm 355(1–2):38–44. https://doi.org/10.1016/j.ijpharm.2008.01.029

Bagad M, Khan ZA (2015) Poly (n-butylcyanoacrylate) nanoparticles for oral delivery of quercetin: preparation, characterization, and pharmacokinetics and biodistribution studies in Wistar rats. Int J Nanomedicine 10:3921. https://doi.org/10.2147/IJN.S80706

Baird AW, Taylor CT, Brayden DJ (1997) Non-antibiotic anti-diarrhoeal drugs: factors affecting oral bioavailability of berberine and loperamide in intestinal tissue. Adv Drug Deliv Rev 23(1–3):111–120. https://doi.org/10.1016/S0169-409X(96)00429-2

Bano G, Amla V, Raina R, Zutshi U, Chopra C (1987) The effect of piperine on pharmacokinetics of phenytoin in healthy volunteers. Planta Med 53(06):568–569. https://doi.org/10.1055/s-2006-962814

Bano G, Raina R, Zutshi U, Bedi K, Johri R, Sharma S (1991) Effect of piperine on bioavailability and pharmacokinetics of propranolol and theophylline in healthy volunteers. Eur J Clin
Pharmacol 41(6):615–617. https://doi.org/10.1007/bf00314996

Bapat P, Ghadi R, Chaudhari D, Katiyar SS, Jain S (2019) Tocophersolan stabilized lipid nanocapsules with high drug loading to improve the permeability and oral bioavailability of curcumin.
Int J Pharm 560:219–227. https://doi.org/10.1016/j.ijpharm.2019.02.013

Barragan-Montero V, Winum J-Y, Molès J-P, Juan E, Clavel C, Montero J-L (2005) Synthesis
and properties of isocannabinoid and cholesterol derivatized rhamnosurfactants: application to
liposomal targeting of keratinocytes and skin. Eur J Med Chem 40(10):1022–1029. https://doi.
org/10.1016/j.ejmech.2005.04.009

Belcaro G, Cesarone M, Dugall M, Pellegrini L, Ledda A, Grossi M, Togni S, Appendino G
(2010a) Product-evaluation registry of Meriva®, a curcumin-phosphatidylcholine complex, for
the complementary management of osteoarthritis. Panminerva Med 52(2 Suppl 1):55–62

Belcaro G, Cesarone MR, Dugall M, Pellegrini L, Ledda A, Grossi MG, Togni S, Appendino
G (2010b) Efficacy and safety of Meriva®, a curcumin-phosphatidylcholine complex, during
extended administration in osteoarthritis patients. Altern Med Rev 15(4):337–344

Bhandari R, Kaur IP (2013) Pharmacokinetics, tissue distribution and relative bioavailability of
isoniazid-solid lipid nanoparticles. Int J Pharm 441(1):202–212. https://doi.org/10.1016/j.
ijpharm.2012.11.042

Bhattaram VA, Graefe U, Kohlert C, Veit M, Derendorf H (2002) Pharmacokinetics
and bioavailability of herbal medicinal products. Phytomedicine 9:1–33. https://doi.
org/10.1078/1433-187x-00210

Brewster ME, Loftsson T (2007) Cyclodextrins as pharmaceutical solubilizers. Adv Drug Deliv
Rev 59(7):645–666. https://doi.org/10.1016/j.addr.2007.05.012

Cai X, Fang Z, Dou J, Yu A, Zhai G (2013) Bioavailability of quercetin: problems and promises.
Curr Med Chem 20(20):2572–2582. https://doi.org/10.2174/09298673113209990120

Chen H, Wu J, Sun M, Guo C, Yu A, Cao F, Zhao L, Tan Q, Zhai G (2012) N-trimethyl chitosan
chloride-coated liposomes for the oral delivery of curcumin. J Liposome Res 22(2):100–109.
https://doi.org/10.3109/08982104.2011.621127

Cheng X, Zeng X, Zheng Y, Wang X, Tang R (2019) Surface-fluorinated and pH-sensitive carboxymethyl chitosan nanoparticles to overcome biological barriers for improved drug delivery
in vivo. Carbohydr Polym 208:59–69. https://doi.org/10.1016/j.carbpol.2018.12.063

Chiappetta DA, Hocht C, Taira C, Sosnik A (2010) Efavirenz-loaded polymeric micelles for pedi-
atric anti-HIV pharmacotherapy with significantly higher oral bioavailability. Nanomedicine
5(1):11–23. https://doi.org/10.2217/nnm.09.90

Cho H-J, Park JW, Yoon I-S, Kim D-D (2014) Surface-modified solid lipid nanoparticles for oral
delivery of docetaxel: enhanced intestinal absorption and lymphatic uptake. Int J Nanomedicine
9:495. https://doi.org/10.2147/IJN.S56648

Chuah AM, Jacob B, Jie Z, Ramesh S, Mandal S, Puthan JK, Deshpande P, Vaidyanathan
VV, Gelling RW, Patel G (2014) Enhanced bioavailability and bioefficacy of an amorphous solid dispersion of curcumin. Food Chem 156:227–233. https://doi.org/10.1016/j.
foodchem.2014.01.108

Dang Y-J, Zhu C-Y (2013) Oral bioavailability of cantharidin-loaded solid lipid nanoparticles.
Chin Med 8(1):1–1. https://doi.org/10.1186/1749-8546-8-1

Dubey RD, Alam N, Saneja A, Khare V, Kumar A, Vaidh S, Mahajan G, Sharma PR, Singh SK,
Mondhe DM (2015) Development and evaluation of folate functionalized albumin nanoparticles for targeted delivery of gemcitabine. Int J Pharm 492(1–2):80–91. https://doi.org/10.1016/j.
ijpharm.2015.07.012

Dubey RD, Saneja A, Qayum A, Singh A, Mahajan G, Chashoo G, Kumar A, Andotra SS, Singh
SK, Singh G (2016) PLGA nanoparticles augmented the anticancer potential of pentacy-
clic triterpenediol in vivo in mice. RSC Adv 6(78):74586–74597. https://doi.org/10.1039/
C6RA14929D

Dutta PK, Sharma R, Kumari S, Dubey RD, Sarkar S, Paulraj J, Vijaykumar G, Pandey M, Sravanti L, Samarla M (2019) A safe and efficacious Pt (ii) anticancer prodrug: design, synthesis, in vitro efficacy, the role of carrier ligands and in vivo tumour growth inhibition. Chem Commun 55(12):1718–1721. https://doi.org/10.1039/C8CC06586A

El-Samaligy M, Afifi N, Mahmoud E (2006a) Increasing bioavailability of silymarin using a buccal liposomal delivery system: preparation and experimental design investigation. Int J Pharm 308(1–2):140–148. https://doi.org/10.1016/j.ijpharm.2005.11.006

El-Samaligy MS, Afifi NN, Mahmoud EA (2006b) Evaluation of hybrid liposomes-encapsulated silymarin regarding physical stability and in vivo performance. Int J Pharm 319(1–2):121–129. https://doi.org/10.1016/j.ijpharm.2006.04.023

Elsheikh MA, Elnaggar YS, Hamdy DA, Abdallah OY (2018) Novel cremochylomicrons for improved oral bioavailability of the antineoplastic phytomedicine berberine chloride: optimization and pharmacokinetics. Int J Pharm 535(1–2):316–324. https://doi.org/10.1016/j.ijpharm.2017.11.023

Fu PP, Xia Q, Hwang H-M, Ray PC, Yu H (2014) Mechanisms of nanotoxicity: generation of reactive oxygen species. J Food Drug Anal 22(1):64–75. https://doi.org/10.1016/j.jfda.2014.01.005

Gabetta B, Bombardelli E, Pifferi G (1988) Complexes of flavanolignans with phospholipids, preparation thereof and associated pharmaceutical compositions. United States patent US 4,764,508

Gaur PK, Mishra S, Bajpai M, Mishra A (2014) Enhanced oral bioavailability of efavirenz by solid lipid nanoparticles: in vitro drug release and pharmacokinetics studies. Biomed Res Int 2014:9. https://doi.org/10.1155/2014/363404

Gera S, Talluri S, Rangaraj N, Sampathi S (2017) Formulation and evaluation of naringenin nanosuspensions for bioavailability enhancement. AAPS PharmSciTech 18(8):3151–3162. https://doi.org/10.1208/s12249-017-0790-5

Giori A, Franceschi F (2009) Phospholipid complexes of curcumin having improved bioavailability. United States patent application US 12/281,994

Girotra P, Singh SK (2017) Multivariate optimization of Rizatriptan benzoate-loaded solid lipid nanoparticles for brain targeting and migraine management. AAPS PharmSciTech 18(2):517–528. https://doi.org/10.1208/s12249-016-0532-0

Godugu C, Patel AR, Doddapaneni R, Somagoni J, Singh M (2014) Approaches to improve the oral bioavailability and effects of novel anticancer drugs berberine and betulinic acid. PLoS One 9(3):e89919. https://doi.org/10.1371/journal.pone.0089919

Gupta PN, Jain S, Nehate C, Alam N, Khare V, Dubey RD, Saneja A, Kour S, Singh SK (2014) Development and evaluation of paclitaxel loaded PLGA: poloxamer blend nanoparticles for cancer chemotherapy. Int J Biol Macromol 69:393–399. https://doi.org/10.1016/j.ijbiomac.2014.05.067

Gursoy RN, Benita S (2004) Self-emulsifying drug delivery systems (SEDDS) for improved oral delivery of lipophilic drugs. Biomed Pharmacother 58(3):173–182. https://doi.org/10.1016/j.biopha.2004.02.001

Hansraj GP, Singh SK, Kumar P (2015) Sumatriptan succinate loaded chitosan solid lipid nanoparticles for enhanced anti-migraine potential. Int J Biol Macromol 81:467–476. https://doi.org/10.1016/j.ijbiomac.2015.08.035

Hartono SB, Hadisoewignyo L, Yang Y, Meka AK, Yu C (2016) Amine functionalized cubic mesoporous silica nanoparticles as an oral delivery system for curcumin bioavailability enhancement. Nanotechnology 27(50):505605. https://doi.org/10.1088/09574484/27/50/505605

Hashem FM, Nasr M, Khairy A (2014) In vitro cytotoxicity and bioavailability of solid lipid nanoparticles containing tamoxifen citrate. Pharm Dev Technol 19(7):824–832. https://doi.org/10.3109/10837450.2013.836218

Hendrikx JJ, Lagas JS, Rosing H, Schellens JH, Beijnen JH, Schinkel AH (2013) P-glycoprotein and cytochrome P450 3A act together in restricting the oral bioavailability of paclitaxel. Int J Cancer 132(10):2439–2447. https://doi.org/10.1002/ijc.27912

Hennenfent K, Govindan R (2005) Novel formulations of taxanes: a review. Old wine in a new bottle? Ann Oncol 17(5):735–749. https://doi.org/10.1093/annonc/mdj100

Hung L-M, Chen J-K, Huang S-S, Lee R-S, Su M-J (2000) Cardioprotective effect of resveratrol, a natural antioxidant derived from grapes. Cardiovasc Res 47(3):549–555. https://doi.org/10.1016/s0008-6363(00)00102-4

Hunter AC, Elsom J, Wibroe PP, Moghimi SM (2012) Polymeric particulate technologies for oral drug delivery and targeting: a pathophysiological perspective. Maturitas 73(1):5–18. https://doi.org/10.1016/j.maturitas.2012.05.014

Hussein A, El-Menshawe S, Afouna M (2012) Enhancement of the in-vitro dissolution and in-vivo oral bioavailability of silymarin from liquid-filled hard gelatin capsules of semisolid dispersion using Gelucire 44/14 as a carrier. Pharmazie 67(3):209–214. https://doi.org/10.1691/ph.2012.1109

Ibrahim WM, AlOmrani AH, Yassin AEB (2014) Novel sulpiride-loaded solid lipid nanoparticles with enhanced intestinal permeability. Int J Nanomedicine 9:129. https://doi.org/10.2147/IJN.S54413

Ingle SG, Pai RV, Monpara JD, Vavia PR (2018) Liposils: an effective strategy for stabilizing paclitaxel loaded liposomes by surface coating with silica. Eur J Pharm Sci 122:51–63. https://doi.org/10.1016/j.ejps.2018.06.025

Jang M, Cai L, Udeani GO, Slowing KV, Thomas CF, Beecher CW, Fong HH, Farnsworth NR, Kinghorn AD, Mehta RG (1997) Cancer chemopreventive activity of resveratrol, a natural product derived from grapes. Science 275(5297):218–220. https://doi.org/10.1126/science.275.5297.218

Ji H, Tang J, Li M, Ren J, Zheng N, Wu L (2016) Curcumin-loaded solid lipid nanoparticles with Brij78 and TPGS improved in vivo oral bioavailability and in situ intestinal absorption of curcumin. Drug Deliv 23(2):459–470. https://doi.org/10.3109/10717544.2014.918677

Johri R, Zutshi U (1992) An Ayurvedic formulation 'Trikatu' and its constituents. J Ethnopharmacol 37(2):85–91. https://doi.org/10.1016/0378-8741(92)90067-2

Jounela A, Pentikäinen P, Sothmann A (1975) Effect of particle size on the bioavailability of digoxin. Eur J Clin Pharmacol 8(5):365–370. https://doi.org/10.1007/bf00562664

Kakkar V, Kaur IP (2011) Evaluating potential of curcumin loaded solid lipid nanoparticles in aluminium induced behavioural, biochemical and histopathological alterations in mice brain. Food Chem Toxicol 49(11):2906–2913. https://doi.org/10.1016/j.fct.2011.08.006

Kakkar V, Mishra AK, Chuttani K, Kaur IP (2013a) Proof of concept studies to confirm the delivery of curcumin loaded solid lipid nanoparticles (C-SLNs) to brain. Int J Pharm 448(2):354–359. https://doi.org/10.1016/j.ijpharm.2013.03.046

Kakkar V, Muppu SK, Chopra K, Kaur IP (2013b) Curcumin loaded solid lipid nanoparticles: an efficient formulation approach for cerebral ischemic reperfusion injury in rats. Eur J Pharm Biopharm 85(3 Part A):339–345. https://doi.org/10.1016/j.ejpb.2013.02.005

Kalepu S, Nekkanti V (2015) Insoluble drug delivery strategies: review of recent advances and business prospects. Acta Pharm Sin B 5(5):442–453. https://doi.org/10.1016/j.apsb.2015.07.003

Kawakami K, Oda N, Miyoshi K, Funaki T, Ida Y (2006) Solubilization behavior of a poorly soluble drug under combined use of surfactants and cosolvents. Eur J Pharm Sci 28(1–2):7–14. https://doi.org/10.1016/j.ejps.2005.11.012

Kesarwani K, Gupta R, Mukerjee A (2013) Bioavailability enhancers of herbal origin: an overview. Asian Pac J Trop Biomed 3(4):253–266. https://doi.org/10.1016/S2221-1691(13)60060-X

Khadka P, Ro J, Kim H, Kim I, Kim JT, Kim H, Cho JM, Yun G, Lee J (2014) Pharmaceutical particle technologies: an approach to improve drug solubility, dissolution and bioavailability. Asian J Pharm Sci 9(6):304–316. https://doi.org/10.1016/j.ajps.2014.05.005

Khalid M, El-Sawy HS (2017) Polymeric nanoparticles: promising platform for drug delivery. Int J Pharm 528(1–2):675–691. https://doi.org/10.1016/j.ijpharm.2017.06.052

Kidd PM (2009) Bioavailability and activity of phytosome complexes from botanical polyphenols: the silymarin, curcumin, green tea, and grape seed extracts. Altern Med Rev 14(3):226–246. http://europepmc.org/abstract/MED/16164374

Kidd P, Head K (2005) A review of the bioavailability and clinical efficacy of milk thistle phytosome: a silybin-phosphatidylcholine complex (Siliphos). Altern Med Rev 10(3):193–203. http://europepmc.org/abstract/MED/16164374

King RE, Bomser JA, Min DB (2006) Bioactivity of resveratrol. Compr Rev Food Sci Food Saf 5(3):65–70. https://doi.org/10.1111/j.1541-4337.2006.00001.x

Kohli K, Chopra S, Dhar D, Arora S, Khar RK (2010) Self-emulsifying drug delivery systems: an approach to enhance oral bioavailability. Drug Discov Today 15(21):958–965. https://doi.org/10.1016/j.drudis.2010.08.007

Kumar S, Randhawa JK (2013) High melting lipid based approach for drug delivery: solid lipid nanoparticles. Mater Sci Eng C 33(4):1842–1852. https://doi.org/10.1016/j.msec.2013.01.037

Kumar P, Singh AK, Raj V, Rai A, Keshari AK, Kumar D, Maity B, Prakash A, Maiti S, Saha S (2018) Poly (lactic-co-glycolic acid)-loaded nanoparticles of betulinic acid for improved treatment of hepatic cancer: characterization, in vitro and in vivo evaluations. Int J Nanomedicine 13:975. https://doi.org/10.2147/IJN.S157391

Lan J, Zhao Y, Dong F, Yan Z, Zheng W, Fan J, Sun G (2015) Meta-analysis of the effect and safety of berberine in the treatment of type 2 diabetes mellitus, hyperlipemia and hypertension. J Ethnopharmacol 161:69–81. https://doi.org/10.1016/j.jep.2014.09.049

Lavelle F, Bissery M, Combeau C, Riou J, Vrignaud P, Andre S (1995) Preclinical evaluation of docetaxel (Taxotere). Semin Oncol 22:3–16

Lee W-H, Loo C-Y, Bebawy M, Luk F, Mason RS, Rohanizadeh R (2013) Curcumin and its derivatives: their application in neuropharmacology and neuroscience in the 21st century. Curr Neuropharmacol 11(4):338–378. https://doi.org/10.2174/1570159X11311040002

Leleux J, Williams RO (2014) Recent advancements in mechanical reduction methods: particulate systems. Drug Dev Ind Pharm 40(3):289–300. https://doi.org/10.3109/03639045.2013.828217

Li Y, Zheng J, Xiao H, McClements DJ (2012) Nanoemulsion-based delivery systems for poorly water-soluble bioactive compounds: influence of formulation parameters on polymethoxyflavone crystallization. Food Hydrocoll 27(2):517–528. https://doi.org/10.1016/j.foodhyd.2011.08.017

Lin C-H, Chen C-H, Lin Z-C, Fang J-Y (2017) Recent advances in oral delivery of drugs and bioactive natural products using solid lipid nanoparticles as the carriers. J Food Drug Anal 25(2):219–234. https://doi.org/10.1016/j.jfda.2017.02.001

Liu A, Lou H, Zhao L, Fan P (2006) Validated LC/MS/MS assay for curcumin and tetrahydrocurcumin in rat plasma and application to pharmacokinetic study of phospholipid complex of curcumin. J Pharm Biomed Anal 40(3):720–727. https://doi.org/10.1016/j.jpba.2005.09.032

Liu Y, Wang P, Sun C, Zhao J, Du Y, Shi F, Feng N (2011) Bioadhesion and enhanced bioavailability by wheat germ agglutinin-grafted lipid nanoparticles for oral delivery of poorly water-soluble drug bufalin. Int J Pharm 419(1):260–265. https://doi.org/10.1016/j.ijpharm.2011.07.019

Liu C, Yang X, Wu W, Long Z, Xiao H, Luo F, Shen Y, Lin Q (2018) Elaboration of curcumin-loaded rice bran albumin nanoparticles formulation with increased in vitro bioactivity and in vivo bioavailability. Food Hydrocoll 77:834–842. https://doi.org/10.1016/j.foodhyd.2017.11.027

Lu Y, Liu S, Zhao Y, Zhu L, Yu S (2014) Complexation of Z-ligustilide with hydroxypropyl-β-cyclodextrin to improve stability and oral bioavailability. Acta Pharma 64(2):211–222. https://doi.org/10.2478/acph-2014-0012

Lu Q, Dun J, Chen J-M, Liu S, Sun CC (2019) Improving solid-state properties of berberine chloride through forming a salt cocrystal with citric acid. Int J Pharm 554:14–20. https://doi.org/10.1016/j.ijpharm.2018.10.062

Maiti K, Mukherjee K, Gantait A, Saha BP, Mukherjee PK (2006) Enhanced therapeutic potential of naringenin-phospholipid complex in rats. J Pharm Pharmacol 58(9):1227–1233. https://doi.org/10.1211/jpp.58.9.0009

Maiti K, Mukherjee K, Gantait A, Saha BP, Mukherjee PK (2007) Curcumin-phospholipid complex: preparation, therapeutic evaluation and pharmacokinetic study in rats. Int J Pharm 330(1–2):155–163. https://doi.org/10.1016/j.ijpharm.2006.09.025

Maiti K, Mukherjee K, Murugan V, Saha BP, Mukherjee PK (2009) Exploring the effect of Hesperetin-HSPC complex – a novel drug delivery system on the in vitro release, therapeutic efficacy and pharmacokinetics. AAPS PharmSciTech 10(3):943–950. https://doi.org/10.1208/s12249-009-9282-6

Maiti K, Mukherjee K, Murugan V, Saha BP, Mukherjee PK (2010) Enhancing bioavailability and hepatoprotective activity of andrographolide from Andrographis paniculata, a well-known medicinal food, through its herbosome. J Sci Food Agric 90(1):43–51. https://doi.org/10.1002/jsfa.3777

Mauri P, Simonetti P, Gardana C, Minoggio M, Morazzoni P, Bombardelli E, Pietta P (2001) Liquid chromatography/atmospheric pressure chemical ionization mass spectrometry of terpene lactones in plasma of volunteers dosed with Ginkgo biloba L. extracts. Rapid Commun Mass Spectrom 15(12):929–934. https://doi.org/10.1002/rcm.316

Md. Makhlaquer R, Ranjit H, Mohd. Aamir M, Sarfaraj H, Arshad H (2011) Oral Lipid Based Drug Delivery System (LBDDS): formulation, characterization and application: a review. Curr Drug Deliv 8(4):330–345. https://doi.org/10.2174/156720111795767906

Minassi A, Sánchez-Duffhues G, Collado JA, Muñoz E, Appendino G (2013) Dissecting the pharmacophore of curcumin. Which structural element is critical for which action? J Nat Prod 76(6):1105–1112. https://doi.org/10.1021/np400148e

Miyake K, Arima H, Hirayama F, Yamamoto M, Horikawa T, Sumiyoshi H, Noda S, Uekama K (2000) Improvement of solubility and oral bioavailability of rutin by complexation with 2-hydroxypropyl-β-cyclodextrin. Pharm Dev Technol 5(3):399–407. https://doi.org/10.1081/PDT-100100556

Moghaddam HK, Baluchnejadmojarad T, Roghani M, Goshadrou F, Ronaghi A (2013) Berberine chloride improved synaptic plasticity in STZ induced diabetic rats. Metab Brain Dis 28(3):421–428. https://doi.org/10.1007/s11011-013-9411-5

Mukherjee K, Venkatesh M, Venkatesh P, Saha B, Mukherjee PK (2011) Effect of soy phosphatidyl choline on the bioavailability and nutritional health benefits of resveratrol. Food Res Int 44(4):1088–1093. https://doi.org/10.1016/j.foodres.2011.03.034

Mukherjee PK, Harwansh RK, Bhattacharyya S (2015) Bioavailability of herbal products: approach toward improved pharmacokinetics evidence-based validation of herbal medicine. Elsevier, pp 217–245. https://doi.org/10.1016/B978-0-12-800874-4.00010-6

Munjal B, Pawar YB, Patel SB, Bansal AK (2011) Comparative oral bioavailability advantage from curcumin formulations. Drug Deliv Transl Res 1(4):322–331. https://doi.org/10.1007/s13346-011-0033-3

Murugan V, Mukherjee K, Maiti K, Mukherjee PK (2009) Enhanced oral bioavailability and antioxidant profile of ellagic acid by phospholipids. J Agric Food Chem 57(11):4559–4565. https://doi.org/10.1021/jf8037105

Nakagawa H, Miyata T, Mohri K, Sugimoto I, Manabe H (1978) Water of crystallization of berberine chloride (author's transl). Yakugaku zasshi. J Pharm Soc Jpn 98(8):981. https://doi.org/10.1248/yakushi1947.98.8_981

Narala A, Veerabrahma K (2013) Preparation, characterization and evaluation of quetiapine fumarate solid lipid nanoparticles to improve the oral bioavailability. J Pharm 2013:1–7. https://doi.org/10.1155/2013/265741

Negi JS, Chattopadhyay P, Sharma AK, Ram V (2014) Development and evaluation of glyceryl behenate based solid lipid nanoparticles (SLNs) using hot self-nanoemulsification (SNE) technique. Arch Pharm Res 37(3):361–370. https://doi.org/10.1007/s12272-013-0154-y

Nehate C, Jain S, Saneja A, Khare V, Alam N, Dhar Dubey R, N Gupta P (2014) Paclitaxel formulations: challenges and novel delivery options. Curr Drug Deliv 11(6):666–686. https://doi.org/10.2174/1567201811666140609154949

Oerlemans C, Bult W, Bos M, Storm G, Nijsen JFW, Hennink WE (2010) Polymeric micelles in anticancer therapy: targeting, imaging and triggered release. Pharm Res 27(12):2569–2589. https://doi.org/10.1007/s11095-010-0233-4

Pandita D, Kumar S, Poonia N, Lather V (2014) Solid lipid nanoparticles enhance oral bioavailability of resveratrol, a natural polyphenol. Food Res Int 62:1165–1174. https://doi.org/10.1016/j.foodres.2014.05.059

Pangeni R, Kang S-W, Oak M, Park EY, Park JW (2017) Oral delivery of quercetin in oil-in-water nanoemulsion: in vitro characterization and in vivo anti-obesity efficacy in mice. J Funct Foods 38:571–581. https://doi.org/10.1016/j.jff.2017.09.059

Pathak D, Pathak K, Singla A (1991) Flavonoids as medicinal agents-recent advances. Fitoterapia 62(5):371–389. https://doi.org/10.2174/157489210790936261

Patro NM, Devi K, Pai RS, Suresh S (2013) Evaluation of bioavailability, efficacy, and safety profile of doxorubicin-loaded solid lipid nanoparticles. J Nanopart Res 15(12):2124. https://doi.org/10.1007/s11051-013-2124-1

Pisha E, Chai H, Lee I-S, Chagwedera TE, Farnsworth NR, Cordell GA, Beecher CW, Fong HH, Kinghorn AD, Brown DM (1995) Discovery of betulinic acid as a selective inhibitor of human melanoma that functions by induction of apoptosis. Nat Med 1(10):1046. https://doi.org/10.1038/nm1095-1046

Pooja D, Kulhari H, Kuncha M, Rachamalla SS, Adams DJ, Bansal V, Sistla R (2016) Improving efficacy, oral bioavailability, and delivery of paclitaxel using protein-grafted solid lipid nanoparticles. Mol Pharm 13(11):3903–3912. https://doi.org/10.1021/acs.molpharmaceut.6b00691

Prasad S, Tyagi AK, Aggarwal BB (2014) Recent developments in delivery, bioavailability, absorption and metabolism of curcumin: the golden pigment from golden spice. Cancer Res Treat 46(1):2. https://doi.org/10.4143/crt.2014.46.1.2

Rahman A, Hussain A, Iqbal Z, Kumar Harwansh R, Ratnakar Singh L, Ahmad S (2013) Nanosuspension: a potential nanoformulation for improved delivery of poorly bioavailable drug. Micro and Nanosyst 5(4):273–287. https://doi.org/10.2174/1876402905041311271216
25

Ramalingam P, Ko YT (2015) Enhanced oral delivery of Curcumin from N-trimethyl chitosan surface-modified solid lipid nanoparticles: pharmacokinetic and brain distribution evaluations. Pharm Res 32(2):389–402. https://doi.org/10.1007/s11095014-1469-1

Ramalingam P, Ko YT (2016) Improved oral delivery of resveratrol from N-trimethyl chitosan-g-palmitic acid surface-modified solid lipid nanoparticles. Colloids Surf B Biointerfaces 139:52–61. https://doi.org/10.1016/j.colsurfb.2015.11.050

Ramalingam P, Yoo SW, Ko YT (2016) Nanodelivery systems based on mucoadhesive polymer coated solid lipid nanoparticles to improve the oral intake of food curcumin. Food Res Int 84:113–119. https://doi.org/10.1016/j.foodres.2016.03.031

Rasenack N, Müller BW (2004) Micron-size drug particles: common and novel micronization techniques. Pharm Dev Technol 9(1):1–13. https://doi.org/10.1081/PDT-120027417

Saneja A, Kumar R, Singh A, Dubey RD, Mintoo MJ, Singh G, Mondhe DM, Panda AK, Gupta PN (2017a) Development and evaluation of long-circulating nanoparticles loaded with betulinic acid for improved anti-tumor efficacy. Int J Pharm 531(1):153–166. https://doi.org/10.1016/j.ijpharm.2017.08.076

Saneja A, Sharma L, Dubey RD, Mintoo MJ, Singh A, Kumar A, Sangwan PL, Tasaduq SA, Singh G, Mondhe DM, Gupta PN (2017b) Synthesis, characterization and augmented anti-cancer potential of PEG-betulinic acid conjugate. Mater Sci Eng C 73:616–626. https://doi.org/10.1016/j.msec.2016.12.109

Saneja A, Kumar R, Mintoo MJ, Dubey RD, Sangwan PL, Mondhe DM, Panda AK, Gupta PN (2019) Gemcitabine and betulinic acid co-encapsulated PLGA– PEG polymer nanoparticles for improved efficacy of cancer chemotherapy. Mater Sci Eng C 98:764–771. https://doi.org/10.1016/j.msec.2019.01.026

Semalty A, Semalty M, Rawat MS, Franceschi F (2010) Supramolecular phospholipids-polyphenolics interactions: the PHYTOSOME strategy to improve the bioavailability of phytochemicals. Fitoterapia 81(5):306–314. https://doi.org/10.1016/j.fitote.2009.11.001

Serajuddin AT (2007) Salt formation to improve drug solubility. Adv Drug Deliv Rev 59(7):603–616. https://doi.org/10.1016/j.addr.2007.05.010

Shanmugam S, Im HT, Sohn YT, Kim Y-I, Park J-H, Park E-S, Woo JS (2015) Enhanced oral bioavailability of paclitaxel by solid dispersion granulation. Drug Dev Ind Pharm 41(11):1864–1876. https://doi.org/10.3109/03639045.2015.1018275

Sharma S, Kulkarni SK, Chopra K (2007) Effect of resveratrol, a polyphenolic phytoalexin, on thermal hyperalgesia in a mouse model of diabetic neuropathic pain. Fundam Clin Pharmacol 21(1):89–94. https://doi.org/10.1111/j.14728206.2006.00455.x

Shulman M, Cohen M, Soto-Gutierrez A, Yagi H, Wang H, Goldwasser J, Lee-Parsons CW, Benny-Ratsaby O, Yarmush ML, Nahmias Y (2011) Enhancement of naringenin bioavailability by complexation with hydroxypropoyl-β-cyclodextrin. PLoS One 6(4):e18033. https://doi.org/10.1371/journal.pone.0018033

Singh B, Bandopadhyay S, Kapil R, Singh R, Katare OP (2009) Self-emulsifying drug delivery systems (SEDDS): formulation development, characterization, and applications. Crit Rev Ther Drug Carrier Syst 26(5). https://doi.org/10.1615/CritRevTherDrugCarrierSyst.v26.i5.10

Singh A, Worku ZA, Van den Mooter G (2011) Oral formulation strategies to improve solubility of poorly water-soluble drugs. Expert Opin Drug Deliv 8(10):1361–1378. https://doi.org/10.1517/17425247.2011.606808

Song Y, Zhuang J, Guo J, Xiao Y, Ping Q (2008) Preparation and properties of a silybin-phospholipid complex. Pharmazie 63(1):35–42. https://doi.org/10.1691/ph.2008.7132

Song Y, Gao H, Zhang S, Zhang Y, Jin X, Sun J (2017) Prescription optimization and oral bioavailability study of Salvianolic acid extracts W/O/W multiple emulsion. Biol Pharm Bull 40(12):2081–2087. https://doi.org/10.1248/bpb.b17-00162

Sun M, Zhao L, Guo C, Cao F, Chen H, Zhao L, Tan Q, Zhu X, Zhu F, Ding T, Zhai Y, Zhai G (2012) Evaluation of an oral carrier system in rats: bioavailability and gastrointestinal absorption properties of curcumin encapsulated PBCA nanoparticles. J Nanopart Res 14(2):705. https://doi.org/10.1007/s11051-011-0705-4

Takahashi M, Uechi S, Takara K, Asikin Y, Wada K (2009) Evaluation of an oral carrier system in rats: bioavailability and antioxidant properties of liposome-encapsulated Curcumin. J Agric Food Chem 57(19):9141–9146. https://doi.org/10.1021/jf9013923

ten Tije AJ, Verweij J, Loos WJ, Sparreboom A (2003) Pharmacological effects of formulation vehicles: implications for cancer chemotherapy. Clin Pharmacokinet 42(7):665–685. https://doi.org/10.2165/00003088-200342070-00005

Thakkar A, Chenreddy S, Wang J, Prabhu S (2015) Ferulic acid combined with aspirin demonstrates chemopreventive potential towards pancreatic cancer when delivered using chitosan-coated solid-lipid nanoparticles. Cell Biosci 5(1):46. https://doi.org/10.1186/s13578-015-0041-y

Tran TH, Guo Y, Song D, Bruno RS, Lu X (2014a) Quercetin-containing self-nanoemulsifying drug delivery system for improving oral bioavailability. J Pharm Sci 103(3):840–852. https://doi.org/10.1002/jps.23858

Tran TH, Ramasamy T, Truong DH, Shin BS, Choi H-G, Yong CS, Kim JO (2014b) Development of Vorinostat-loaded solid lipid nanoparticles to enhance pharmacokinetics and efficacy against multidrug-resistant cancer cells. Pharm Res 31(8):1978–1988. https://doi.org/10.1007/s11095-014-1300-z

Tsai M-J, Huang Y-B, Wu P-C, Fu Y-S, Kao Y-R, Fang J-Y, Tsai Y-H (2011) Oral apomorphine delivery from solid lipid nanoparticles with different monostearate emulsifiers: pharmacokinetic and behavioral evaluations. J Pharm Sci 100(2):547–557. https://doi.org/10.1002/jps.22285

Uekama K, Fujinaga T, Otagiri M, Seo H, Tsuruoka M (1981) Enhanced bioavailability of digoxin by gamma-cyclodextrin complexation. J Pharmacobiodyn 4(9):735–737. https://doi.org/10.1248/bpb1978.4.735

Vasconcelos T, Sarmento B, Costa P (2007) Solid dispersions as strategy to improve oral bioavailability of poor water soluble drugs. Drug Discov Today 12(23):1068–1075. https://doi.org/10.1016/j.drudis.2007.09.005

Vedagiri A, Thangarajan S (2016) Mitigating effect of chrysin loaded solid lipid nanoparticles against amyloid β25–35 induced oxidative stress in rat hippocampal region: an efficient formulation approach for Alzheimer's disease. Neuropeptides 58:111–125. https://doi.org/10.1016/j.npep.2016.03.002

Wan S, Sun Y, Qi X, Tan F (2012) Improved bioavailability of poorly water-soluble drug Curcumin in cellulose acetate solid dispersion. AAPS PharmSciTech 13(1):159–166. https://doi.org/10.1208/s12249-011-9732-9

Wang T, Wang N, Song H, Xi X, Wang J, Hao A, Li T (2011) Preparation of an anhydrous reverse micelle delivery system to enhance oral bioavailability and anti-diabetic efficacy of berberine. Eur J Pharm Sci 44(1):127–135. https://doi.org/10.1016/j.ejps.2011.06.015

Wang W, Kang Q, Liu N, Zhang Q, Zhang Y, Li H, Zhao B, Chen Y, Lan Y, Ma Q, Wu Q (2015) Enhanced dissolution rate and oral bioavailability of Ginkgo biloba extract by preparing solid dispersion via hot-melt extrusion. Fitoterapia 102:189–197. https://doi.org/10.1016/j.fitote.2014.10.004

Williams HD, Trevaskis NL, Charman SA, Shanker RM, Charman WN, Pouton CW, Porter CJ (2013) Strategies to address low drug solubility in discovery and development. Pharmacol Rev 65(1):315–499. https://doi.org/10.1124/pr.112.005660

Wong M-Y, Chiu GN (2011) Liposome formulation of co-encapsulated vincristine and quercetin enhanced antitumor activity in a trastuzumab-insensitive breast tumor xenograft model. Nanomedicine 7(6):834–840. https://doi.org/10.1016/j.nano.2011.02.001

Wong J, Yuen K (2001) Improved oral bioavailability of artemisinin through inclusion complexation with β-and γ-cyclodextrins. Int J Pharm 227(1–2):177–185. https://doi.org/10.1016/S0378-5173(01)00796-7

Xu Y, Meng H (2016) Paclitaxel-loaded stealth liposomes: development, characterization, pharmacokinetics, and biodistribution. Artificial Cells Nanomed Biotechnol 44(1):350–355. https://doi.org/10.3109/21691401.2014.951722

Yoshimatsu K, Nakabayashi S, Ogaki J, Kimura M, Horikoshi I (1981) Dielectric study on the water of crystallization of berberine chloride (author's transl). Yakugaku zasshi. J Pharm Soc Jpn 101(12):1143. https://doi.org/10.1248/yakushi1947.101.12_1143

Youssouf L, Bhaw-Luximon A, Diotel N, Catan A, Giraud P, Gimié F, Koshel D, Casale S, Bénard S, Meneyrol V, Lallemand L, Meilhac O, Lefebvre D'Hellencourt C, Jhurry D, Couprie J (2019) Enhanced effects of curcumin encapsulated in polycaprolactone-grafted oligocarrageenan nanomicelles, a novel nanoparticle drug delivery system. Carbohydr Polym 217:35–45. https://doi.org/10.1016/j.carbpol.2019.04.014

Yu F, Ao M, Zheng X, Li N, Xia J, Li Y, Li D, Hou Z, Qi Z, Chen XD (2017) PEG–lipid–PLGA hybrid nanoparticles loaded with berberine–phospholipid complex to facilitate the oral delivery efficiency. Drug Deliv 24(1):825–833. https://doi.org/10.1080/10717544.2017.1321062

Yuan H, Chen C-Y, Chai G-h, Du Y-Z, Hu F-Q (2013) Improved transport and absorption through gastrointestinal tract by PEGylated solid lipid nanoparticles. Mol Pharm 10(5):1865–1873. https://doi.org/10.1021/mp300649z

Zafar N, Fessi H, Elaissari A (2014) Cyclodextrin containing biodegradable particles: from preparation to drug delivery applications. Int J Pharm 461(1):351–366. https://doi.org/10.1016/j.ijpharm.2013.12.004

Zhang H, Wei J, Xue R, Wu J-D, Zhao W, Wang Z-Z, Wang S-K, Zhou Z-X, Song D-Q, Wang Y-M (2010) Berberine lowers blood glucose in type 2 diabetes mellitus patients through increasing insulin receptor expression. Metabolism 59(2):285–292. https://doi.org/10.1016/j.metabol.2009.07.029

Zhang Y, Li Z, Zhang K, Yang G, Wang Z, Zhao J, Hu R, Feng N (2016) Ethyl oleate-containing nanostructured lipid carriers improve oral bioavailability of trans-ferulic acid ascompared with conventional solid lipid nanoparticles. Int J Pharm 511(1):57–64. https://doi.org/10.1016/j.ijpharm.2016.06.131

Zhao L, Feng SS (2010) Enhanced oral bioavailability of paclitaxel formulated in vitamin E-TPGS emulsified nanoparticles of biodegradable polymers: in vitro and in vivo studies. J Pharm Sci 99(8):3552–3560. https://doi.org/10.1002/jps.22113

Zhao X, Deng Y, Zhang Y, Zu Y, Lian B, Wu M, Zu C, Wu W (2016) Silymarin nanoparticles through emulsion solvent evaporation method for oral delivery with high antioxidant activities, bioavailability, and absorption in the liver. RSC Adv 6(95):93137–93146. https://doi.org/10.1039/C6RA12896C

Zhou Y, Zhang G, Rao Z, Yang Y, Zhou Q, Qin H, Wei Y, Wu X (2015) Increased brain uptake of venlafaxine loaded solid lipid nanoparticles by overcoming the efflux function and expression of P-gp. Arch Pharm Res 38(7):1325–1335. https://doi.org/10.1007/s12272-014-0539-6

Chapter 2
Solubility Enhancement Techniques for Natural Product Delivery

Harsha Jain and Naveen Chella

Abstract The products extracted from the natural source are of great importance as they are useful for the treatment of a variety of ailment and thus they are an important source of pharmaceuticals. The major drawbacks with the use of these natural products are their isolation from the source in pure form, the variability in the content of samples, poor aqueous solubility, and decreased oral bioavailability. Poor aqueous solubility of the natural product is the major obstacle involved with the development of any formulation and causes of low bioavailability and thus the delivery of drugs at a sub-optimal level. Poorly soluble components require a high amount of them to be incorporated into the formulation to get the required concentration in the body to elicit a pharmacological effect.

Different strategies are being used for the enhancement of their aqueous solubility of an active product of natural origin. In the current chapter major focus was on particle engineering technologies, dispersions, complexation based technologies, and nanotechnology related aspects were discussed as these already proven their efficiency in terms of solubility improvement and commercialization with synthetic molecules. Nanotechnology offers great advantages for solubility improvement as surface to volume ratio increases. Particle engineering includes particle size reduction that improves their surface area and thus enhances solubility. Complexation using different complexing agent improve solubility as well as stability of natural products. Nanosuspension or Nanocrystals are emerging technologies where, higher dose drugs also can be used with utilizing lesser quantity of stabilizer. Selection of an appropriate method for solubility improving also results in improved bioavailability and a reduction in dosage frequency as well as patient improved compliance.

Keywords Natural products · Solubility · Particle engineering · Complexation · Nanotechnology · Liposomes · Emulsion

H. Jain · N. Chella (✉)
Department of Pharmaceutics, National Institute of Pharmaceutical Education and Research – Hyderabad, Hyderabad, Telangana, India
e-mail: naveen.niperhyd@gov.in

A. Saneja et al. (eds.), *Sustainable Agriculture Reviews 43*, Sustainable
Agriculture Reviews 43, https://doi.org/10.1007/978-3-030-41838-0_2

2.1 Introduction

Natural products derived from the living organism such as plant, animal and micro-organisms are a great source of a variety of pharmaceuticals. They exert nutritional and health benefits and thus used in the pursuit of health and well-being (Cragg and Newman 2013). They cover a variety of therapeutic compounds and thus are useful in the treatment of many diseases since very old time. Because of the diversity in their structure and steric properties, these products are of great importance as they offer selectivity as well as an affinity towards molecular receptors (Guo 2017; Hong 2011). During the last few decades, the use of the natural product in pharmaceutical industries has increased tremendously due to their safety over the synthetic mole-cules (Siddiqui et al. 2014). Studies have shown that compounds generally obtained from natural sources such flavonoids, terpenoids and others like some natural oils possess variety of function that to treat many chronic and severe conditions. Few examples include, artemisinin for antimalarial activity (Klayman 1985), agathosma betulina for antihypertensive activity (Etuk 2006), Taxol derivatives for anticancer activity (Cragg and Newman 2005), ranunculus ternatus for anti-tuberculosis (Deng et al. 2013) and curcumin for anti-infective, analgesic and anti-inflammatory activ-ity (Garg 2005). Besides a wide range of utility of natural products, there are certain limitations like structural complexity, isolation or extraction of pure active mole-cule, poor aqueous solubility, poor stability in *in vitro* and *in vivo* conditions and uncontrolled-release which are hindering their development as clinical candidate.

The present chapter mainly emphasized on the significance of solubility in dos-age form design, and about the different methods or strategies such as particle engi-neering, complexation, and nanotechnology for enhancement of aqueous solubility of these natural substances.

2.2 Solubility

Solubility is the property of any substance or solute molecule to get dissolved into the solvent and plays a significant function in achieving the desired concentration in the blood or in plasma as to have high bioavailability (Lachman et al. 1976; Martin 1993). Solubility is a key parameter in the journey of drug compound from discov-ery to development phase at various stages (Krishnaiah 2010; Sharma et al. 2009). In the initial stages of drug discovery from natural products, if the drug has poor aqueous solubility, it may limit the extraction efficiency of the natural molecules as water being the universal solvent. Later during preclinical development, making formulations suitable for administration to selected animal models also creates problem. Solubility is the rate limiting parameter for the drug which are adminis-tered orally to achieve required concentration in blood plasma to elicit a therapeutic or pharmacological response (Wen et al. 2015). Another problem with these poorly soluble drugs is that, they frequently require high amount of dose to attend thera-peutic concentrations in blood plasma upon administration.

The slower dissolution and slower absorption guided by poor aqueous solubility leads to lower bioavailability and reduced therapeutic efficacy. Lack of sufficient bioavailability leads to the withdrawal of their development from pipeline (Kalepu and Nekkanti 2015). Hence, pharmaceutical scientists were working on various technologies and strategies for enhancing the solubility and thereby enhancing their bioavailability, therapeutic potential and clinical application (Kumari et al. 2012; Lipinski et al. 1997; Savjani et al. 2012).

2.2.1   Factors Influencing the Solubility of Substances

The rate and amount of solute to be dissolved in a particular solvent relies on many factors, such as physicochemical properties of the substance being dissolved (including chemical nature, type of surface functional groups present, purity of the substance, melting point, crystallinity) temperature, pressure, pH of the media, polarity of the solvent (Recharla et al. 2017).

2.2.1.1   Effect of Particle Size and Shape

Smaller particle sizes possess higher effective surface area and increased surface to volume ratio which may improve the solubility. The reduced particle size shows only higher dissolution rate. However, the maximum amount of compound dissolved will be same (Morales et al. 2016). Furthermore, symmetrical molecules exerts low solubility when compared with asymmetrical (Merisko-Liversidge et al. 2003; Mosharraf and Nyström 1995).

2.2.1.2   Effect of Molecular Weight

In general, substances with higher molecular weight are considered as poorly soluble compounds. While in other cases like organic compounds, the solubility depends on branching of carbon atoms. The more will be the branching of carbon atom higher will be the solubility of that compound. Therefore, the branched polymer shows higher solubility than that of the linear polymer having the identical molecular weight (Ravve 2013; Small 1953).

2.2.1.3   Effect of Temperature

Solubility of many substance increases as temperature increases (either liquid or solid). The reason is that the kinetic energy of particles increases with the rise in temperature and this elevated temperature allows the molecules of solvent to break the solute molecules more effectively which are held together by intermolecular forces of attraction (Chaudhary et al. 2012; Kadam et al. 2009).

2.2.1.4 Effect of Molecular Polarity

Polarity of a molecule is related to its intermolecular interactions which intern effects the other properties like melting point, boiling point and solubility of the compound. As a rule of thumb, like dissolves like polar compounds will be solubilized by polar solvents and vice versa. As water is more polar in nature, behaves as good solvent for most of the polar molecules.

2.2.1.5 Effect of Solid State Forms

Amorphous forms of natural bioactive compounds possess greater aqueous solubility when compared with crystalline form. In case of crystalline molecules also, different polymorphic forms possess different solubilities. The structural arrangements into the crystal lattice can potentially affect the various physicochemical properties of any bioactive molecules (Chawla and Bansal 2008; Huang and Tong 2004). Similarly, there is variation in solubility of molecule from hydrates to anhydrous form.

2.2.1.6 Effect of pH Condition of the Medium

Solubility can also be affected by the pH condition of solvent present in surrounding depending on the pKa of the compound. Various compounds either hydrophilic or lipophilic will show difference in solubilities at different pH conditions (Hörter and Dressman 2001).

2.2.1.7 Presence of Excipients

Stabilizers also named as surfactants or emulsifiers are used in many pharmaceutical dosage forms to maintain the stability. The surfactants contain both hydrophilic and lipophilic group in their structure and these are used to reduce the surface or interfacial tension that exist between air water or water and oil interface to enhance the solubility and stability of the compounds (Park and Choi 2006).

2.3 Strategies for Solubility Enhancement

There a large number of existing methodologies are used for improving solubility of poorly aqueous soluble drugs. The techniques for improving aqueous solubility are selected based on different factors such as various physiochemical properties related to drug, nature of excipients, and type of dosage form (Vemula et al. 2010). Some of the strategies for improving the water solubility of natural products were discussed in Table 2.1 with advantages and limitations.

Table 2.1 Methods of solubility enhancement of natural products

Method	Advantages	Limitations	References
Particle engineering (Micronization, Nano suspension and Nano crystals)	No need to alter the chemical structure to increase solubility of drug substances	Agglomeration due to developed charges	Sutradhar et al. (2013)
	Scale up feasibility	Decreased wetting	
Complexation	Enhanced solubility and stability	Not suitable for large dose drugs as it may increase bulk	Loftsson and Brewster (1996)
	Odor and taste masking		
	Reduction in irritation		
	Conversion to dry powders		
Solid dispersion	Improved wettability	Physical instability	Chiou and Riegelman (1971)
	Increased surface area rapid absorption and enhanced bioavailability		
Nanotechnology (Solid lipid nanoparticles, liposomes, polymeric micelles, SNEDDS)	Improved surface to volume ratio	Cost and stability	Müller et al. (2000a)
	Possibility of targeted delivery		
	Scale up feasibility		
	Reduction in side effects along with improved solubility		
Micellar solubilization	Thermodynamic stability	Use of higher surfactant concentration may be toxic	Rangel-Yagui et al. (2005)
	Easy and reproducible to produce in large scale		
	Provides longer circulation time		
	Also provides enhanced penetration due to use of surfactants		
	Possibility of site specific drug delivery		

2.3.1 Particle Engineering

Particle engineering mainly refers to obtaining the optimal particle size and size distribution, along with morphological and surface characteristics. This technique is mainly utilized to improve solubility, dissolution rate and thereby bioavailability (Blagden et al. 2007). Micronization or conventional size reduction with the help of high pressure homogenization or milling is an age old technique engaged in pharmaceutical industry for solubility and dissolution improvement. This augmented surface area as a result of particle size reduction will increase dissolution as per Noyes–Whitney equation (Khadka et al. 2014). But, it has very limited effect on solubility which is more dependent on solid state characteristics of drug substance.

However, reducing the size of particles further to nanometer range has showed improvement in solubility (especially kinetic solubility) as per Ostwald-Freundlich equation (Sun et al. 2012; Williams et al. 2013). Further, use of polymers and surfactants to stabilize the system will also help in improving the solubility. Nanosuspensions/nanocrystals are the well reported technologies in the literature for improving the solubility of many natural products such as curcumin (Gao et al. 2011), Zerumbone (rhizomes available from the Zinigiber zerumbet) (Md et al. 2018), resveratrol (Hao et al. 2014), quercetin (Sun et al. 2010).

Top down or bottom up approaches are used to prepare nanosuspension or nanocrystals. This was discussed in detail in section "Preparation of Nanosuspension".

2.3.2 Complexation

Complexation is one of the most constructive method for enhancing the aqueous solubility, dissolution and bioavailability of poorly soluble drugs. Complexation is an intermolecular association of substrate (drug) and ligand molecules or ions either by strong coordinate covalent bonds or by relatively weak non-covalent forces like hydrophobic interactions or dipole forces, hydrogen bonding, electrostatic interactions and Van der Waals forces. Complexes are diversified based on the type of substrate and ligand involved. However, inclusion complexes in which hydrophobic drug molecules (guest) are encapsulated or included into the hydrophobic internal cavity of host molecule gained much interest in pharmaceutical industry due to their ease of preparation, scalability and successful commercialization. Cyclodextrin, cyclosophoraoses, and chitooligosaccharides are the few examples of host molecules used in improving the solubility of natural products (Chaudhary et al. 2012; Kim et al. 2008; Kumar et al. 2013b).

2.3.2.1 Cyclodextrin Based Complexes

Cyclodextrins are the oldest compounds discovered (100 years ago) and most used material for improving the solubility, dissolution and thereby bioavailability of many hydrophobic drugs. The hydrophilic functional group on the surface makes them water soluble whereas hydrophobic cavity allows the incorporation of either compete structure or part of the structure of non-polar compound enabling the improvement in solubility (Singh et al. 2002). Cyclodextrins are majorly composed of monomers of glucose organized themselves in the shape like donut ring having a hydrophilic part outside and hydrophobic cavity inside as shown in Fig. 2.1.

Further, detailed discussion on the structural properties, types of cyclodextrin, modified derivatives of cyclodextrin and their relevant physicochemical properties along with applications in drug delivery were discussed by many authors (Challa et al. 2005; Loftsson and Duchêne 2007; Loftsson et al. 2005; Tiwari et al. 2010; Zhang and Ma 2013). There are about 35 products (either solids or liquid dosage form) in the market based on cyclodextrin complexation for solubility and bioavailability improvement throughout the world (Brewster and Loftsson 2007; Jansook et al. 2018). Modified cyclodextrins like hydroxypropyl derivatives of β cyclodextrin and γ cyclodextrin, methylated β-cyclodextrin, maltosyl β-cyclodextrin, and the sulfobutylether β cyclodextrin with improved physicochemical properties were also found to have applications in drug delivery (Brewster and Loftsson 2007).

Cyclodextrins also behave as permeation enhancers as they carry the drug molecules across the aqueous membrane barrier that exists before the non-aqueous biological membranes (Del Valle 2004; Loftsson and Brewster 1996; Loftsson and Duchene 2007; Stella and Rajewski 1992). The present chapter focuses on the preparation methods and discussion of few case studies with respect to natural products. Natural cyclodextrins like, α- cyclodextrin, β-cyclodextrin and γ-cyclodextrin were included in the generally regarded as safe (GRAS) list by FDA and other modified derivatives such as hydroxypropyl β- cyclodextrin and sulfobutylether β- cyclodextrin were included in the inactive ingredient database of FDA.

Various methodologies were reported in the literature for preparing the cyclodextrin complexes with drug molecules. Few techniques that are mostly cited in the literature and have commercial feasibility were discussed here.

Fig. 2.1 Structure of cyclodextrin

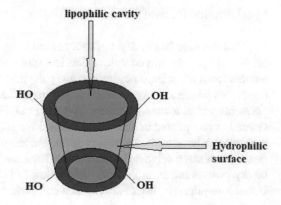

Physical Mixture or Blending Technique

A physical mixture of cyclodextrin and hydrophobic molecule can be prepared by simple mechanical mixing (Del Valle 2004).

Kneading Method

Kneading involves preparing a paste of cyclodextrin by macerating with small quantity of water or hydro alcoholic solutions. The hydrophobic drug is then added to the prepared paste and kneading is done for a specific period. The resultant mixture is dried and passed through a sieve if necessary to convert it into dry powder. At the laboratory scale, mortar pestle can be used for kneading. But at industrial scale, kneading is performed on extruders. Kneading is the most widely preferred and simple technique used for the preparation of inclusion complexes (Loftsson and Duchene 2007; Singh et al. 2010).

Microwave Irradiation Method

Microwave irradiation method can be used for industrial scalability purpose as the method is novel, requires smaller reaction times, and gives higher practical yield. In general, equimolar ratio of complexing agents and drug are taken in a round bottom flask containing water, organic solvent or mixture of both. This mixture is kept in microwave oven at 60 °C for the completion of reaction. After this, free drug and complexing agent are removed from the reaction mixture by using suitable solvent. The resulting product is separated by using Whatman filter paper and dried at 40 °C (Nacsa et al. 2008).

Lyophilization/Freeze-Drying Method

Lyophilization or freeze-drying is accounted as an appropriate method for the preparation of cyclodextrin complex. This technique allows the absolute amputation of solvent from the solution during primary drying. This method is more suitable for molecules which are thermo labile in nature as the drying takes place at lower temperatures and at lowered pressure (Savjani et al. 2012). This technique can be considered as an optional or alternative method for solvent evaporation. It also involves molecular mixing of the hydrophobic drug along with hydrophilic carrier i.e. cyclodextrin in a same solvent (Loftsson and Duchene 2007). The resultant product will be dry, porous and amorphous in nature. Some of the main limitation of this method is that it requires the use of specialized equipment and time-consuming process.

Supercritical Anti-solvent Precipitation Process

In this method, the drug and cyclodextrin are dissolved in a mixture of organic solvents depending on solubility. Then, supercritical CO_2 and the drug cyclodextrin solution sprayed into vessel simultaneously at a constant rate. During this, solvent will be extracted by supercritical CO_2 and complex will be precipitated as dry powder.

Borghetti GS et al., reported about 4.6 folds and 2.2 folds improvement in solubility of natural anti-oxidants such as quercetin after complexation with β-cyclodextrin by spray drying and physical mixing methods respectively. The authors also considered the effect of operating conditions on the complexation efficiency using factorial design. Quercetin- β-cyclodextrin complex was prepared in 1:1 molar ratios per phase solubility studies and the yield was found to be 77% in case of spray drying (Borghetti et al. 2009). Zhang et al. studied the impact of complexation on the water solubility of astilbin with α, β, and γ-cyclodextrin determined through phase solubility study and UV – visible spectral analysis. The formation constant (Ka) of astilbin complexes with all three cyclodextrin was calculated. They observed that, complexation with cyclodextrin increased its aqueous solubility. The solubility of astilbin in complex with β-cyclodextrin prepared by the freeze-drying technique was improved by 122.1 fold, and there is also improvement in their dissolution profile. The stability of astilbin complexes was found in increased order of α-cyclodextrin < γ-cyclodextrin < β-cyclodextrin (Zhang et al. 2012). In another study, Trollope L et al., showed 44 folds enhancement in solubility of resveratrol after complexation with hydroxypropyl β-cyclodextrin and 63-fold after complexation with randomly methylated β-cyclodextrin (Trollope et al. 2014).

Another study done by Upreti M et al., revealed significant improvement in the aqueous solubility of Steviol glycosides after complexation with γ- cyclodextrin. The solubility study indicated, the concentration of Baudioside (reb) A, C and D in water was observed as 1.43%, 2% and 4% in presence of cyclodextrin against 0.8% and 0.1–0.2% in absence of cyclodextrin respectively. Thermal and spectroscopic analysis confirmed the stable interaction between glycosides and cyclodextrin (Upreti et al. 2011).de Lima Petito et al., formulated red bell pepper extracted carotenoids into 2-hydroxypropyl-β-cyclodextrin complexes to improve the aqueous solubility of carotenoids and thereby their use in food. The pure carotenoid extract was completely insoluble in water whereas, carotenoid after complexation with cyclodextrin showed solubility of 8.00 ± 2.6 mg/mL for the sample 1:4 compared to 3.53 mg/mL with physical mixture in same ratio (de Lima Petito et al. 2016). Nanringenin shows greater potential in the treatment of hyperlipidemia but its efficacy is hampered owing to poor aqueous solubility. Shulman and others tried to enhance the aqueous solubility of Naringenin by preparing inclusion complex of naringenin with hydroxypropyl-β-cyclodextrin. The resulting complex showed about 400 times enhancement in the solubility of naringenin (Shulman et al. 2011).

2.3.2.2 Cyclosophoraoses Based Complexes

Cyclosophoraoses are the category of unbranched cyclic oligosaccharides produced by all the members belong to rhizobiaceae, mainly separated from Rhizobium leguminous arum biovartrifolii (Abe et al. 1982). Cyclosophoraoses comprises of unbranched cyclic β-(1,2)-D-glucans, having 17 to 25 ring size range in their degree of polymerization (Breedveld and Miller 1994). Initially, these are grown in periplasmic as well as extracellular media of fast-growing soil bacteria. Generally, cyclosophoraoses are synthesized in periplasmic space and transferred to the extracellular space. These play a key role in maintaining the osmolarity, in the formation of root nodules during nitrogen fixation and also provide attachment for bacterial cells to the host (Lee et al. 2004; Miller et al. 1986). Cyclosophoraoses gained much attention as they have ability to form inclusion complexes with a range of hydrophobic molecules. Besides, these molecules are also used to encapsulate a range of molecules and act as solubility enhancers for the poorly soluble molecules. They can be utilized in naive form or by modifying their functional groups such as succinyl, sulfonyl, methyl, butyryl, and carboxymethyl groups.

Kim et al. demonstrated enhancement in the aqueous solubility of atrazine by 3.69 times at 20 mM concentration of cyclosophoraoses after formation inclusion complex. They also suggested the potentiality of biological and environmental applications of cyclosophoraoses in removing very less soluble unhealthy materials from aqueous solutions due to presence of a large number of residues of glucose extracted from rhizobial species (Kim et al. 2019). Jeong, D et al., investigated the effect of complexation on the solubility of fisetin using novel cyclosophoroase dimer and compared with that of β-cyclodextrin fisetin complex. Cyclosophoraose dimer was synthesized by modification of cyclosophoraose produced from Rhizobium species. Nuclear magnetic resonance and Fourier-transform infrared spectroscopy studies confirmed the structural modification was successful. Solubility studies indicated, about 6.5-fold raise in solubility of fisetin compared to pure drug and about 2.4 fold improvement compared to β-cyclodextrin fisetin complex. Fisetin cyclosophoraose complex also showed higher cytotoxicity against HeLa cells compared to its counterpart indicating enhanced bioavailability. The authors concluded that, the synthesized cyclosophoraose dimer can become an effective alternate to cyclodextrin for the successful delivery of fisetin (Jeong et al. 2013). Piao et al. also explored the effect of hydroxypropyl cyclosophoraoses complex on the solubility of α-naphthoflavone and suggested that, hydroxypropyl cyclosophoraose act as a novel host carrier with the ability to potentially interact with different polyaromatic compounds and also with α-naphthoflavone in improving the solubility (Piao et al. 2014). Luteolin is a bio flavonoid exuded from alfalfa reported to possess anti-cancer effects. However, luteolin use was precluded by its low water solubility. The solubility issue of this compound was addressed by making complex with cyclosophoraoses oligosaccharide by Sanghoo Lee and others (Lee et al. 2003).

The cyclosophoraoses complexes were prepared by solvent evaporation method similar to cyclodextrin complexation.

2.3.2.3 Chitooligosaccharide Based Complexes

Chitooligosaccharide, also called as chitosan oligomers, is obtained by either chemical or enzymatic hydrolysis of chitosan. Readers can refer to the literature for detailed description of various synthetic procedures for preparation of chitooligosaccharide from chitin (Aam et al. 2010; Lodhi et al. 2014). Compared to chitosan, these oligosaccharides have less molecular weight, low viscosity, and better solubility that offers the advantage of complex development with natural products for their solubility enhancement (Xia et al. 2011).

Zhou and co-workers prepared a complex of hesperidin with chitooligosaccharides by spray-drying method. The prepared complex showed an interaction between aromatic group of the hesperidin and chitooligosaccharide via hydrogen bond. Because of such interaction between hesperidin and host molecule, the resulting complex showed significant improvement in aqueous solubility (3.8 fold at 1:10 ratio) and better antioxidant activity than native hesperidin. The authors also reported that, the solubility was increased with increasing carrier to drug ratio (Cao et al. 2018).

Cao, R et al., studied about the complexation of naringin with chitooligosaccharide at different mole ratios (1:1, 1:5, 1:10) and also evaluated the effect of these complexes on the solubility of naringin. Complexes were prepared by spray drying method and characterized by scanning electron microscopy, ^{1}H nuclear magnetic resonance (NMR), fourier-transform infrared spectroscopy analysis (FT-IR). Further, the authors also determined their solubility, antioxidant activity and antibacterial property in comparison to pure naringin. FT-IR and NMR studies confirmed the formation of complexes and involvement of hydrogen bond between naringin and chitooligosaccharide. The complex showed significant improvement in solubility and antioxidant activity with superior anti-bacterial activity (Cao et al. 2019).

2.3.2.4 Phospholipid Complex

Phospholipid complexation results in formation of hydrogen bond between drug and phospholipid. The resulting complex will self-assemble in aqueous media. The amphiphilic nature of the phospholipid helps in enhancing the solubility of drug molecule in both aqueous and non-aqueous media leading to improved solubility, enhanced permeation and thereby improvement in bioavailability (Loftsson 2017). Phospholipids are amphiphilic molecules derived from soyabean, milk, egg yolk and marine organisms and comprise of a glycerol backbone esterified with fatty acids at 1 and 2 position and with phosphate in position 3 (van Hoogevest 2017; Zhou and Rakariyatham 2019). Based on the backbone present these are classified into Glycero phospholipids, Sphingo phospholipids and sterols (Lu et al. 2019; Singh et al. 2017). Phospholipid complexes are prepared in different stoichiometric ratios of drug (natural product) to lipid (1:1, 1:2 and 1:3). These are prepared using solvent evaporation, co-grinding method, mechanical dispersion, super critical fluid process, co-solvent lyophilization and anti-solvent precipitation (Kuche et al. 2019).

Zhang, K et al., studied the effect of phospholipid complexation on the aqueous solubility, bioavailability and hepato protective activity of quercetin. The authors prepared quercetin-phospholipid complex and analyzed by Fourier-transform infrared spectroscopy, Differential scanning calorimetry, X-ray diffraction and water/n-octanol solubility. Oral bioavailability and hepatoprotective activity was studied in SD rats. The resulting quercetin-phospholipid complex showed about 13 folds improvement in aqueous solubility and 1.6 folds enhancement in n-octanol solubility (Zhang et al. 2016a). Saoji et al., showed about 12 folds improvement in the solubility of standardized Centella extract compared to pure extract and physical mixture of extract and phospholipid. The complexation also resulted in significant increase in dissolution and permeability that resulted in improved efficacy of the extract as tested in animal models (Saoji et al. 2016). Li et al., reported enhancement in the solubility of ferulic acid using phospholipid complexation. Complex was prepared in different ratios of drug to phospholipid (2:1, 1:1, 1:2, 1:3, and 1:4) using Soy lecithin by rotaevaporation method. The resulting complex showed significant improvement in solubility of ferulic acid from complex (1.68 ± 0.01 mg/ml) to the pure drug (0.67 ± 0.02 mg/ml). The enhanced solubility resulted in improved bioavailability and cellular melanogenesis inhibition activity (Li et al. 2017).

2.3.3 Solid Dispersion

Solid dispersion is one of the renowned systems for the improvement of solubility, dissolution and thus bioavailability of hydrophobic drugs. Natural products being poorly soluble in aqueous solvent are less bioavailable also. Thus, solid dispersion may serve as one of the approach to augment the water solubility, dissolution and bioavailability of natural actives with poor aqueous solubility (Kumar et al. 2013a). Solid dispersion can be described as "the dispersion system containing single or more active ingredients in an inert matrix of carrier that can be prepared by the melting or fusion method, solvent or melting-solvent method" (Aggarwal et al. 2010; Dhirendra et al. 2009). Formulating solid dispersions for natural products with poor aqueous-solubility reduces the issue related to solubility and increases their bioavailability also. Solid dispersions majorly increases the dissolution rate and few carriers may improve the solubility which may be ascribed to size reduction, superior wetting, change in drug crystallinity and presence of water soluble carriers (Craig 2002).

2.3.3.1 Methods of Preparation of Solid Dispersion

Various methods like fusion method, kneading, spray drying, melt extrusion, solvent evaporation; lyophilization and co-precipitation were reported in the literature for preparing solid dispersions. Only methods which have commercial application such as spray drying and co-precipitation and hot melt extrusion (Huang and Dai 2014) were discussed in the following section.

Hot Melt Extrusion

Hot melt extrusion technology is a well-established technology used in the food and plastic industry since 1930s. Recently, it found application in pharmaceutical industry to prepare dispersions. Essential elements of hot melt extruder include barrels that enclose single or twin screws for the transport of material through a die. This barrel is subjected to heating to a temperature which is equivalent to the melting point of drug and carrier. Due to the heat and force applied on the material by the screws, the polymer will be plasticized and the drug will be molecularly dispersed or dissolved depending on the nature and extruded from die cavity. The molten mass will be cooled to room temperature, crushed, sieved and stored in desiccators for further use (Shah et al. 2013).

Spray Drying

Spray-drying is considered as one of the most frequently employed as a solvent evaporation technique to formulate solid dispersions. It includes dissolving of the drug substance and carrier material, followed by spraying it into spraying chamber with a stream of heated air flow to evaporate the solvent. Spray drying completes in three steps namely, atomization, drying, and collection of the powder. During atomization, a fine mist with a large surface area is sprayed into a spraying chamber. Heat is transferred and immediate removal of the solvent occurs because of the formation of fine droplets. After atomization of liquid, the droplets enter in the drying chamber in which hot air evaporates the solvent. After drying, the air stream carries particles out of the drying chamber and into a separation device. Centrifugal force is used in a cyclone in order to separate powder from the air stream and the resultant powder product is collected in the collector (Arunachalam et al. 2010; Baghel et al. 2016).

Co-precipitation

In co-precipitation method, drug and polymer are dissolved in a common solvent and vortexed if necessary to get a transparent solution. Then, anti-solvent (mostly water) was added drop wise to the organic solvent to induce precipitation. The precipitate was filtered, dried and stored in dessicator (Tran et al. 2019).

Kanaze and coworkers developed a solid dispersion system for flavonone aglycone drugs naringenin and hesperetin by solvent evaporation method using polyvinylpyrrolidone as carrier material. The prepared dispersions of polyvinylpyrollidone and naringenin-hesperedin (80/20 w/w) were compressed into tablets and filled as such in capsules for further use. These are evaluated for drug release and their stability under accelerated condition (45 °C and 75% RH) for 3 months. The authors reported faster release of drug from capsules compared to tablets and both the formulations are stable up to 3 months without any precipitation (Kanaze et al. 2010). Teixeira and coworkers prepared ternary solid dispersions of curcumin containing

Aerosil®-Gelucire®50/13 using spray drying method and studied the impact on the solubility and oral bioavailability of curcumin. The results shown that there was an approximately 3600 times enhancement in aqueous solubility of curcumin. Furthermore, the anti-inflammatory activity due to increase in gastrointestinal absorption (Teixeira et al. 2016). Khan et al., prepared the solid dispersions of naringenin by solvent evaporation and kneading methods with different carriers such as soluplus and inulin in various ratios of drug to the carrier (1:4, 3:7, 2:3 and 1:1). The aqueous solubility of naringenin was increased significantly from solid dispersion compared to pure drug (390 and 270 times with soluplus and inulin respectively). *In vitro* drug release indicated complete release of drug from SDs in 2 h compared to only 15% release at the end of 8 h. *In vivo* pharmacokinetic studies also indicated significant improvement in area under the curve compared to pure drug (Khan et al. 2015a). Other natural compounds reported in literature, whose solubility was enhanced using solid dispersion technology include ellagic acid (Li et al. 2013a), quercetin (Li et al. 2013b) and curcumin (Li et al. 2013c; Seo et al. 2012).

2.3.4 Nanotechnology

Nanotechnology is being used to develop nano-based system for the delivery of various therapeutic compounds or drugs for medicinal purposes. Therapy based on the nanoparticulate drug delivery system is widely explored in the management of many ailments by modifying pharmacokinetic and pharmacodynamic properties. Based on the type of material used, nanoparticulate drug delivery system has been differentiated into polymer based, lipid based and hybrid drug delivery system. Nanotechnology can be used for the problem associated with some natural compounds like poor aqueous solubility and bioavailability (Ansari and Farha Islam 2012; Bhadoriya et al. 2011; Watkins et al. 2015). Nanoparticles are described as the dispersions of particulate or solid particles possess size range of 10–100 nm. The drug in such a system may be entrapped, encapsulated, dissolved, or adsorbed on to nanoparticle matrix. Different strategies can be used through which nanoparticles for poorly aqueous soluble molecules can be formulated (Gunasekaran et al. 2014).

The major advantages of nanoparticulate drug delivery system include:

- Increased effective surface area with reduced particle size, thus higher solubility and dissolution.
- Increased chemical or metabolic stability.
- Increased penetration, permeability and thus promotes targeted delivery.

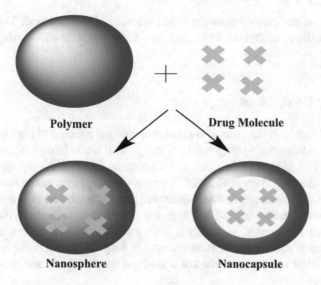

Fig. 2.2 Polymer based nanoparticles nanocapsules and nanospheres

2.3.4.1 Nano Particles

Based upon the type of material used as delivery vehicle in preparation of nanoparticles, they are generally categorized into the following classes.

(a) Polymer nanoparticles
(b) Lipid nanoparticles

Polymer-Based Nanoparticles

These are colloidal particles of polymer having submicron size into which, the drug is completely encapsulated in their matrix or adsorbed or conjugated on polymeric surface. They can be further classified as nanocapsules and nanospheres on the basis of the method used for preparation. Nanocapsules are submicron ranged nanoparticles in which the drug molecule is restricted to a cavity which surrounded by a membrane of polymer, whereas, in nanospheres the drug is uniformly dispersed or distributed within the polymeric matrix system (Fig. 2.2). The polymeric nanoparticles are used for their targeted delivery, enhancement of stability and solubility of poorly water soluble molecules (Jawahar and Meyyanathan 2012; Kumari et al. 2010b). Anwer et al., and coworkers developed PLGA based nanoparticulate system of quercetin, a flavonoid in order to enhance the aqueous solubility and therapeutic effects. The data suggested significant improvement in solubility of quercetin using PLGA nanoparticles and concluded that, these can be used for quercetin delivery (Anwer et al. 2016). Lee et al., improved water solubility of milk thistle silymarin by 7.7-folds using nanocapsules. These nanocapsules were prepared using chitosan and poly-γ-glutamic acid by ionic gelation method. The improved solubility further resulted in enhanced anti-microbial activity of silymarin (Lee et al. 2017).

Lipid Based Nanoparticles

Lipid based nanoparticles are nano-sized structure comprising of core having lipid. Lipid nanoparticles can be prepared by mixing an oil phase into phospholipids as an emulsifier. The oily core is used to incorporate hydrophobic drugs and lipid-conjugated prodrugs (Puri et al. 2009). Hydrophobic drugs are easily loaded into the lipid and after administration these lipids will cause stimulation of bile acids which will act as surfactants and aids in dispersing poorly soluble drugs in physiological fluid resulting in improved solubility (Chella and Shastri 2017; Feeney et al. 2016). Various types of lipid-based carriers are used in drug delivery. Among them, widely used lipid-based carriers systems are solid-lipid nanoparticles and nanostructured lipid carriers.

Solid Lipid Nanoparticles

Solid lipid nanoparticles are an attractive and alternative delivery system to conventional colloidal carriers like liposomes and conventional emulsion based system. Solid lipid nanoparticles are defined as an aqueous colloidal dispersion of particles composed of solid lipids, that are stabilized, in aqueous dispersion by surfactants or polymers (Müller et al. 2000b). These formulation mainly contains solid lipid content (about 0.1–30%) as base material i.e. lipid such as triglycerides, mixtures of complex glyceride or waxes, surfactant or stabilizers such as polysorbate, lecithin, bile salts, polyvinyl alcohol and the drug substances to be incorporated. These may be prepared by different methods which majorly include micro emulsion, solvent evaporation, high-speed homogenization, ultra-sonication, phase inversion, solvent injection and membrane contractor technique (Mehnert and Mäder 2012; Shah et al. 2019) (Fig. 2.3).

Nanostructured Lipid Carriers

Nanostructured lipid carriers are developed to deal with shortcomings of solid lipid nanoparticles. Their preparation can be done by the mixing of both solid lipid as well as liquid lipids. Nanostructured lipid carriers also offers some advantages over the other polymeric nanoparticles that includes reduced toxicity, increased stability of drugs, biodegradability, release of drugs in controlled manner, and avoidance of use of organic solvents during formulation. Furthermore, nanostructured lipid carriers increase the drug loading capacity when compared with solid lipid nanoparticles. These lipid carriers have been used as delivery carriers for hydrophilic and hydrophobic drugs and also for natural products (Tamjidi et al. 2013). Various

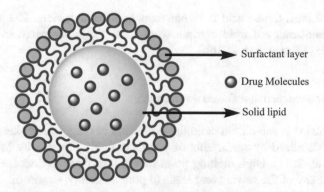

Fig. 2.3 Solid lipid nanoparticles

Fig. 2.4 Nanostructured
lipid carrier

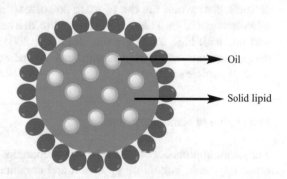

methods used to prepare of nanostructured lipid carriers are mainly high-pressure/shear homogenization, micro emulsion, emulsification sonification, solvent diffusion, phase inversion, solvent evaporation, solvent injection membrane contractor, solvent displacement method (Salvi and Pawar 2019). High-pressure homogenization through which scale up is possible is the widely used method to prepare nanostructured lipid carriers (Fig. 2.4).

Fang and coworkers prepared solid lipid nanoparticles, nanostructured lipid carriers and lipid emulsions of tryptanthrin (a lipophilic alkaloid) to overcome the insolubility issue and improve the targeting. Out of all the lipid nanoparticles, nanostructured lipid carriers produced smaller particles with high partitioning efficiency due to amorphous core. All the lipid particles showed sustained release with zero order kinetics (Fang et al. 2011). Zhang, Y et al., prepared nanostructured lipid carrier of trans-ferulic acid and compared its oral bioavailability with solid lipid nanoparticles. Trans-ferulic acid is phenolic acid with antioxidant, anti-inflammatory and also cardio protective effects. It has very poor solubility resulting in poor oral bioavailability. Nanostructured lipid carriers were prepared using glyceryl behenate as solid lipid and ethyl oleate as liquid lipid by micro emulsion method. Significant improvement in bioavailability of trans-ferulic acid from nanostructured lipid carriers was observed and the authors attributed the improvement was due to increased solubility of trans-ferulic acid from nanostructured lipid carriers. The authors also reported nanostructured lipid carriers were more stable compared to solid lipid nanoparticles (Zhang et al. 2016b).

2.3.4.2 Nanosuspension Based Approach

Nanosuspension is submicron colloidal dispersion of drug having size in the nano range and stabilized by surfactants or polymers firstly reported by Muller et al., (Müller et al. 2011). High melting point makes it difficult to dissolve the crystal structure in any of the solvent and leads to poor solubility, slower dissolution and thereby poor oral bioavailability. Nanosuspension offers various advantages such as higher drug loading (can be used to drugs with high dose), use of low concentrations of surfactant, avoids the use of toxic co-solvents and has scale up feasibility over other nanoparticles. These are also used to deliver insoluble drugs (both in water and oil) with high melting point (Rabinow 2004). Enhancement in solubility of drugs from nanosuspension is attributed to increased surface area as a result of reduced particle size.

———

Preparation of Nanosuspension

The major approaches employed in the preparation of nanosuspension include top-down approach, bottom-up approach and combination technologies. In top-down approach, starting material having larger size are reduced to nanoscale size. Top down approach technology uses either high energy milling like high-pressure homogenization or low energy milling technology like media milling. The bottom-up approach includes self-association of small molecules. These technologies generally include precipitation techniques like anti-solvent precipitation, precipitation under sonication and flash nanoprecipitation method. Combined technologies uses bottom up approach such as precipitation followed by top-down process such as homogenization (Patel and Agrawal 2011; Zhang et al. 2017). There are various other technologies also mentioned in the literature including supercritical fluid technology, emulsion diffusion method emulsification-solvent evaporation technique, melt emulsification method, and Nanojet technology (Ahire et al. 2018; S. Pawar et al. 2017).

Precipitation Method (Bottom-Up Approach)

In precipitation method, the material i.e. active pharmaceutical ingredient is primarily dissolved in suitable solvent to prepare a solution followed by addition of this mixed solution with an anti-solvent along with sufficient quantity of surfactant. Progressive mixing of this drug solution to the anti-solvent leads to supersaturation of solution to form amorphous nanoparticles (Kocbek et al. 2006; Verma et al. 2009).

Homogenization (Top-Down Approach)

High-pressure homogenization is generally carried out with piston-gap homogenizers. Homogenization may be done in aqueous media known as dissocubes technique as well in non-aqueous media also called as Nano pure technology. In aqueous media, bioactive material, water, and surfactants are mixed to prepare coarse suspension. This mixture of coarse suspension is then pushed with small piston through a homogenization gap. The particles having nano size diameter are produced by either shear forces, vacuum forces or may be because of collision between particles. In non-aqueous media, the homogenization is performed using same piston gap in very less amount of water or even completely free from water (Verma et al. 2009).

Media Milling Technique (Top-Down Approach)

In this technique, the particles with nanosize are produced due to collision between excipients and milling media. The size of particles can be reduced by high shear forces (Zhang et al. 2014).

Shen G et al., reported significant improvement in the solubility of herpetospermum caudigerum lignans using nanosupension technology. Herpetrione and herpetin are the most profuse ingredients having anti-hepatitis B virus and hepatoprotective effects. However, they found less clinical application due to their low bioavailability as a result of poor aqueous solubility. Their nanosuspension is prepared by using combination of precipitation and homogenization technology. The resulting nanosuspension showed particle size of 286 nanometers with PDI of 0.215 and solubility was improved by 4–5 folds compared to pure drug at pH 1.2 and 7.4. The authors concluded that, this nanosuspension can be helpful tool to improve the solubility, dissolution rate and thereby bioavailability of poorly soluble drugs with pH dependent solubility (Shen et al. 2016). Improvement in the aqueous solubility of quercetin was observed by karadag and others with the help of nanosuspension prepared using high pressure homogenization combined with spray drying technique (Karadag et al. 2014).

2.3.5 Emulsion-Based Approach

Emulsions are considered as biocompatible carriers that allow the encapsulation of various lipophilic molecules into the oil droplets. This carrier system provides platform to improve the delivery of lipophilic drugs by solubilizing the drug molecule, and by reducing toxicity. Oil in water emulsion can entrap the high amount of lipophilic drugs with high efficiency. Moreover, emulsions can be modified further to improve their specificity and targeting ability. The emulsion-based delivery can further classified into nanoemulsion, micro emulsion, and self-emulsifying delivery system (McClements et al. 2007).

2.3.5.1   Nano Emulsions

For poorly water-soluble lipophilic compounds delivery through nanoemulsion can be a potential strategy to enhance their aqueous solubility (Chen et al. 2011). Nanoemulsion comprising of an oil phase and aqueous phase with some stabilizers also as known as emulsifiers is proven as a promising carrier for the delivery of poorly soluble drugs. The mean droplet size of nano emulsion ranges from 50–100 nm, which generally exist either as oil-in-water form or water-in-oil. Decreased particle size promotes more surface area of particles and thereby improves the aqueous solubility and stability of the incorporated drug. Nanoemulsions can be formulated by employing low energy techniques or high energy techniques (Koroleva and Yurtov 2012).

Li et al. developed a nanoemulsion system using high-pressure homogenization method for polymethoxyflavones to increase solubilization capacity (Li et al. 2012). In order to make soluble formulations of curcumin with improved solubility, bio-availability and photostability, nanoemulsion of curcumin were formulated by Onoue and coworkers (Onoue et al. 2010). In comparison with naturally occurring curcumin, nanoemulsion greatly improved the aqueous solubility. It was found that approximately nine fold enhancement of bioavailability in plasma following oral administration in the rat model when studied. Yen C-C et al., reported improvement in solubility, anti-inflammatory activity and oral bioavailability of andrographolide from nanoemulsion composed of α-tocopherol, cremophor EL, ethanol, and water. Nanoemulsion formulations markedly increased the solubility and permeability of andrographolide with sixfolds improvement in bioavailability. Further, the prepared nanoemulsions also showed stronger anti-inflammatory effect than pure drug and respective AG suspension (Yen et al. 2018).

2.3.5.2   Self-Emulsifying Drug Delivery Systems

Self-emulsifying drug delivery systems is an attractive tool for increasing aqueous solubility of low water soluble drugs. This system has an capability to disperse and form spontaneous colloidal structures when exposed to gastrointestinal fluids. This system comprises a mixture of drug, oil phase, surfactant, and co-surfactant (Sapra et al. 2012). This can be differentiated based on droplet size into two major categories such as self-nano emulsifying drug delivery systems having droplet size in between 100–200 nm and Self-micro emulsifying drug delivery systems having less than 100 nm droplet size (Pouton 1997). Another modification of this system is solidified self-emulsifying drug delivery systems where the liquid material is adsorbed onto carriers for improved stability.

Quercetin, a flavonoid exhibit potential chemo protective action but has low bio-availability due to poor water solubility. Tran and coworkers tried to enhance the solubility of quercetin by developing this delivery system using Cremophore RH 40, castor oil and Tween 80. The data obtained suggested that this formulation can improve the solubility and oral bioavailability of quercetin (Tran et al. 2014). Other natural products whose solubility was improved using self-emulsifying drug delivery systems technology include resveratrol (Balata et al. 2016), naringenin (Khan et al. 2015b), curcumin (Dhumal et al. 2015).

2.3.6   Liposomes Based Approach

Liposomes are vesicles having spherical geometry, mainly composed of one or more phospholipid bilayer membrane that surrounds interior aqueous core. Phospholipid molecules possess a hydrophilic head having affinity for aqueous phases, and a hydrophobic tail end (Brandl 2001). A liposome is a promising carrier to encapsulate both hydrophilic and lipophilic compounds. Liposome exerts large number of applications in drug delivery, cosmetics, and improving the efficacy of natural products. The liposome offers great advantages like improved solubility of drugs, improved stability, controlled and targeted drug delivery, reduced toxicity induced by drug, increased circulation time, and improved therapeutic efficacy (Akbarzadeh et al. 2013) (Fig. 2.5).

Perillaldehyde, is a bioactive compound obtained from Perilla frutescens. However, the potentiality of this compound was reduced due to poor aqueous solubility and delivery to non-specific site. Liposomal preparation of perillaldehyde results in the significant improvement in its aqueous solubility as well relative bioavailability when compared with native free form (Omari Siaw et al. 2016). Resveratrol, potential natural active compounds possess limited therapeutic index because of low aqueous solubility and inadequate bioavailability. Joraholmen and co-workers prepared liposomal formulations of resveratrol coated with chitosan to overcome all these shortcomings and to get better delivery of resveratrol from topical formulation (Jøraholmen et al. 2015).

Coimbra and others studied various natural compounds namely caffeic acid (derivatives), pterostilbene (derivatives), carvacrol (derivatives), N-(3-oxo-dodecanoyl)-l-homoserine lactone and thymol for their anti-inflammatory activity but they are poorly soluble and possess less stability. They encapsulated these

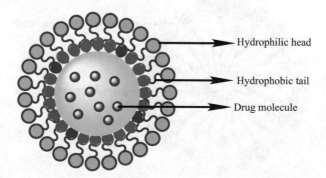

Fig. 2.5 Typical structure of liposomes

natural compounds into liposomes and studied the effect of liposomal formulation on their solubility and stability. Results have shown that these low soluble drugs can be incorporated into liposomes and their solubility can be increased (Coimbra et al. 2011).

2.3.7 Micellar Solubilization

Micelles are spontaneously formed colloidal carriers in nano size range (20–80 nm) composed of hydrophobic core and hydrophilic shell. These are made up of amphiphilic molecules with the unique self-assembling ability and capacity to load hydrophilic and hydrophobic molecules for a wide variety of therapeutic applications in diagnostics and drug delivery (Rangel-Yagui et al. 2005). Formulating micelles can be used as a powerful strategy to improve the solubility and stability of the drugs and other natural compounds. These can be called as polymeric micelle, mixed micelle and lipid –polymer hybrid micelle based on the nature of carrier materials used in the formulation such as polymers, polysaccharides, surfactants are and lipids either alone or in combination respectively. Micelles can be formed in water solution when the concentration of these copolymer reaches more than the critical micelle concentration (Adams et al. 2003) (Fig. 2.6).

Micelles are being used in the delivery of drugs as nanocarrier system for the solubility improvement of several low soluble drugs (Chen et al. 2011). The augmentation in aqueous solubility of quercetin was achieved by formulating nanomicelles with PEGylated lipid (Xu et al. 2015). The bioavailability and stability of vitamin D has been increased developing re-assembled casein micelles encapsulating vitamin D using ultrahigh-pressure homogenization method (Haham et al. 2012). Casein micelles were also used to solubilize the curcumin and used as a nanocarrier for targeting cancer cell in drug delivery resulting in improved aqueous solubility and stability of native curcumin (Esmaili et al. 2011; Sahu et al. 2008) (Table 2.2).

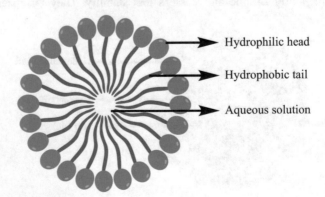

Fig. 2.6 Micellar structure

Table 2.2 List of few natural products along with solubility improvement methods

S. no.	Drug/natural product	Type of carriers	References
1	Curcumin	Chitosan and alginate based polymeric nanocapsules	Hernandez-Patlan et al. (2019)
2	Taxifolin	Polymeric nanoparticles	Zu et al. (2014)
3	Quercetin	Poly-D,L-Lactide nanoparticles	Kumari et al. (2010a)
4	Reseveratrol	Solid lipid nanoparticles	Summerlin et al. (2015)
5	Quercetin	Lipid nanocapsule	Barras et al. (2009)
6	Piperine	Selfemulsifying drug delivery system	Shao et al. (2015)
7	Ferulic acid	Liposome	Qin et al. (2008)
8	Boswellic acids	Selfemulsifying drug delivery system	Anshuman Bhardwaj et al. (2016)
9	Resveratrol	Micelles	Atanacković et al. (2009)
10	Curcumin	Nanocrystal	Rachmawati et al. (2013)
11	Apigenin	Nanocrystal	Al Shaal et al. (2011)
12	Hesperetin	Nanoparticles	Kaur and Kaur (2014)
13	Genistein	Micelles	Kwon et al. (2007)
14	Zerumbone	Nanosuspension	Md et al. (2018)
15	Piperine	Nanosuspension	Zafar et al. (2019)
16	Silymarin	Nanosuspension	Jahan et al. (2016)

2.3.8   Other Technologies

2.3.8.1   Co-solvency

The water solubility of poorly soluble natural bioactive compounds can be enhanced with addition of some solvents that are miscible with water. These solvents are termed as co-solvents. The most common liquids that are used as co-solvents include ethanol, glycofural, glycerin, and (PEG 400) polyethylene glycols. The co-solvent system works based on decreasing the interfacial tension that exist in between the non-polar solute and polar solvent. Many co-solvents are either hydrogen bond acceptor or donor. Their water soluble groups promote water miscibility, whereas hydrophobic groups help in reducing intermolecular attraction of water molecule. The mechanism of improving solubility is the disruption the self-association of water molecules thereby reducing ability of water molecules to remove out non-polar and hydrophobic substances. The solubility of ginger bioactive compounds was improved by addition of 20–50% ethanol as co-solvent (Mulligan et al. 2001).

2.3.8.2   Hydrotrophy

Hydrotrophy utilizes the addition of a large quantity of some derivative solute that ultimately improves the solubility of the less water soluble compounds. In hydrotrophy technique, the solubility of the poorly soluble substance can be improved by using various hydrotropic agents like urea, sodium citrate, sodium benzoate. In hydrotrophy, use of an organic solvent can be diminished by alternative use of hydrotropic. Hydrotropic technique is one of the most potential solubilization methods used for the formulations of poorly aqueous-soluble drugs (Jain et al. 2010). The extraction of limonin, a bioactive compounds that belongs to limonoids obtained from sour orange was done by using sodium salicylate and sodium cumene sulphonate used as hydrotrophic agents (Dandekar et al. 2008; Mangal et al. 2011).

2.3.8.3 Bioconjugation Technique

Bioconjugation is the covalent linkage between two molecules at their side ends. This strategy is also used to improve solubility, stability, and biocompatibility of proteins, peptides and enzymes as well. Water soluble or amphilphilic molecules such as polyethylene glycol, hyaluronic acid and others are being used to overcome the problem of poor solubility of various natural products.

Bioconjugation with Polyethylene Glycol or PEGylation

Hydrophilic polymers like polyethylene glycol are being used to enhance solubility of various poorly soluble compounds. Polyethylene glycol has been conjugated to taxol, which is a potent chemotherapeutic agent employed for the treatment of breast and ovarian cancer. The conjugate has shown approximately six times more solubility than that of the parent drug without losing the taxol activity (Greenwald et al. 1996). Another example of conjugation with polyethylene glycol is camptothecin, which is an active topoisomerase-1 inhibitor increased solubility with better cell uptake and diminished efflux activity (Yu et al. 2005). Dihydroartemisinin, a derivative of artemisinin demonstrates anti-cancer activity. However, its clinical potential is hampered by low solubility and rapid metabolism. The conjugation of dihydroartemisinin with polyethylene glycol significantly improved its solubility by approximately 160 folds compared to pure drug (Dai et al. 2014). Other natural products which are conjugated to polyethylene glycol for improved solubility include curcumin (Kim et al. 2011), camptothecin and Tacrolimus (Li et al. 2013d) (Fig. 2.7).

Bioconjugation with Hyaluronic Acid

Hyluronic acid, a natural polysaccharides can also be used as conjugating agent with many natural compounds to increase their solubility as well as stability. Solubility of curcumin was significantly improved by making bioconjugate with hyaluronic acid. This conjugate improved solubility upto 7.5 mg/ml when compared with free drug with 0.1 mg/ml solubility (Hani and Shivakumar 2014). Another natural products which is conjugated to hyaluronic acid include dihydroartemisinin (Kumar et al. 2019) for improved solubility.

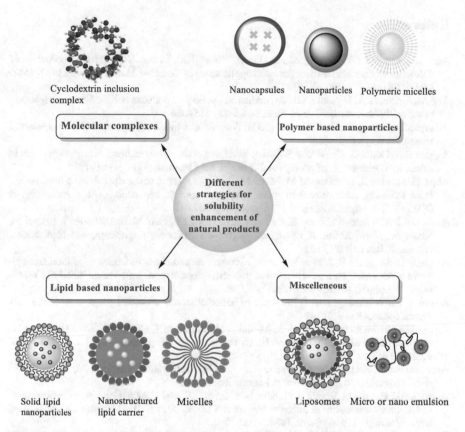

Fig. 2.7 Schematic representations of different strategies of solubility enhancement of natural products

2.4 Conclusion

Natural products are a great source of therapeutics and are being used as alternative medicines since ancient times. These products have been employed for treating the variety of human ailments as they possess many health promoting outcomes. However, there clinical application is limited by their poor aqueous solubility, poor stability followed by poor oral bioavailability. Many formulation strategies including particle engineering, complex formation, dispersion based techniques and nanoparticulate carriers that can improve solubility, stability and hence bioavailability was discussed in the present chapter. Each technique as their own pros and cons. Careful selection of proper formulation strategy based on their physicochemical properties, enhancement of solubility required and final dose and dosage form required will leads to optimum results. Development of suitable delivery system not only improves its efficacy but a delivery system with good scale up feasibility will also lead to increased use of natural products for the benefit of mankind.

References

Aam BB, Heggset EB, Norberg AL, Sørlie M, Vårum KM, Eijsink VGH (2010) Production of chitooligosaccharides and their potential applications in medicine. Mar Drugs 8(5):1482–1517. https://doi.org/10.3390/md8051482

Abe M, Amemura A, Higashi S (1982) Studies on cyclic β-1, 2-glucan obtained from periplasmic space ofRhizobium trifolii cells. Plant Soil 64(3):315–324

Adams ML, Lavasanifar A, Kwon GS (2003) Amphiphilic block copolymers for drug delivery. J Pharm Sci 92(7):1343–1355

Aggarwal S, Gupta G, Chaudhary S (2010) Solid dispersion as an eminent strategic approach in solubility enhancement of poorly soluble drugs. Int J Pharm Sci Res 1(8):1–13

Ahire E, Thakkar S, Darshanwad M, Misra M (2018) Parenteral nanosuspensions: a brief review from solubility enhancement to more novel and specific applications. Acta Pharm Sin B 8(5):733–755. https://doi.org/10.1016/j.apsb.2018.07.011

Akbarzadeh A, Rezaei-Sadabady R, Davaran S, Joo SW, Zarghami N, Hanifehpour Y, Samiei M, Kouhi M, Nejati-Koshki K (2013) Liposome: classification, preparation, and applications. Nanoscale Res Lett 8(1):102

Al Shaal L, Shegokar R, Müller RH (2011) Production and characterization of antioxidant apigenin nanocrystals as a novel UV skin protective formulation. Int J Pharm 420(1):133–140. https://doi.org/10.1016/j.ijpharm.2011.08.018

Ansari S, Farha Islam M (2012) Influence of nanotechnology on herbal drugs: a review. J Adv Pharm Technol Res 3(3):142

Anwer MK, Al-Mansoor MA, Jamil S, Al-Shdefat R, Ansari MN, Shakeel F (2016) Development and evaluation of PLGA polymer based nanoparticles of quercetin. Int J Biol Macromol 92:213–219

Arunachalam A, Karthikeyan M, Konam K, Prasad PH, Sethuraman S, Ashutoshkumar S (2010) Solid dispersions: a review. Current Pharma Res 1(1):82

Atanacković M, Poša M, Heinle H, Gojković-Bukarica L, Cvejić J (2009) Solubilization of resveratrol in micellar solutions of different bile acids. Colloids Surf B: Biointerfaces 72(1):148–154. https://doi.org/10.1016/j.colsurfb.2009.03.029

Baghel S, Cathcart H, O'Reilly NJ (2016) Polymeric amorphous solid dispersions: a review of amorphization, crystallization, stabilization, solid-state characterization, and aqueous solubilization of biopharmaceutical classification system class II drugs. J Pharm Sci 105(9):2527–2544

Balata GF, Essa EA, Shamardl HA, Zaidan SH, Abourehab MA (2016) Self-emulsifying drug delivery systems as a tool to improve solubility and bioavailability of resveratrol. Drug Des Devel Ther 10:117–128. https://doi.org/10.2147/DDDT.S95905

Barras A, Mezzetti A, Richard A, Lazzaroni S, Roux S, Melnyk P, Betbeder D, Monfilliette-Dupont N (2009) Formulation and characterization of polyphenol-loaded lipid nanocapsules. Int J Pharm 379(2):270–277. https://doi.org/10.1016/j.ijpharm.2009.05.054

Bhadoriya SS, Mangal A, Madoriya N, Dixit P (2011) Bioavailability and bioactivity enhancement of herbal drugs by "nanotechnology": a review. J Curr Pharm Res 8:1–7

Bhardwaj A, Dwivedi H, Kymonil KM, Pareek A, Upadhyay SC, Tripathi CB, Sara SA (2016) Solubility enhancement of Boswellia serrata Roxb. Ex Colebr. Extract through a self dispersible lipidic formulation approach. Indian J Nat Prod Resour 7(1):9–18

Blagden N, de Matas M, Gavan PT, York P (2007) Crystal engineering of active pharmaceutical ingredients to improve solubility and dissolution rates. Adv Drug Deliv Rev 59(7):617–630

Borghetti GS, Lula IS, Sinisterra RD, Bassani VL (2009) Quercetin/beta-cyclodextrin solid complexes prepared in aqueous solution followed by spray-drying or by physical mixture. AAPS PharmSciTech 10(1):235–242. https://doi.org/10.1208/s12249-009-9196-3

Brandl M (2001) Liposomes as drug carriers: a technological approach. Biotechnol Annu Rev 7:59–85

Breedveld MW, Miller KJ (1994) Cyclic beta-glucans of members of the family Rhizobiaceae. Microbiol Mol Biol Rev 58(2):145–161

Brewster ME, Loftsson T (2007) Cyclodextrins as pharmaceutical solubilizers. Adv Drug Deliv Rev 59(7):645–666. https://doi.org/10.1016/j.addr.2007.05.012

Cao R, Zhao Y, Zhou Z (2018) Preparation and physicochemical properties of hesperidin-chitooligosaccharide complex. Shipin Kexue/Food Sci 39(10):14–19

Cao R, Li X, Zhou Z, Zhao Z (2019) Synthesis and biophysical analysis of Naringin-Chitooligosaccharide complex. Nat Prod Res:1–7. https://doi.org/10.1080/14786419.2019.1628752

Challa R, Ahuja A, Ali J, Khar RK (2005) Cyclodextrins in drug delivery: an updated review. AAPS PharmSciTech 6(2):E329–E357. https://doi.org/10.1208/pt060243

Chaudhary A, Nagaich U, Gulati N, Sharma V, Khosa R, Partapur M (2012) Enhancement of solubilization and bioavailability of poorly soluble drugs by physical and chemical modifications: a recent review. J Adv Pharm Educ Res 2(1):32–67

Chawla G, Bansal A (2008) Improved dissolution of a poorly water soluble drug in solid dispersions with polymeric and non-polymeric hydrophilic additives. Acta Pharma 58(3):257–274

Chella N, Shastri NR (2017) Lipid carriers: role and applications in Nano drug delivery. In: Jana S, Jana S (eds) Particulate technology for delivery of therapeutics. Springer, Singapore, pp 253–289

Chen H, Khemtong C, Yang X, Chang X, Gao J (2011) Nanonization strategies for poorly water-soluble drugs. Drug Discov Today 16(7–8):354–360

Chiou WL, Riegelman S (1971) Pharmaceutical applications of solid dispersion systems. J Pharm Sci 60(9):1281–1302

Coimbra M, Isacchi B, van Bloois L, Torano JS, Ket A, Wu X, Broere F, Metselaar JM, Rijcken CJ, Storm G (2011) Improving solubility and chemical stability of natural compounds for medicinal use by incorporation into liposomes. Int J Pharm 416(2):433–442

Cragg GM, Newman DJ (2005) Plants as a source of anti-cancer agents. J Ethnopharmacol 100(1–2):72–79

Cragg GM, Newman DJ (2013) Natural products: a continuing source of novel drug leads. Biochim Biophys Acta (BBA)-Gen Subj 1830(6).3670–3695

Craig DQM (2002) The mechanisms of drug release from solid dispersions in water-soluble polymers. Int J Pharm 231(2):131–144. https://doi.org/10.1016/S0378-5173(01)00891-2

Dai L, Wang L, Deng L, Liu J, Lei J, Li D, He J (2014) Novel multiarm polyethylene glycol-dihydroartemisinin conjugates enhancing therapeutic efficacy in non-small-cell lung cancer. Sci Rep 4:5871–5871. https://doi.org/10.1038/srep05871

Dandekar DV, Jayaprakasha G, Patil BS (2008) Hydrotropic extraction of bioactive limonin from sour orange (Citrusaurantium L.) seeds. Food Chem 109(3):515–520

de Lima Petito N, da Silva Dias D, Costa VG, Falcão DQ, de Lima Araujo KG (2016) Increasing solubility of red bell pepper carotenoids by complexation with 2-hydroxypropyl-β-cyclodextrin. Food Chem 208:124–131

Del Valle EM (2004) Cyclodextrins and their uses: a review. Process Biochem 39(9):1033–1046

Deng K-Z, Xiong Y, Zhou B, Guan Y-M, Luo Y-M (2013) Chemical constituents from the roots of Ranunculus ternatus and their inhibitory effects on Mycobacterium tuberculosis. Molecules 18(10):11859–11865

Dhirendra K, Lewis S, Udupa N, Atin K (2009) Solid dispersions: a review. Pak J Pharm Sci 22(2):234–246

Dhumal DM, Kothari PR, Kalhapure RS, Akamanchi KG (2015) Self-microemulsifying drug delivery system of curcumin with enhanced solubility and bioavailability using a new semi-synthetic bicephalous heterolipid: in vitro and in vivo evaluation. RSC Adv 5(110):90295–90306. https://doi.org/10.1039/C5RA18112G

Esmaili M, Ghaffari SM, Moosavi-Movahedi Z, Atri MS, Sharifizadeh A, Farhadi M, Yousefi R, Chobert J-M, Haertlé T, Moosavi-Movahedi AA (2011) Beta casein-micelle as a nano vehicle for solubility enhancement of curcumin; food industry application. LWT-food Sci Technol 44(10):2166–2172

Etuk E (2006) A review of medicinal plants with hypotensive or antihypertensive effects. J Med Sci 6(6):894–900

Fang YP, Lin YK, Su YH, Fang JY (2011) Tryptanthrin-loaded nanoparticles for delivery into cultured human breast cancer cells, MCF7: the effects of solid lipid/liquid lipid ratios in the inner core. Chem Pharm Bull 59(2):266–271. https://doi.org/10.1248/cpb.59.266

Feeney OM, Crum MF, McEvoy CL, Trevaskis NL, Williams HD, Pouton CW, Charman WN, Bergström CAS, Porter CJH (2016) 50years of oral lipid-based formulations: provenance, progress and future perspectives. Adv Drug Deliv Rev 101:167–194. https://doi.org/10.1016/j.addr.2016.04.007

Gao Y, Li Z, Sun M, Guo C, Yu A, Xi Y, Cui J, Lou H, Zhai G (2011) Preparation and characterization of intravenously injectable curcumin nanosuspension. Drug Deliv 18(2):131–142. https://doi.org/10.3109/10717544.2010.520353

Garg, S. C. (2005). Essential oils as therapeutics. Nat Prod Rad 4(1):18–26

Greenwald RB, Gilbert CW, Pendri A, Conover CD, Xia J, Martinez A (1996) Drug delivery systems: water soluble Taxol 2 '-poly (ethylene glycol) Ester prodrugs design and in vivo effectiveness. J Med Chem 39(2):424–431

Gunasekaran T, Haile T, Nigusse T, Dhanaraju MD (2014) Nanotechnology: an effective tool for enhancing bioavailability and bioactivity of phytomedicine. Asian Pac J Trop Biomed 4:S1–S7

Guo Z (2017) The modification of natural products for medical use. Acta Pharm Sin B 7(2):119–136

Haham M, Ish-Shalom S, Nodelman M, Duek I, Segal E, Kustanovich M, Livney YD (2012) Stability and bioavailability of vitamin D nanoencapsulated in casein micelles. Food Funct 3(7):737–744

Hani U, Shivakumar H (2014) Solubility enhancement and delivery systems of curcumin a herbal medicine: a review. Curr Drug Deliv 11(6):792–804

Hao J, Gao Y, Zhao J, Zhang J, Li Q, Zhao Z, Liu J (2014) Preparation and optimization of resveratrol nanosuspensions by antisolvent precipitation using box-Behnken design. AAPS PharmSciTech 16(1):118–128. https://doi.org/10.1208/s12249-014-0211-y

Hernandez-Patlan D, Solis-Cruz B, Cano-Vega MA, Beyssac E, Garrait G, Hernandez-Velasco X, Lopez-Arellano R, Tellez G, Rivera-Rodriguez GR (2019) Development of chitosan and alginate Nanocapsules to increase the solubility, permeability and stability of Curcumin. J Pharm Innov 14(2):132–140. https://doi.org/10.1007/s12247-018-9341-1

Hong J (2011) Role of natural product diversity in chemical biology. Curr Opin Chem Biol 15(3):350–354

Hörter D, Dressman J (2001) Influence of physicochemical properties on dissolution of drugs in the gastrointestinal tract. Adv Drug Deliv Rev 46(1–3):75–87

Huang Y, Dai W-G (2014) Fundamental aspects of solid dispersion technology for poorly soluble drugs. Acta Pharm Sin B 4(1):18–25. https://doi.org/10.1016/j.apsb.2013.11.001

Huang L-F, Tong W-QT (2004) Impact of solid state properties on developability assessment of drug candidates. Adv Drug Deliv Rev 56(3):321–334

Jahan N, Aslam S, Rahman K, Fazal T, Anwar F, Saher R (2016) Formulation and characterisation of nanosuspension of herbal extracts for enhanced antiradical potential. J Exp Nanosci 11(1):72–80. https://doi.org/10.1080/17458080.2015.1025303

Jain P, Goel A, Sharma S, Parmar M (2010) Solubility enhancement techniques with special emphasis on hydrotrophy. Int J Pharm Professional's Res 1(1):34–45

Jansook P, Ogawa N, Loftsson T (2018) Cyclodextrins: structure, physicochemical properties and pharmaceutical applications. Int J Pharm 535(1):272–284. https://doi.org/10.1016/j.ijpharm.2017.11.018

Jawahar N, Meyyanathan S (2012) Polymeric nanoparticles for drug delivery and targeting: a comprehensive review. Int J Health Allied Sci 1(4):217

Jeong D, Choi JM, Choi Y, Jeong K, Cho E, Jung S (2013) Complexation of fisetin with novel cyclosophoroase dimer to improve solubility and bioavailability. Carbohydr Polym 97(1):196–202

Jøraholmen MW, Škalko-Basnet N, Acharya G, Basnet P (2015) Resveratrol-loaded liposomes for topical treatment of the vaginal inflammation and infections. Eur J Pharm Sci 79:112–121

Kadam Y, Yerramilli U, Bahadur A (2009) Solubilization of poorly water-soluble drug carbamezapine in Pluronic® micelles: effect of molecular characteristics, temperature and added salt on the solubilizing capacity. Colloids Surf B: Biointerfaces 72(1):141–147

Kalepu S, Nekkanti V (2015) Insoluble drug delivery strategies: review of recent advances and business prospects. Acta Pharm Sin B 5(5):442–453

Kanaze FI, Kokkalou E, Niopas I, Barmpalexis P, Georgarakis E, Bikiaris D (2010) Dissolution rate and stability study of flavanone aglycones, naringenin and hesperetin, by drug delivery systems based on polyvinylpyrrolidone (PVP) nanodispersions. Drug Dev Ind Pharm 36(3):292–301. https://doi.org/10.3109/03639040903140589

Karadag A, Ozcelik B, Huang Q (2014) Quercetin nanosuspensions produced by high-pressure homogenization. J Agric Food Chem 62(8):1852–1859

Kaur H, Kaur G (2014) A critical appraisal of solubility enhancement techniques of polyphenols. J Pharm 2014:14. https://doi.org/10.1155/2014/180845

Khadka P, Ro J, Kim H, Kim I, Kim JT, Kim H, Cho JM, Yun G, Lee J (2014) Pharmaceutical particle technologies: an approach to improve drug solubility, dissolution and bioavailability. Asian J Pharm Sci 9(6):304–316. https://doi.org/10.1016/j.ajps.2014.05.005

Khan AW, Kotta S, Ansari SH, Sharma RK, Ali J (2015a) Enhanced dissolution and bioavailability of grapefruit flavonoid Naringenin by solid dispersion utilizing fourth generation carrier. Drug Dev Ind Pharm 41(5):772–779

Khan AW, Kotta S, Ansari SH, Sharma RK, Ali J (2015b) Self-nanoemulsifying drug delivery system (SNEDDS) of the poorly water-soluble grapefruit flavonoid Naringenin: design, characterization, in vitro and in vivo evaluation. Drug Deliv 22(4):552–561. https://doi.org/10.310 9/10717544.2013.878003

Kim H-M, Kim H-W, Jung S-H (2008) Aqueous solubility enhancement of some flavones by complexation with cyclodextrins. Bull Kor Chem Soc 29(3):590–594

Kim CY, Bordenave N, Ferruzzi MG, Safavy A, Kim KH (2011) Modification of curcumin with polyethylene glycol enhances the delivery of curcumin in preadipocytes and its antiadipogenic property. J Agric Food Chem 59(3):1012–1019. https://doi.org/10.1021/jf103873k

Kim Y, Shinde VV, Jeong D, Choi Y, Jung S (2019) Solubility enhancement of atrazine by complexation with Cyclosophoraose isolated from rhizobium leguminosarum biovar trifolii TA-1. Polymers 11(3):474

Klayman DL (1985) Qinghaosu (artemisinin): an antimalarial drug from China. Science 228(4703):1049–1055

Kocbek P, Baumgartner S, Kristl J (2006) Preparation and evaluation of nanosuspensions for enhancing the dissolution of poorly soluble drugs. Int J Pharm 312(1–2):179–186

Koroleva MY, Yurtov EV (2012) Nanoemulsions: the properties, methods of preparation and promising applications. Russ Chem Rev 81(1):21

Krishnaiah YS (2010) Pharmaceutical technologies for enhancing oral bioavailability of poorly soluble drugs. J Bioequiv Availab 2(2):28–36

Kuche K, Bhargavi N, Dora CP, Jain S (2019) Drug-phospholipid complex—a go through strategy for enhanced Oral bioavailability. AAPS PharmSciTech 20(2):43. https://doi.org/10.1208/ s12249-018-1252-4

Kumar S, Bhargava D, Thakkar A, Arora S (2013a) Drug carrier systems for solubility enhancement of BCS class II drugs: a critical review. Crit Rev Ther Drug Carrier Syst 30(3):217–256

Kumar SK, Sushma M, Raju PY (2013b) Dissolution enhancement of poorly soluble drugs by using complexation technique-a review. J Pharm Sci Res 5(5):120

Kumar R, Singh M, Meena J, Singhvi P, Thiyagarajan D, Saneja A, Panda AK (2019) Hyaluronic acid - dihydroartemisinin conjugate: synthesis, characterization and in vitro evaluation in lung cancer cells. Int J Biol Macromol 133:495–502. https://doi.org/10.1016/j.ijbiomac.2019.04.124

Kumari A, Yadav SK, Pakade YB, Singh B, Yadav SC (2010a) Development of biodegradable nanoparticles for delivery of quercetin. Colloids Surf B: Biointerfaces 80(2):184–192. https:// doi.org/10.1016/j.colsurfb.2010.06.002

Kumari A, Yadav SK, Yadav SC (2010b) Biodegradable polymeric nanoparticles based drug delivery systems. Colloids Surf B: Biointerfaces 75(1):1–18

Kumari A, Kumar V, Yadav S (2012) Nanotechnology: a tool to enhance therapeutic values of natural plant products. Trends Med Res 7(2):34–42

Kwon SH, Kim SY, Ha KW, Kang MJ, Huh JS, Tae Jong I, Kim YM, Park YM, Kang KH, Lee S, Chang JY, Lee J, Choi YW (2007) Pharmaceutical evaluation of genistein-loaded pluronic micelles for oral delivery. Arch Pharm Res 30(9):1138–1143. https://doi.org/10.1007/bf02980249

Lachman L, Lieberman HA, Kanig JL (1976) The theory and practice of industrial pharmacy. Lea & Febiger Philadelphia

Lee S, D-h S, H-l P, Choi Y, Jung S (2003) Solubility enhancement of a hydrophobic flavonoid, luteolin by the complexation with cyclosophoraoses isolated from rhizobium meliloti. Antonie Van Leeuwenhoek 84(3):201

Lee S, Park H, Seo D, Choi Y, Jung S (2004) Synthesis and characterization of carboxymethylated cyclosophoraose, and its inclusion complexation behavior. Carbohydr Res 339(3):519–527

Lee J-S, Hong DY, Kim ES, Lee HG (2017) Improving the water solubility and antimicrobial activity of silymarin by nanoencapsulation. Colloids Surf B: Biointerfaces 154:171–177. https://doi.org/10.1016/j.colsurfb.2017.03.004

Li Y, Zheng J, Xiao H, McClements DJ (2012) Nanoemulsion-based delivery systems for poorly water-soluble bioactive compounds: influence of formulation parameters on polymethoxyflavone crystallization. Food Hydrocoll 27(2):517–528

Li B, Harich K, Wegiel L, Taylor LS, Edgar KJ (2013a) Stability and solubility enhancement of ellagic acid in cellulose ester solid dispersions. Carbohydr Polym 92(2):1443–1450. https://doi.org/10.1016/j.carbpol.2012.10.051

Li B, Konecke S, Harich K, Wegiel L, Taylor LS, Edgar KJ (2013b) Solid dispersion of quercetin in cellulose derivative matrices influences both solubility and stability. Carbohydr Polym 92(2):2033–2040. https://doi.org/10.1016/j.carbpol.2012.11.073

Li B, Konecke S, Wegiel LA, Taylor LS, Edgar KJ (2013c) Both solubility and chemical stability of curcumin are enhanced by solid dispersion in cellulose derivative matrices. Carbohydr Polym 98(1):1108–1116. https://doi.org/10.1016/j.carbpol.2013.07.017

Li W, Zhan P, De Clercq E, Lou H, Liu X (2013d) Current drug research on PEGylation with small molecular agents. Prog Polym Sci 38(3):421–444. https://doi.org/10.1016/j.progpolymsci.2012.07.006

Li L, Liu Y, Xue Y, Zhu J, Wang X, Dong Y (2017) Preparation of a ferulic acid-phospholipid complex to improve solubility, dissolution, and B16F10 cellular melanogenesis inhibition activity. Chem Cent J 11(1):26–26. https://doi.org/10.1186/s13065-017-0254-8

Lipinski CA, Lombardo F, Dominy BW, Feeney PJ (1997) Experimental and computational approaches to estimate solubility and permeability in drug discovery and development settings. Adv Drug Deliv Rev 23(1–3):3–25

Lodhi G, Kim Y-S, Hwang J-W, Kim S-K, Jeon Y-J, Je J-Y, Ahn C-B, Moon S-H, Jeon B-T, Park P-J (2014) Chitooligosaccharide and its derivatives: preparation and biological applications. Biomed Res Int 2014:13. https://doi.org/10.1155/2014/654913

Loftsson T (2017) Drug solubilization by complexation. Int J Pharm 531(1):276–280. https://doi.org/10.1016/j.ijpharm.2017.08.087

Loftsson T, Brewster ME (1996) Pharmaceutical applications of cyclodextrins. 1. Drug solubilization and stabilization. J Pharm Sci 85(10):1017–1025

Loftsson T, Duchene D (2007) Cyclodextrins and their pharmaceutical applications. Int J Pharm 329(1–2):1–11

Loftsson T, Duchêne D (2007) Cyclodextrins and their pharmaceutical applications. Int J Pharm 329(1):1–11. https://doi.org/10.1016/j.ijpharm.2006.10.044

Loftsson T, Jarho P, Másson M, Järvinen T (2005) Cyclodextrins in drug delivery. Expert Opin Drug Deliv 2(2):335–351. https://doi.org/10.1517/17425247.2.1.335

Lu M, Qiu Q, Luo X, Liu X, Sun J, Wang C, Lin X, Deng Y, Song Y (2019) Phyto-phospholipid complexes (phytosomes): a novel strategy to improve the bioavailability of active constituents. Asian J Pharm Sci 14(3):265–274. https://doi.org/10.1016/j.ajps.2018.05.011

Mangal A, Bhadoriya SS, Joshi S, Agrawal G, Gupta A, Mandoria N (2011) Extraction of herbal drugs by using hydrotropy solublization phenomenon. Drugs 5:24. Pharm. App Sci., 2012; 2(1):63–74

Martin A (1993) *Physical pharmacy: physical chemical principles in the pharmaceutical sciences.* BI Waverly. Pvt Ltd.

McClements D, Decker E, Weiss J (2007) Emulsion-based delivery systems for lipophilic bioactive components. J Food Sci 72(8):R109–R124

Md S, Kit B, Jagdish S, David D, Pandey M, Chatterjee L (2018) Development and in vitro evaluation of a Zerumbone loaded Nanosuspension drug delivery system. Crystals 8(7):286. https://doi.org/10.3390/cryst8070286

Mehnert W, Mäder K (2012) Solid lipid nanoparticles: Production, characterization and applications. Adv Drug Deliv Rev 64:83–101. https://doi.org/10.1016/j.addr.2012.09.021

Merisko-Liversidge E, Liversidge GG, Cooper ER (2003) Nanosizing: a formulation approach for poorly-water-soluble compounds. Eur J Pharm Sci 18(2):113–120

Miller KJ, Kennedy EP, Reinhold VN (1986) Osmotic adaptation by gram-negative bacteria: possible role for periplasmic oligosaccharides. Science 231(4733):48–51

Morales JO, Watts AB, McConville JT (2016) Mechanical particle-size reduction techniques. In Formulating Poorly Water Soluble Drugs (pp. 165–213). Springer, Cham.

Mosharraf M, Nyström C (1995) The effect of particle size and shape on the surface specific dissolution rate of microsized practically insoluble drugs. Int J Pharm 122(1–2):35–47

Müller RH, Mäder K, Gohla S (2000a) Solid lipid nanoparticles (SLN) for controlled drug delivery–a review of the state of the art. Eur J Pharm Biopharm 50(1):161–177

Müller RH, Mäder K, Gohla S (2000b) Solid lipid nanoparticles (SLN) for controlled drug delivery – a review of the state of the art. Eur J Pharm Biopharm 50(1):161–177. https://doi.org/10.1016/S0939-6411(00)00087-4

Müller RH, Gohla S, Keck CM (2011) State of the art of nanocrystals – special features, production, nanotoxicology aspects and intracellular delivery. Eur J Pharm Biopharm 78(1):1–9 https://doi.org/10.1016/j.ejpb.2011.01.007

Mulligan C, Yong R, Gibbs B (2001) Surfactant-enhanced remediation of contaminated soil: a review. Eng Geol 60(1–4):371–380

Nacsa A, Ambrus R, Berkesi O, Szabo-Revesz P, Aigner Z (2008) Water-soluble loratadine inclusion complex: analytical control of the preparation by microwave irradiation. J Pharm Biomed Anal 48(3):1020–1023

Omari-Siaw E, Wang Q, Sun C, Gu Z, Zhu Y, Cao X, Firempong CK, Agyare R, Xu X, Yu J (2016) Tissue distribution and enhanced in vivo anti-hyperlipidemic-antioxidant effects of perillaldehyde loaded liposomal nanoformulation against Poloxamer 407-induced hyperlipidemia. Int J Pharm 513(1–2):68–77

Onoue S, Takahashi H, Kawabata Y, Seto Y, Hatanaka J, Timmermann B, Yamada S (2010) Formulation design and photochemical studies on nanocrystal solid dispersion of curcumin with improved oral bioavailability. J Pharm Sci 99(4):1871–1881

Park S-H, Choi H-K (2006) The effects of surfactants on the dissolution profiles of poorly water-soluble acidic drugs. Int J Pharm 321(1–2):35–41

Patel VR, Agrawal YK (2011) Nanosuspension: an approach to enhance solubility of drugs. J Adv Pharm Technol Res 2(2):81–87. https://doi.org/10.4103/2231-4040.82950

Pawar SS, Dahifale BR, Nagargoje SP, Shendge RS (2017) Nanosuspension technologies for delivery of drugs. Nanosci Nanotechnol Res 4(2):59–66. http://pubs.sciepub.com/nnr/4/2/4

Piao J, Jang A, Choi Y, Tahir MN, Kim Y, Park S, Cho E, Jung S (2014) Solubility enhancement of α-naphthoflavone by synthesized hydroxypropyl cyclic-(1→ 2)-β-d-glucans (cyclosophoroases). Carbohydr Polym 101:733–740

Pouton CW (1997) Formulation of self-emulsifying drug delivery systems. Adv Drug Deliv Rev 25(1):47–58

Puri A, Loomis K, Smith B, Lee J-H, Yavlovich A, Heldman E, Blumenthal R (2009) Lipid-based nanoparticles as pharmaceutical drug carriers: from concepts to clinic. Crit Rev Ther Drug Carrier Syst 26(6):523

Qin J, Chen D, Lu W, Xu H, Yan C, Hu H, Chen B, Qiao M, Zhao X (2008) Preparation, characterization, and evaluation of liposomal ferulic acid in vitro and in vivo. Drug Dev Ind Pharm 34(6):602–608. https://doi.org/10.1080/03639040701833559

Rabinow BE (2004) Nanosuspensions in drug delivery. Nat Rev Drug Discov 3(9):785–796. https://doi.org/10.1038/nrd1494

Rachmawati H, Al Shaal L, Muller RH, Keck CM (2013) Development of curcumin nanocrystal: physical aspects. J Pharm Sci 102(1):204–214. https://doi.org/10.1002/jps.23335

Rangel-Yagui CO, Pessoa A Jr, Tavares LC (2005) Micellar solubilization of drugs. J Pharm Pharm Sci 8(2):147–163

Ravve A (2013) Principles of polymer chemistry. Springer Science & Business Media.

Recharla N, Riaz M, Ko S, Park S (2017) Novel technologies to enhance solubility of food-derived bioactive compounds: a review. J Funct Foods 39:63–73

Sahu A, Kasoju N, Bora U (2008) Fluorescence study of the curcumin– casein micelle complexation and its application as a drug nanocarrier to cancer cells. Biomacromolecules 9(10):2905–2912

Salvi VR, Pawar P (2019) Nanostructured lipid carriers (NLC) system: a novel drug targeting carrier. J Drug Delivery Sci Technol 51:255–267. https://doi.org/10.1016/j.jddst.2019.02.017

Saoji SD, Raut NA, Dhore PW, Borkar CD, Popielarczyk M, Dave VS (2016) Preparation and evaluation of phospholipid-based complex of standardized Centella extract (SCE) for the enhanced delivery of Phytoconstituents. AAPS J 18(1):102–114. https://doi.org/10.1208/s12248-015-9837-2

Sapra K, Sapra A, Singh S, Kakkar S (2012) Self emulsifying drug delivery system: a tool in solubility enhancement of poorly soluble drugs. Indo Global J Pharm 2(3):313–332

Savjani KT, Gajjar AK, Savjani JK (2012) Drug solubility: importance and enhancement techniques. ISRN Pharm 2012

Seo S-W, Han H-K, Chun M-K, Choi H-K (2012) Preparation and pharmacokinetic evaluation of curcumin solid dispersion using Solutol® HS15 as a carrier. Int J Pharm 424(1):18–25. https://doi.org/10.1016/j.ijpharm.2011.12.051

Shah S, Maddineni S, Lu J, Repka MA (2013) Melt extrusion with poorly soluble drugs. Int J Pharm 453(1):233–252. https://doi.org/10.1016/j.ijpharm.2012.11.001

Shah MK, Khatri P, Vora N, Patel NK, Jain S, Lin S, A.M. (Ed.) (2019) Chapter 5 – lipid nanocarriers: preparation, characterization and absorption mechanism and applications to improve oral bioavailability of poorly water-soluble drugs. In: Grumezescu AM (ed) Biomedical applications of nanoparticles. William Andrew Publishing, pp 117–147

Shao B, Cui C, Ji H, Tang J, Wang Z, Liu H, Qin M, Li X, Wu L (2015) Enhanced oral bioavailability of piperine by self-emulsifying drug delivery systems: in vitro, in vivo and in situ intestinal permeability studies. Drug Deliv 22(6):740–747. https://doi.org/10.3109/10717544.2014.898109

Sharma D, Soni M, Kumar S, Gupta G (2009) Solubility enhancement–eminent role in poorly soluble drugs. Res J Pharm Technol 2(2):220–224

Shen G, Cheng L, Wang LQ, Zhang LH, Shen BD, Liao WB, Li JJ, Zheng J, Xu R, Yuan HL (2016) Formulation of dried lignans nanosuspension with high redispersibility to enhance stability, dissolution, and oral bioavailability. Chin J Nat Med 14(10):757–768. https://doi.org/10.1016/S1875-5364(16)30090-5

Shulman M, Cohen M, Soto-Gutierrez A, Yagi H, Wang H, Goldwasser J, Lee-Parsons CW, Benny-Ratsaby O, Yarmush ML, Nahmias Y (2011) Enhancement of naringenin bioavailability by complexation with hydroxypropoyl-β-cyclodextrin. PLoS One 6(4):e18033

Siddiqui AA, Iram F, Siddiqui S, Sahu K (2014) Role of natural products in drug discovery process. Int J Drug Dev Res 6(2):172–204

Singh M, Sharma R, Banerjee U (2002) Biotechnological applications of cyclodextrins. Biotechnol Adv 20(5–6):341–359

Singh R, Bharti N, Madan J, Hiremath S (2010) Characterization of cyclodextrin inclusion complexes—a review. J Pharm Sci Technol 2(3):171–183

Singh RP, Gangadharappa HV, Mruthunjaya K (2017) Phospholipids: unique carriers for drug delivery systems. J Drug Deliv Sci Technol 39:166–179. https://doi.org/10.1016/j.jddst.2017.03.027

Small P (1953) Some factors affecting the solubility of polymers. J Appl Chem 3(2):71–80

Stella V, Rajewski R (1992) Derivatives of cyclodextrins exhibiting enhanced aqueous solubility and the use thereof. Google Patentsdoi

Summerlin N, Soo E, Thakur S, Qu Z, Jambhrunkar S, Popat A (2015) Resveratrol nanoformulations: challenges and opportunities. Int J Pharm 479(2):282–290. https://doi.org/10.1016/j.ijpharm.2015.01.003

Sun M, Gao Y, Pei Y, Guo C, Li H, Cao F, Yu A, Zhai G (2010) Development of nanosuspension formulation for oral delivery of quercetin. J Biomed Nanotechnol 6(4):325–332

Sun J, Wang F, Sui Y, She Z, Zhai W, Wang C, Deng Y (2012) Effect of particle size on solubility, dissolution rate, and oral bioavailability: evaluation using coenzyme Q_{10} as naked nanocrystals. Int J Nanomedicine 7:5733–5744. https://doi.org/10.2147/IJN.S34365

Sutradhar KB, Khatun S, Luna IP (2013) Increasing possibilities of Nanosuspension. J Nanotechnol 2013:12. https://doi.org/10.1155/2013/346581

Tamjidi F, Shahedi M, Varshosaz J, Nasirpour A (2013) Nanostructured lipid carriers (NLC): a potential delivery system for bioactive food molecules. Innovative Food Sci Emerg Technol 19:29–43

Teixeira C, Mendonca L, Bergamaschi M, Queiroz R, Souza G, Antunes L, Freitas L (2016) Microparticles containing curcumin solid dispersion: stability, bioavailability and anti-inflammatory activity. AAPS PharmSciTech 17(2):252–261

Tiwari G, Tiwari R, Rai AK (2010) Cyclodextrins in delivery systems: applications. J Pharm Bioallied Sci 2(2):72–79. https://doi.org/10.4103/0975-7406.67003

Tran TH, Guo Y, Song D, Bruno RS, Lu X (2014) Quercetin-containing self-nanoemulsifying drug delivery system for improving oral bioavailability. J Pharm Sci 103(3):840–852

Tran P, Pyo Y-C, Kim D-H, Lee S-E, Kim J-K, Park J-S (2019) Overview of the manufacturing methods of solid dispersion Technology for Improving the solubility of poorly water-soluble drugs and application to anticancer drugs. Pharmaceutics 11(3):132. https://doi.org/10.3390/pharmaceutics11030132

Trollope L, Cruickshank DL, Noonan T, Bourne SA, Sorrenti M, Catenacci L, Caira MR (2014) Inclusion of trans-resveratrol in methylated cyclodextrins: synthesis and solid-state structures. Beilstein J Org Chem 10:3136–3151. https://doi.org/10.3762/bjoc.10.331

Upreti M, Strassburger K, Chen YL, Wu S, Prakash I (2011) Solubility enhancement of steviol glycosides and characterization of their inclusion complexes with gamma-cyclodextrin. Int J Mol Sci 12(11):7529–7553. https://doi.org/10.3390/ijms12117529

van Hoogevest P (2017) Review – an update on the use of oral phospholipid excipients. Eur J Pharm Sci 108:1–12. https://doi.org/10.1016/j.ejps.2017.07.008

Vemula VR, Lagishetty V, Lingala S (2010) Solubility enhancement techniques. Int J Pharm Sci Rev Res 5(1):41–51

Verma S, Gokhale R, Burgess DJ (2009) A comparative study of top-down and bottom-up approaches for the preparation of micro/nanosuspensions. Int J Pharm 380(1–2):216–222

Watkins R, Wu L, Zhang C, Davis RM, Xu B (2015) Natural product-based nanomedicine: recent advances and issues. Int J Nanomedicine 10:6055

Wen H, Jung H, Li X (2015) Drug delivery approaches in addressing clinical pharmacology-related issues: opportunities and challenges. AAPS J 17(6):1327–1340

Williams HD, Trevaskis NL, Charman SA, Shanker RM, Charman WN, Pouton CW, Porter CJH (2013) Strategies to address low drug solubility in discovery and development. Pharmacol Rev 65(1):315–499. https://doi.org/10.1124/pr.112.005660

Xia W, Liu P, Zhang J, Chen J (2011) Biological activities of chitosan and chitooligosaccharides. Food Hydrocoll 25(2):170–179

Xu G, Shi H, Ren L, Gou H, Gong D, Gao X, Huang N (2015) Enhancing the anti-colon cancer activity of quercetin by self-assembled micelles. Int J Nanomedicine 10:2051

Yen C-C, Chen Y-C, Wu M-T, Wang C-C, Wu Y-T (2018) Nanoemulsion as a strategy for improving the oral bioavailability and anti-inflammatory activity of andrographolide. Int J Nanomedicine 13:669–680. https://doi.org/10.2147/IJN.S154824

Yu D, Peng P, Dharap SS, Wang Y, Mehlig M, Chandna P, Zhao H, Filpula D, Yang K, Borowski V, Borchard G, Zhang Z, Minko T (2005) Antitumor activity of poly(ethylene glycol)–camptothecin conjugate: the inhibition of tumor growth in vivo. J Control Release 110(1):90–102. https://doi.org/10.1016/j.jconrel.2005.09.050

Zafar F, Jahan N, Khalil Ur R, Bhatti HN (2019) Increased Oral bioavailability of Piperine from an optimized Piper nigrum Nanosuspension. Planta Med 85(03):249–257. https://doi.org/10.1055/a-0759-2208

Zhang J, Ma PX (2013) Cyclodextrin-based supramolecular systems for drug delivery: recent progress and future perspective. Adv Drug Deliv Rev 65(9):1215–1233. https://doi.org/10.1016/j.addr.2013.05.001

Zhang Q-F, Nie H-C, Shangguang X-C, Yin Z-P, Zheng G-D, Chen J-G (2012) Aqueous solubility and stability enhancement of astilbin through complexation with cyclodextrins. J Agric Food Chem 61(1):151–156

Zhang X, Chiu Li L, Mao S (2014) Nanosuspensions of poorly water soluble drugs prepared by top-down technologies. Curr Pharm Des 20(3):388–407

Zhang K, Zhang M, Liu Z, Zhang Y, Gu L, Hu G, Chen X, Jia J (2016a) Development of quercetin-phospholipid complex to improve the bioavailability and protection effects against carbon tetrachloride-induced hepatotoxicity in SD rats. Fitoterapia 113:102–109. https://doi.org/10.1016/j.fitote.2016.07.008

Zhang Y, Li Z, Zhang K, Yang G, Wang Z, Zhao J, Hu R, Feng N (2016b) Ethyl oleate-containing nanostructured lipid carriers improve oral bioavailability of trans-ferulic acid ascompared with conventional solid lipid nanoparticles. Int J Pharm 511(1):57–64. https://doi.org/10.1016/j.ijpharm.2016.06.131

Zhang J, Xie Z, Zhang N, Zhong J (2017) Nanosuspension drug delivery system: preparation, characterization, postproduction processing, dosage form, and application. In Nanostructures for Drug Delivery (pp. 413–443). Elsevier.

Zhou D-Y, Rakariyatham K (2019) Phospholipids. In: Melton L, Shahidi F, Varelis P (eds) Encyclopedia of food chemistry. Academic, Oxford, pp 546–549

Zu Y, Wu W, Zhao X, Li Y, Wang W, Zhong C, Zhang Y, Zhao X (2014) Enhancement of solubility, antioxidant ability and bioavailability of taxifolin nanoparticles by liquid antisolvent precipitation technique. Int J Pharm 471(1–2):366–376. https://doi.org/10.1016/j.ijpharm.2014.05.049

Chapter 3
Delivery of Natural Products Using Polymeric Particles for Cancer Chemotherapeutics

Rahul Ahuja, Neha Panwar, Jairam Meena, Debi P. Sarkar, and Amulya K. Panda

Abstract Cancer is a major cause of deaths throughout the world. According to World Health Organization, cancer lead to the loss of about 9.6 million lives in the year 2018 with low and middle income countries accounting for 70% of these deaths. With its ability to grow in an uncontrolled manner and spreading out to different sites in the body, cancer is seen as that dreaded disease with no cure or treatment. Several promising cancer chemotherapeutic advancements have emerged but these came along with unavoidable limitations like toxicity, limits on the administered dosage as well as affordability for the people who most need them. Natural products with anticancer potential have also emerged as an immensely attractive option to treat cancer. The use of natural products has been credited to doubling of human life span in twentieth century. Furthermore, of all the anti-cancer agents approved and pre-new drug application candidates till date, more than 60% are of natural origin or a derivative thereof. However, an estimated 4 in 10 of the new chemical entity proposed against cancer turns out to be lipophilic. Problems of low aqueous solubility, low bioavailability and non-specific action still pose as major stumbling blocks in their advancement. Recently, polymeric nanocarriers have been used to deliver cancer chemotherapeutics to circumvent the associated limitations and improve their efficacy. Polymers like polylactic acid/poly lactic-co-glycolic acid are have a long history of use in medical field. These are safe and biodegradable with about 20 products approved by the Food and Drug Administration and European Medicine Agency for use in humans. The polymeric nanocarriers with

R. Ahuja · J. Meena · A. K. Panda (✉)
Product Development Cell-II, National Institute of Immunology, New Delhi, India
e-mail: amulya@nii.ac.in

N. Panwar
Product Development Cell-II, National Institute of Immunology, New Delhi, India

Department of Biochemistry, University of Delhi South Campus, New Delhi, India

D. P. Sarkar
Department of Biochemistry, University of Delhi South Campus, New Delhi, India

© The Editor(s) (if applicable) and The Author(s), under exclusive license to
Springer Nature Switzerland AG 2020
A. Saneja et al. (eds.), *Sustainable Agriculture Reviews 43*, Sustainable
Agriculture Reviews 43, https://doi.org/10.1007/978-3-030-41838-0_3

their versatility offer solution to many of the hurdles of natural product delivery for cancer chemotherapy.

In this chapter, we briefly discuss about the characteristics acquired by a cell progressing towards carcinogenesis. We then explore some of the conventional chemotherapeutic strategies that have been in use for the treatment of cancer. Moving further we discuss about the emergence of natural products in the cancer chemotherapy field. We examine how their non-toxic nature and ability to target multiple mechanisms has garnered attention. We conclude by reviewing about the use of polymeric nanocarriers to deliver natural products. We discuss some of the advantages offered by them and how they can be used to circumvent the limitations associated with the delivery of natural products for cancer chemotherapy. With this chapter we hope to shed light on how delivering anticancer natural products via polymeric nanocarriers can offer a safe and patient compliant strategy for the treatment of cancer.

Keywords Cancer · Chemotherapy · Biodegradable polymer · Nanocarrier · Nanoparticle · Poly (lactic acid)-poly (Lactide-co-glycolic acid)

Abbreviations

DNA   deoxyribonucleic acid
PLGA   poly lactic-co-glycolic acid
RNA   ribonucleic acid

3.1 Introduction

Cancer has been the one of the major fatal disease claiming about 1 in every 6 deaths occurring globally. In the year 2018 an estimated 9.6 million lives were lost due to cancer. About 70% of these deaths occurred in low and middle income countries (WHO n.d.). It is marked by deregulated cell growth forming a tumoral mass. The capability of new blood vessel formation not only provides them with continued oxygen and nutrient supply but also grant cancer its dreaded metastastic potential leading it to further spread in other sites of the body ultimately culminating in death. Chemotherapy has been the most successful choice for treatment of cancers. It is based on targeting abnormal cancer cells but unfortunately also damages healthy cells of the body (Danhier et al. 2012; Perez-Herrero and Fernandez-Medarde 2015). As an alternative to conventional chemotherapeutics, the use of natural products with anticancer properties has gained much appreciation. Their natural origin, relatively non-toxic nature and pleiotropic effects have been largely responsible for this attention. A wide variety of natural products like epigallocate-chin gallate, curcumin, resveratrol and many others have been investigated for their anticancer action. About 6 in 10 drugs approved and pre-new drug application can-didates against cancer are from a natural product or derived from them (Demain and Vaishnav 2011). However, most of them face challenges due to low solubility, sta-bility, poor permeability and with approximately 40% of the new chemical entity emerging against cancer being lipophilic, they face major hurdles in their further development (Arora and Jaglan 2016; Siddiqui and Sanna 2016).

The application of synthetic polymeric nanocarriers for cancer chemotherapy has drawn a lot of attention in the scientific community. Polymers like poly lactic acid or poly lactide-co-glycolide have a long been used in biomedical applications (Frazza and Schmitt 1971; O'Hagan and Singh 2003). They are Food and Drug Administration (FDA) as well as European Medical Agency approved (Danhier et al. 2012). They offer many advantages due to their unique properties like tunable particle size, increased stability, high surface area to volume ratio, possibility of surface functionalization in addition to being biocompatible and biodegradable. Roughly about 20 products based on these polymers are in market for various indi-cations like periodontal disease, diabetes, schizophrenia including cancer (Schoubben et al. 2019). The particulate delivery systems prepared using these polymers can be fine-tuned to enable targeted delivery of the bioactive molecules. Moreover, the well-established production methods offer uniform preparation and scalability.

Over the recent years many promising anticancer products have been delivered using polymeric nanocarriers to improve upon their potential. The use of polymeric nanocoarriers to deliver these natural products can circumvent the limitations asso-ciated with them. They can gain access to the leaky tumor vasculature due to mere small size or can be surface decorated for targeted delivery. They can be used to formulate particles with different size, shape and porosity. Different molecular weight polymers can be formulated to prepare particles with desirable release pro-files (Anderson and Shive 2012). Various molecules can be decorated on surface to enable selective delivery of bioactive molecules. Molecules to be grafted are selected so as to bind to specific overexpressed receptors on tumor cells. This reduces off-target effects as well as the dose required to achieve the desired effect. Apart from safety, delivery using poly lactic acid or poly lactide-co-glycolide based particulate systems offers other advantages like protection of therapeutic agents from cellular enzymes, better stability during storage and lower toxicity. Thus, the use of polymeric nanocarriers for delivering anticancer natural products seems to offer a safe and patient compliant method for cancer treatment.

In this review, we briefly discuss certain characteristics associated with cancer moving on to the various chemotherapy options available for cancer treatment. We then discuss the emergence of natural products in cancer chemotherapy field and the possible mechanisms that have been elucidated for few compounds. Finally, we end our discussion with the impact of polymeric nanocarriers on chemotherapy field and how they have been used to improve upon the various limitations offering the opti-mal and safest solution to cancer.

3.2   Derailment of Normal Functioning in Cancer

Cancer is a major health complication affecting millions of lives every year. One in six deaths globally occur due to cancer. World Health Organization (WHO) ranks cancer as the second leading cause of death with an estimation of 9.6 million deaths in 2018 (WHO n.d.). Low and middle income countries account for the major proportion of cancer deaths being as high as 70%. According to WHO (2014) as of year 2010, estimation of the total annual economic cost for cancer stands at around $1.16 trillion. Cancers of lung, colorectal, stomach, liver and breasts constitute the leading cause of deaths (WHO 2014, WHO n.d.).

When normal regulatory mechanisms of a cell breakdown and it start to grow in an uncontrolled manner, a cancerous transformation is said to have occur. Malignant neoplasm, a subset of neoplasms is often simply referred to as cancer. Cancerous cell can be defined as having certain key attributes which are discussed below (Table 3.1).

Table 3.1 Features acquired by cell progressing towards carcinogenesis

Capability acquired	Description of process	Example	Reference
Uncontrolled proliferation of cells	Enhanced growth factor release, receptors sensing growth signals up regulated, activation of cell survival pathways	Mutations affecting certain serine/threonine kinases like B-Raf	Davies and Samuels (2010) and Hanahan and Weinberg (2011)
Evading growth suppressors	Response to regulatory anti-growth signalling lost, cells continue to grow and divide, loss of contact inhibition	Defects in retinoblastoma protein	Amin et al. (2015)
Resisting apoptosis	Overexpression of anti-apoptotic proteins, down regulation of pro-apoptotic proteins, cell survival	Defects in protein p53	Hanahan and Weinberg (2011) and Joerger and Fersht (2016)
Immortal replicative capacity	Expression of enzymes telomerase, indefinite cell division	Mutations in *hTERT* promoter	Jafri et al. (2016)
Nutrient and oxygen maintenance	Formation of new blood vessels, increased pro-angiogenic factors, cell survival	Up regulation of VEGF	Kessenbrock et al. (2010)
Invasion and metastasis	Increased cell motility, cells escape primary site, cause secondary carcinoma	Loss of E-cadherin	Mendonsa et al. (2018)

B-Raf Rapidly Accelerated Fibrosarcoma Homolog B, *p53* transformation-related protein 53, *hTERT* human telomerase reverse transcriptase, *VEGF* vascular endothelial growth factor, *E-cadherin* epithelial cadherin

3.2.1   Uncontrolled Proliferation of Cells

The growth and division in a normal cell is a tightly regulated phenomenon. The positive growth promoting signals are carefully balanced by negative feedback mechanisms. This is of extreme importance in order to ensure a coordinated growth of a healthy organism as a whole. When a cell breaks free of these controls and divides uncontrollably, it is said to have turned cancerous.

Sustained chronic growth is probably the basic foundational trait of a cancer cell. Cancerous cells may start producing growth factors themselves, up regulate receptors responding to growth signals or may even instruct tumor associated stroma to release growth factors (Bhowmick et al. 2004; Cheng et al. 2008; Lemmon and Schlessinger 2010). They may then constitutively activate downstream signaling pathways independent of growth signals (Hanahan and Weinberg 2011). For example, mutations affecting a serine/threonine-protein kinase named B-Raf, can result in constitutive signalling via mitogen activated protein kinase pathway and are known to occur in a significant proportion of human melanomas. Phosphoinositide 3-kinase (PI3-kinase) mutants activating the PI3-kinase signaling have been detected in a wide variety of cancers (Yuan and Cantley 2008; Jiang and Liu 2009; Davies and Samuels 2010).

The negative feedback mechanisms are equally important in a cell to counter the positive growth inducing signals to maintain the healthy state. Defects in these, axiomatically predispose the cells to an uncontrolled growth. For instance, certain mutations in *ras* genes result in decreased hydrolysis of guanosine triphosphate to guanosine diphosphate thereby keeping Ras proteins in constitutively active from (Amit et al. 2007; Hanahan and Weinberg 2011). Similarly, defects in *phosphatase and tensin* homolog protein, via abolishment of its function result in phosphoinositide 3-kinase signal amplification and thus tumorigenesis (Yuan and Cantley 2008; Jiang and Liu 2009).

3.2.2   Evading Growth Suppressors

There exist a number of mechanisms in a cell which function as a regulatory break to cell division. Various tumor suppressors function to limit uncontrolled cell proliferation and hence tumor development. These play an important role in determining whether a cell should proceed through division or not. Any defect which obliterates their function leaves the cell without any negative control and uncontrolled proliferation results. For example, retinoblastoma gene encodes a protein that functions to majorly transduce extracellular inhibitory signals. Defects in this gene are responsible for many cancers. It also serves as target of DNA oncoviruses like simian virus 40, adenovirus and human papillomavirus which act to inhibit its activity (Burkhart and Sage 2008). Another important tumor suppressor named 'tumor protein p53' largely responds to various intracellular stress signals. It can cause cell cycle arrest due to stresses like DNA damage giving time for repair or cause apoptosis if the damage is irreparable. Any alterations in this protein are the responsible for a wide range of cancers. (Sherr and McCormick 2002; Deshpande et al. 2005).

Contact inhibition is another important feature of normal cells which enables coordinated growth and development. In simple terms, it can be described as cells when they come in contact based on the density of cells around them are inhibited from further division in culture conditions, thus forming monolayers. Loss of this key property results in tumor formation. Merlin for instance, a NF2 gene product, seems to have a role in controlling contact dependent inhibition as it functions by coupling cell-surface adhesion molecules to the transmembrane receptor tyrosine kinases. Defects in Merlin have shown to be mitogenic and promote tumorigenesis. (Okada et al. 2005; Curto et al. 2007). Similarly, a protein in liver named liver kinase B1 functions to inhibit uncontrolled proliferation organizing the epithelial structure and maintaining tissue integrity. Defects in the function of this protein can destabilize epithelial integrity and make them susceptible to the effects of oncogene overexpression (Hezel and Bardeesy 2008; Partanen et al. 2009).

Cytokine signaling alterations like those of can also lead to cancer development. For example, a cytokine named 'Transforming growth factor beta 1' is unique as it can have either tumor suppressive or tumor promoter functions. It is affected not only by mechanisms operating within the tumor but also by tumor-stroma interaction. It is known to activate epithelial to mesenchymal transition resulting in the problematic malignancy that cancer cells possess (Massague 2008; Ikushima and Miyazono 2010).

3.2.3   Resisting Apoptosis

Normal cells undergo regulated programmed cell death as a mechanism to balance cell division and also as a defense strategy to remove damaged/dangerous cells. It plays an important role during development by getting rid of unwanted cells in tissues. Apoptosis and autophagy, both are forms of programmed cell death.

At times, a situation arises when the cell decides commit suicide for the broader benefit of the organism. For, instance apoptosis is responsible for the morphological development like the formation of digits during embryonic development. Also if a cell is infected by foreign organism which can hijack its machinery and spread to other cells, in order to prevent that spread, the infected cell may undergo apoptosis. The cell signaling machinery working to activate apoptosis can be extrinsic pathway and intrinsic pathway. Extrinsic pathway senses signals from outside the cell while intrinsic pathway senses signals originating from within the cell activating different caspases (cysteine-dependent aspartate-directed proteases). Both of them converge to activate effector caspase named 'caspase 3' leading to classic apoptosis characteristics like membrane blebbing, nucleic acid fragmentation, chromatin condensation and eventual consumption of cell (Lowe et al. 2004; Adams and Cory 2007; Hanahan and Weinberg 2011).

Apoptosis is controlled by regulatory proteins belonging to B-cell lymphoma 2 family. They may be pro-apoptotic or anti-apoptotic. In a normal cell, pro-apoptotic proteins are blocked by interactions with anti-apoptotic proteins. However, when pro-apoptotic proteins are activated they impede the anti-apoptotic members, thus tipping the balance in favor of pro-apoptotic members and the cell undergoes apoptosis (Willis and Adams 2005). Various stresses encountered by cancer cells like up regulation of oncogenes, down regulation of tumor suppressors and DNA damage can induce apoptosis. Thus, it acts as an intrinsic obstacle to cancer development (Adams and Cory 2007; Hanahan and Weinberg 2011). In tumorigenesis, the cells amass strategies to inhibit apoptosis, rendering their continued survival no matter the danger. Loss of function of tumor protein p53 which acts as a critical damage sensor is one mechanism to block apoptosis. Similarly, overexpression of anti-apoptotic proteins or down regulation of pro-apoptotic proteins can circumvent apoptosis in cancer cells. Several studies have shown impaired apoptosis in tumors progressing to high grade malignancy (Lowe et al. 2004; Adams and Cory 2007).

Autophagy is another mechanism which must be overcome during tumor development. It works to maintain the overall health by degrading cellular organelles and rechanneling catabolites to be used for biosynthesis in case of stresses like nutrient deprivation. The organelles are engulfed into vesicles called autophagosome which then fuse with lysosome and are degraded. Recent studies show that autophagy and apoptosis are interconnected via their regulatory circuits. For instance survival signals transducing their effect via pathways regulating cell cycle function to inhibit autophagy in addition also lead to impeding apoptosis (Levine and Kroemer 2008). White and Dipola reported that an important interconnection is the Beclin-1 protein (member of apoptotic regulatory proteins) which serves to induce autophagy. *Beclin-1* gene, if inactivated results in suppressed autophagy and increased susceptibility to tumorigenesis (White and DiPaola 2009).

Necrosis though undervalued for its role in cancer can also be a contributory factor. The cells undergoing necrosis release pro-inflammatory signals and interleukin-1 alpha recruiting inflammatory cells and stimulating neighboring cell proliferation. Immune inflammatory cells can help tumor development if they acquire capabilities of angiogenesis and proliferation (Galluzzi and Kroemer 2008; Grivennikov et al. 2010).

3.2.4 Immortal Replicative Capacity

In normal condition, a cell is able to pass through a limited number of cell division cycles. Cell senescence is the non-proliferative but viable state of a cell while crisis is characterized by cell death. Both of them act as natural barriers to proliferation (Greenberg 2005). In contrast to normal cells, cancer cells acquire uncontrolled replicative potential giving rise to macroscopic tumors.

Tandem hexanucleotide repeats called telomeres at the ends of chromosomes shorten with subsequent cell divisions and slowly lose the ability to protect chromosomal ends ultimately compromising cell viability. The length of the telomere dictates how many cell division a cell can pass through before losing protective function and leading to cell death. Consequently, if somehow a cell is able to maintain the length of telomeres, it could undergo divisions indefinitely thus giving rise to cancer. Telomerase, an enzyme capable of adding telomere repeats to the ends of chromosome, is expressed in cancers granting them the expansive replicative potential. It has been seen that transient telomere deficiency seems to facilitate tumor progression. Absence of telomerase in early stages may allow the acquisition of tumor promoting mutations and their expression later may confer them with limitless replicative capacity leading to cancers (Chin et al. 2004; Raynaud et al. 2010).

3.2.5 Nutrient and Oxygen Maintenance

Angiogenesis is simply described as the development of new blood vessels. In a normal adult, the development of vessel system enters in quiescent state with exception to processes like wound healing and menstrual cycle in females. It is regulated by a variety of proteins that bind to cell surface receptors either stimulating or impeding vessel formation.

Angiogenesis is a very crucial part of cancer progression as the new vessels serve to provide cells with continuous supply of nutrients and oxygen to stay alive and proceed to tumor development. In absence of this supply, cancer cells are bound to die off. For example, vascular endothelial growth factor (VEGF) acts on endothelial cells and is known to promote not only angiogenesis but also cell migration. Both hypoxia and oncogene signaling which exhibited by cancerous cells can up regulate its expression. Further, sequestration of VEGF in extracellular matrix keeps them in latent form which can later be activated and released by matrix degrading proteases (Carmeliet 2005; Kessenbrock et al. 2010). Additionally, chronic activation of fibroblast growth factor family members have been shown to induce tumor angiogenesis in a variety of cancers (Ronca et al. 2015).

On other hand, thrombospondin-1 protein works to counter the pro-angiogenic effects by inducing endothelial cell apoptosis and impeding cell migration and proliferation. The loss of thrombospondin-1 is seen as the trigger tipping the balance towards pro-angiogenic factors in tumor progression (Huang et al. 2017). Thus, loss of angiogenesis inhibitors can also result in increased growth of tumors whereas increased levels of anti-angiogenic factors can impede the growth of tumors. This has been the basis for the design of various angiogenesis inhibitors in the market.

Furthermore, a diverse population of cells originating in bone marrow is now known to contribute to the process of tumor angiogenesis. Cells of immune system can gain access to premalignant or progressed tumors and help to induce and sustain angiogenesis enabling further tumor growth and invasion. They may also protect these tumors from drugs targeting endothelial signaling. Particularly, tumor associated macrophages form a notable part of these infiltrating immune cells and not only promote angiogenesis in a variety of cancers but are also linked to resistance to cancer therapies (Qian and Pollard 2010; Goswami et al. 2017).

3.2.6 Invasion and Metastasis

The major complication in cancer which makes it arduous to treat is its capability of invasion and metastasis. This enables it to reach distant sites and causing cancer in sites far away from the primary sites. Invasion and metastasis is responsible for the progression carcinomas to high grades of malignancy and is the major cause for cancer mortality. Some of the cells in tumor mass tend to exhibit alteration in their shape and attachment to nearby cells or extracellular matrix enabling them to gain access to nearby vessels. These then travel via lymphatics and blood vessels escaping the primary sites and lodge themselves into distant sites. Here they again start to proliferate and form secondary tumors. This multistep process is often termed as invasion-metastasis cascade (Hanahan and Weinberg 2011).

E-cadherin molecule is responsible for the maintenance of stable cell junctions. They form adherens junctions among adjacent epithelial cells. Loss of this molecule is associated with most cancers. In fact reconstitution of E-cadherin is suppressive to invasive phenotype of many cancers (Cavallaro and Christofori 2004; Berx and van Roy 2009). Additionally, molecules linked to cell migration are often found to be up regulated in many cancers. In many invasive carcinomas, down regulation of E-cadherin is accompanied by increased expression of a transmembrane protein called N-cadherin which induces pro-migratory and invasive signals via fibroblast growth factor receptor interaction (Cavallaro and Christofori 2004).

Cancer cells are extremely smart in embodying a process involved during embryogenesis and wound healing. This empowers them with features enabling invasion and metastasis. The regulatory mechanism by which transformed epithelial cells acquire capabilities to invade and disseminate to distant sites is called 'epithelial-mesenchymal transition' (EMT) (Klymkowsky and Savagner 2009; Thiery et al. 2009). This involves the loss of cell-cell adhesive interaction and cell polarity and acquiring migratory properties similar to those in mesenchymal stem cells. Several regulatory transcription factors like Slug, Snail, Zeb1/2, and Twist which are involved in EMT are also known to play similar roles during embryogenesis. They are overexpressed in a number of cancers and even elicit metastasis if ectopically expressed. These factors also evoke signals leading to repression of E-cadherins thus disrupting adherens junctions, altering the cell morphology and increasing motility (Micalizzi et al. 2010; Taube et al. 2010). As in early development process, the signals inducing their expression comes from the neighboring cells, in cancer cells too, the interaction with the associated stromal cells can induce the EMT inducing transcription factors (Sleeboom et al. 2018; Karnoub et al. 2007; Micalizzi et al. 2010).

3.3   The Advent of Chemotherapy

Up until mid-1900s, surgery and radiation therapy were the only available options for cancer treatment. But these were only effective against local cancers and ineffective against metastatic cancers. As it was realized that even a single surviving cancer cell in the body could give rise to recurrent cancers so drugs which could reach all the cells of the body became the focus of research in finding solution to cancer. This gave rise to the field of chemotherapy.

Chemotherapy is a major field devoted to the treatment of cancer. Usage drugs for the treatment of cancer is referred to as 'chemotherapy'. These drugs are cytotoxic and act on rapidly dividing cells. Most chemotherapy treatments are not too specific to cancer cells and target all the rapidly proliferating cells. As a consequence they harm the tissues bearing high replication rate like those of bone marrow, gut intestinal lining, oral mucosa and hair follicles. Due to the associated side effects and limitations on dose that can be administered in a single sitting, chemotherapy regimens are given in repeated steps killing a fraction of cells each time.

Despite the limitations associated with chemotherapy, it still remains the indispensable choice for cancer treatment even today. With concerted efforts of the scientific community, there is hope to find new anticancer drugs with improved efficacy and limited toxicity (He et al. 2017).

Various types of chemotherapeutic agents effective against cancer have been devised. Some of them are discussed below (Table 3.2).

3.3.1   Alkylating Agents

They are the oldest class of agents used against cancers and date back to 1949 FDA approval of mechlorethamine. As an option against refractory tumors and capability to treat some malignancy, they still occupy a significant place (Puyo et al. 2014). Most of the alkylating agents are cell-cycle specific with their effect most pronounced during the resting phase of cells.

The cytotoxic effects are exerted mainly by alkylation of DNA bases and impairing processes like replication and transcription. Depending on the nature of alkylating agent, alkylation may occur at different sites in DNA and most give rise to a mixture of adducts. Treatment with nitrosoureas can also lead to the production of isocyanate derivatives which can alkylate proteins and lead to their inactivation. Thus, proteins may also serve as targets of alkylation (Puyo et al. 2014). They include nitrogen mustards, ethylene imines, nitrosoureas, hydrazines, alkyl alkane sulfonates and platinum compounds.

Table 3.2 Types of chemotherapeutic agents for cancer therapy

Chemotherapeutic agent	Example	Description	Reference
Alkylating agents	Nitrogen mustards, ethylene imines, alkyl alkane sulfonates	Alkylate DNA bases impairing crucial cell processes like replication and transcription	Puyo et al. (2014)
Antimetabolites	Fluorouracil, mercatopurine	Mimic natural metabolites and disrupt cell metabolism	Mehrmohamadi et al. (2017)
Antimitotic agents	Taxanes, vinca alkaloids	Inhibit cell division by stabilizing or destabilizing microtubules	Cortazar et al. (2012) and van Vuuren et al. (2015)
Topoisomerase inhibitors	Doxorubicin, ironotecan	Inhibit topoisomerase enzyme leading to single or double strand breaks in DNA	Mordente et al. (2017)
Ribonucleotide reductase inhibitors	Hydroxyurea, gemcitabine, clofarabine	Limit the supply of deoxyribonucleotide triphosphates required for DNA synthesis	Mannargudi and Deb (2017)
Oncogene inhibitors	Panitumumab, Erlotinib	Repress the overexpressed oncogene proteins	Pento (2017) and Yang et al. (2017b)
Angiogenesis inhibitors	Bevacizumab, sunitinib	Inhibit the formation of new blood vessels limiting the supply of food and oxygen	El-Kenawi and El-Remessy (2013) and Pento (2017)

DNA deoxyribonucleic acid

3.3.2 Antimetabolites

Probably, the earliest rationally designed drugs against cancer were those developed against nucleic acids. These analogs of nucleic acids (DNA or RNA) are referred to as 'antimetabolites'. Being similar to natural metabolites, these act as mimics and get into the metabolism of the cell, thereby functioning to disrupt it. They act at specific cell stage in cell cycle i.e. synthesis (S) phase.

Methotrexate (aminopterin) was the first purine analog approved in 1953. In 1962, first pyrimidine analog, 5-Fluorouracil saw approval. Since then several purine and pyrimidine analogs like Pemetrexed, Clofarabine, Pralatrexate, Hydroxyurea, gemcitabine and others have been developed for use in cancer treatment. (Peters 2014; Balboni et al. 2019). Drugs such as 6-mercaptopurine and 6-thioguanine are purine analogs whereas drugs like 5-fluorouracil, floxuridine, cytarabine and gemcitabine are pyrimidine analogs. This category also includes adenosine deaminase inhibitors like cladribine, nelarabine and pentostatin as well as folic acid antagonist like methotrexate.

3.3.3   Antimitotic Agents

These are another class of cell cycle specific inhibitors used in cancer chemotherapy and act at the mitotic (M) phase of cell cycle. As the name suggests, these agents work by inhibiting the mitosis of cancer cells. They act by targeting the microtubules which are important for aligning the chromosomes during cell division. By resisting this process, these activate the spindle assembly checkpoint (SAC) which ceases the cell division and the cell subsequently undergoes apoptosis (van Vuuren et al. 2015).

These include drugs that stabilize microtubules like taxanes and epothilones and those that destabilize microtubules like vinca alkaloids (Cortazar et al. 2012; van Vuuren et al. 2015).

3.3.4   Topoisomerase Inhibitors

During the process of replication or transcription of DNA, it may become overwound or under wound. Enzymes that catalyze the cycle of breakage and rejoining of DNA backbone and correct these topological problems are called 'topoisomerases'. Thus, these are crucial in maintaining the stability of genome.

Inhibiting this process may result in single stranded or double stranded breaks compromising genome integrity. This stress is sensed by the cell and it eventually leads to apoptosis. Topoisomerases serve as important targets for various drugs against cancer. These include anthracycline, doxorubicin, etoposide, ironotecan and topotecan (Mordente et al. 2017).

3.3.5   Ribonucleotide Reductase Inhibitors

Continuous supply of deoxyribonucleotides is essential for maintenance of nucleic acid integrity within a cell with their levels being altered depending on the stage of cell cycle. Alterations in their supply can result in hampered genome integrity if they are not present in appropriate levels. The enzyme ribonucleotide reductase catalyzes the reduction of ribonucleotides to deoxyribonucleotides thus acting to supply the building blocks required for DNA synthesis. It is involved in the rate limiting step of de novo pathway of deoxyribonucleotide triphosphates biosynthesis (Aye et al. 2015).

Uncontrolled proliferation of cancer cells must be supported by appropriate deoxyribonucleotide triphosphate supply and in order to achieve this many cancer cells are known to acquire increased ribonucleotide reductase expression. Ribonucleotide reductase inhibitors work by limiting the supply of DNA building blocks thereby inhibiting the cancer cell growth. These include agents like hydroxyurea, gemcitabine and clofarabine (Aye et al. 2015; Mannargudi and Deb 2017).

3.3.6   Oncogene Inhibitors

As earlier described, activation of various oncogenes can be the cause of cancer. Targeting these oncogenic proteins to repress their function is another strategy to mitigate cancer. The inception of this strategy began with the usage of retinoic acid, a metabolite of vitamin A, to treat acute promyelocyte leukemia (APL). APL involves reciprocal translocation between the promyelocyte leukemia gene (PML) and Retinoic acid receptor alpha gene (RARalpha). The resulting PML/RARalpha protein acts to repress cell differentiation. Administration of retinoic acid induces differentiation of cells by inactivating the fusion protein (Uray et al. 2016).

Monoclonal antibodies targeting various oncogenic receptors have also been developed. For example, Panitumumab targets epidermal growth factor receptor (EGFR) expressed on colon cancer cells. Cetuximab is another EGFR targeting antibody. Trastuzumab is used to treat human epidermal growth factor receptor 2 (HER2) positive breast cancer and other cancers expressing HER-2 protein. It acts at extracellular HER-2 protein impeding the EGFR activity (Pento 2017).

Small molecule inhibitors like gefitinib and erlotinib targeting overexpressed EGFR have also been in use in treatment of cancers like non-small cell lung cancer (NSCLC) (Yang et al. 2017b; Rawluk and Waller 2018).

3.3.7   Angiogenesis Inhibitors

When cancer cells start growing in uncontrolled manner, their demand for nutrients and oxygen soar. This exorbitant call is then supplied by the new sprouting vessels that are induced by various pro angiogenic mediators like growth factors, cytokines and matrix degrading enzymes. Conversely, inhibiting the growth of vessels can starve cancer cells and impede their unlimited growth.

This class includes agents like bevacizumab which target vascular endothelial growth factor (VEGF) and inactivates the VEGF receptor thereby inhibiting blood vessel formation and restricting their growth (Pento 2017). Another example is an epidermal growth factor receptor (EGFR) Tyrosine Kinase Inhibitor (TKI), Iressa, capable of impeding pro-angiogenic factor expression in tumors (El-Kenawi and El-Remessy 2013). Sunitinib is another example which is a multiple receptor TKI which prevents the response of tumors to VEGF by inactivating receptors for VEGF, platelet-derived growth factor and stem cell factor (Kit) (Roskoski 2007; El-Kenawi and El-Remessy 2013).

Despite many chemotherapeutic treatment options available against cancer, even today most of the cancer treatment therapies are not only expensive and toxic but many a times also have very low effectiveness. In addition repeated administrations due to dose considerations and side effects add to the patient discomfort. In such a scenario investing in natural compounds is a very attractive recourse against cancer which offers higher efficacy at a low cost and fewer side effects (Barr et al. 2016;

Granja et al. 2017). In the following section we explore various natural products that have proven to be promising alternative chemotherapeutics and how they can be bettered in providing the optimal and patient compliant cancer treatment.

3.4 Emergence of Natural Products in Cancer Chemotherapy

Natural products have intrigued the cancer chemotherapy field and captured its attention in the recent time. Their natural origins, non-toxic nature as well as their ability to act via multiple targets are some of the attributes that have been largely responsible for this interest. Co-administration of natural products with conventional chemotherapeutics has also been implicated in alleviating toxicity and reducing dosage of conventional drugs (Conklin 2005; Lee et al. 2011). As conventional agents directed against single targets often prove to be unsuccessful, natural products are emerging as the ideal candidates against the war on cancer. They are increasingly proving to be successful in impeding proliferation of tumor cells, inducing their apoptosis and cell cycle arrest as well as suppressing angiogenesis and metastasis thereby attacking cancer from multiple sides in order to attain better treatment against carcinogenesis (Arora and Jaglan 2016). Moreover, greater than 50% of chemotherapeutic drugs are from natural source or its derivatives (Watkins et al. 2015; Saneja et al. 2017b).

Green tea, for instance, is a popular beverage and is proclaimed to have many health benefits. A polyphenol, epigallocatechin gallate (EGCG), occurring in green tea is largely credited for the beneficial effects with anticancer effect drawing a huge amount of attention. EGCG have been shown to possess anticancer properties in many studies. It also shows induction of apoptosis, cell cycle arrest and inhibition of proliferation. It has been implicated in affect multiple signaling pathways involved in survival and proliferation of tumor cells as well as reducing the expression of invasion promoting proteins (Gan et al. 2018).

Curcumin, another nutracuetical, is popular in South and Southeast Asia and used as a food flavoring agent. This bright yellow compound from turmeric has been investigated in the past decades for its anticancer potential. It has been shown to control cancer by acting on a multitude of targets ranging from various transcription factors, protein kinases, growth factors to inflammatory cytokines and oncogenic proteins. The results of an inverse molecular docking study showed that curcumin is capable of binding to multitude of protein targets ranging from metalloproteinase-2, folate receptor β, mitogen-activated protein kinase 9, apoptosis-inducing factor 1 to epidermal growth factor receptor. Furthermore, several potentially new targets were identified which included deoxycytidine kinase, tyrosine-protein phosphatase non-receptor type 11, macrophage colony-stimulating factor 1 receptor and death-associated protein kinase 2 (Furlan and Konc 2018).

Resveratrol, a polyphenolic phytoalexin antioxidant present in red grapes is another compound that has been emerging due to its anticancer properties. It like other natural compounds has also been shown to be effective in a wide range of cancers like breast, prostate and colon to name a few by its effects on a variety of transcription factors, proteins or receptors involved multiple processes like apoptosis, cell survival, cell cycle arrest or angiogenesis (Rauf et al. 2018). In one study resveratrol treatment of tongue and liver carcinoma cells led to a reduction of blood vessel promoting factors like hypoxia-inducible factor-1 alpha and vascular endothelial growth factor expression. Another study concluded that resveratrol was capable of inducing apoptosis in renal carcinoma cells by down regulating anti-apoptotic factors like B-cell lymphoma 2 (Bcl-2) and up regulating pro-apoptotic factors like Bcl-2 associated X protein (Bax) (Zhang et al. 2005; Liu et al. 2018b).

Though how natural products exert their anticancer properties is not fully elucidated, we try to shed light on the possible ways by which some of the products have been reported to act on the system to put a brake on carcinogenesis.

3.4.1 Curcumin

Curcumin, a principal curcuminoid, is a polyphenolic compound derived from *Curcumin longa*. Its use has been described in ancient Ayurvedic medicine but its anti-inflammatory and anti-cancer potential has been in investigation for the past few decades. It is speculated to act in controlling cancer by targeting multiple properties and acting negatively on various growth factors, transcription factors, protein kinases, inflammatory cytokines, and oncogenic proteins associated with cancerous cells thereby impeding initiation, progression and metastasis (Shanmugam et al. 2015). Probably first reported study was that by Kuttan in 1987 where patients with external cancerous lesions experienced symptomatic relief on treatment with curcumin (Kuttan et al. 1987). Since curcumin has shown to be effective in inhibiting a wide variety of cancer cells like those of digestive, lymphatic and immune, reproductive, urinary, nervous, pulmonary, skeletal systems, and the skin.

Curcumin is known to act in a multifaceted manner targeting many mechanisms at a time.

3.4.1.1 Effect on Transcription Factors

Several signal transduction pathways like those involving mitogen-activated protein kinase (MAPK), nuclear factor kappa light chain enhancer of activated B cells (NF-kB) and activator protein 1 (AP-1) to name a few are altered in cancerous cells. Curcumin can help in restoring these pathways to their unaltered state. It leads to induction of apoptosis in tumor cells by modulation of various signal transduction pathways thereby impeding carcinogenesis (Shanmugam et al. 2015). Inverse molecular docking of curcumin with human protein structures showed that it binds to a variety of protein targets like metalloproteinase-2, folate receptor β, DNA (cytosine-5)-methyltransferase 3A, mitogen-activated protein kinase 9, apoptosis-inducing factor 1 and epidermal growth factor receptor. It also identified deoxycytidine kinase, tyrosine-protein phosphatase non-receptor type 11, core histone macro-H2A.1, macrophage colony-stimulating factor 1 receptor, tryptophan-tRNA ligase and death-associated protein kinase 2 as potentially new targets through which it may exert its anticancer effects (Furlan and Konc 2018).

A study exploring the effects of curcumin along with irradiation of esophageal squamous-cell carcinoma (ESCC) observed that pro-apoptotic effects of irradiation were enhanced in the presence of curcumin. Also, a significant sensitization of ESCC cells to irradiation (IR) occurred on pretreatment with curcumin. ESCC xenograft mice survived longer on IR therapy when curcumin was co-administered. It was also reported that these effects were probably due to inhibition of NF-kB signaling. Thus, synergistic effects of curcumin and irradiation were evident (Liu et al. 2018a). Another study evaluating the potential synergism of curcumin and paclitaxel in breast cancer showed induction of apoptosis and necrosis by both. Enhanced apoptosis was observed with combined treatment of curcumin and paclitaxel (Calaf et al. 2018).

3.4.1.2 Effect on Apoptosis

As discussed earlier apoptosis functions to remove damaged or unwanted cells from the system. Cancer cells acquire mechanisms to evade apoptosis for their survival. Curcumin shows anticancer effects by promoting apoptosis. In a study, mouse myeloma cells were used to look into the anticancer effects of curcumin. The expression of genes associated with Notch3-p53 signaling axis were evaluated. Apoptosis was shown by curcumin treated cells and an enhanced expression of Notch-3 responsive genes was observed. A significant down regulation of B-cell lymphoma-2 (Bcl-2) at both mRNA and protein level with up regulation of Bcl-2-associated X was also reported. Thus, the anticancer effects of curcumin on mouse myeloma cells were attributed to its effects on Notch3-p53 axis leading to apoptosis of cancer cells (Zhang et al. 2019). A separate study investigating the effects of curcumin on gastric cancer cells found inhibition of cell proliferation, induction of apoptosis and autophagy in cancer cells. The study also probed the effects on p53 and phosphoinositide 3-kinase (PI3K) signalling pathways. An activation of p53 signalling along with inhibition of PI3K signalling was reported. Activation of p53 pathway was associated with enhanced p53 and p21, while inhibition of PI3K pathway was associated with decreased PI3K phosphorylation of protein kinase B (PKB), and mammalian target of rapamycin (mTOR) (Fu et al. 2018). Curcumin has also been reported to exert anticancer effects against colon cancer. A study investigating its effects on Dukes' type C colon cancer examined cellular viability, proliferation, cytotoxicity, apoptosis and cell cycle distribution. A significant inhibition of cellular viability and proliferation in a dose and time dependent manner with no cytotoxic effects was observed on curcumin treatment. Also, cell cycle arrest at Gap1 phase, reduction of cell population in Synthesis phase with induction of apoptosis in colon adenocarcinoma cells was also reported as the effects of curcumin treatment (Dasiram et al. 2017).

The effects of curcumin on cancer stem cells (CSCs) have also been investigated. Sonic hedgehog (Shh) and Wnt/β-catenin pathways are known to play a crucial role in the maintenance of stemness of CSCs. A study examining the inhibition of breast CSCs on curcumin treatment reported a decreased activity of breast CSCs. This was supported by the observed inhibition of tumor sphere formation, decrease in breast CSCs markers like P-glycoprotein 1, Nanog, Oct4 as well as inhibition of proliferation and induction of apoptosis. They also showed that Shh and Wnt/β-catenin down regulation lead to inhibition of breast CSCs. They concluded that the inhibitory effect exerted by curcumin on breast CSCs was by suppression of both Shh and Wnt/β-catenin pathways (Li et al. 2018a). Another study investigating the effects of curcumin on prostate cancer cells observed that survival and proliferation of prostate cancer cells were inhibited on curcumin treatment. It concluded that Notch-1 signal pathway played a role in the curcumin mediated inhibition of survival and metastasis in prostate cancer cells (Yang et al. 2017a).

When acute myeloid leukemia or burkitt lymphoma cells were subjected to curcumin treatment, the CSC associated markers were reduced. Additionally, colony formation decreased and several CSC-associated proteins like Notch-1, Gli-1 and Cyclin D1 were down regulated. This implicated that curcumin mediated decrease of CSC markers in lymphoma/leukemia cells was potentially through inhibition of self-renewal (Li et al. 2018b). Inhibition of proliferation and induction of apoptosis on curcumin administration was also shown on human colorectal carcinoma cells. P53 mutations and pre-mRNA processing factor 4B (Prp4B) overexpression have been reported as characteristics of cancer cells. Curcumin administration activated caspase-3 and lead to a reduction of p53 and Prp4B expression in human colon cancer cells (Shehzad et al. 2013). Curcumin is also known to mediate its inhibitory effects on primary tumors occurring in central nervous system called gliomas. A notable inhibition of growth in gliomas on curcumin treatment was observed in a study. The same study also reported that a low dose of curcumin lead to the induction of G2/M cell cycle arrest while a high dose lead to the induction of G2/M cell cycle arrest along with S phase arrest. It also showed that expression of inhibitor of growth protein 4 and p53 was up regulated during cell cycle arrest (Liu et al. 2007).

3.4.2 Epigallocatechin-3-Gallate

It has been stipulated that some of the polyphenols occurring in green tea have anti-cancer activities. Epigallocatechin-3-gallate (EGCG), for instance, has been shown to inhibit cell growth cancer cells. A study by Du and Zhang compared the chemopreventive potentials of some of the polyphenols and found that most potent antiproliferative effects on colon cancer cells were shown by EGCG. G1 phase cell cycle arrest and apoptosis were also significantly induced. The gallic acid group seemed to enhance the anticancer property (Du et al. 2012). Catechin, a plant secondary metabolite belonging to the family of flavonoids is a naturally occurring antioxidant. It's isomer in cis configuration forming an ester with gallic acid is called epigallocatechin gallate (EGCG) which is the most abundant catechin found in tea. EGCG has been shown to have the capability to inhibit all stages of carcinogenesis from initiation and promotion to progression. Though its mechanism is still not conclusively known. It possess antioxidant property but this property alone cannot explain its antiproliferative effects. However, its pro-oxidant property has also been shown by Farhan, Khan, et al. It has hypothesized that EGCG is able to mobilize copper bound to chromatin leading to intranucleuosomal DNA breakage. Interaction of EGCG with copper bound to DNA and localized generation of hydroxyl radicals causes catechin-copper induced DNA damage. The reduction of copper gives rise to oxidized species. Hence, copper is a key factor that shifts EGCG from anti-oxidant to pro-oxidant action. As cancer cells have elevated levels of copper, thus, they are more likely to generate reactive oxygen species (ROS). ROS stress above a threshold then results in apoptosis. Hence, EGCG can also promote pro-oxidant cancer cell death, by mobilizing copper (Farhan et al. 2016; Granja et al. 2017).

EGCG also acts to modulate multiple cell signalling pathways related to proliferation, survival or death, and also down regulating proteins involved in invasion thus, helping to manage cancer.

3.4.2.1 Effect on Tumor Suppressors

The guardian of genome, p53, is altered or inactivated in a variety of tumors. As we know from earlier discussion tumor suppressor alteration leaves the cell without any regulatory check resulting in uncontrolled proliferation and hence, tumorigenesis. Enhancement of transcriptional activity of p53 gene related to cell cycle arrest and apoptosis is known to occur via acetylation of carboxy-terminal lysine. The acetylation of genes is counterbalance by histone deacetylases (HDACs). In a study of human prostate cancer cells it was demonstrated that epigallocatechin-3-gallate (EGCG) impede HDACs and activated p53 via acetylation at lysine residues (Lys 373, Lys 382). Additionally, this inhibition was dose and time dependent and its interruption lead to a loss of p53 acetylation. Also, increased expression of proteins p21 and Bax occurred which are associated with the induction of apoptosis (Thakur et al. 2012). Another study by Qin J showed p53 dependent apoptosis of cancer cells by EGCG. Using mouse epithelial cells, they showed that EGCG down regulated protein serine/threonine phosphatase-2A (PP-2A), hyperphosphorylated p53 and up regulated Bak which induced apoptosis. They also demonstrated that RNA interference mediated knockdown of p53 and Bak caused substantial inhibition of EGCG induced apoptosis. Overexpression PP-2A catalytic subunit down regulated p53 phosphorylation at Ser15, and attenuated expression of Bak, thus antagonizing EGCG induced apoptosis whereas inhibition of PP-2A promoted EGCG induced apoptosis (Qin et al. 2008).

The possibility of enhanced efficacy and overcoming of resistance to conventional therapies via co-administration with natural products is an active area of research. Combinatorial treatment with EGCG and sulforaphane (SFN) along with cisplatin therapy on ovarian cancer cells reduced cancer cell viability in a time and dose dependent manner. It also enhanced G2/M arrest and apoptosis in addition to up regulating p21 expression (Chen et al. 2013). Antifolates are known to disturb the folic acid metabolism, inhibiting DNA and RNA synthesis. They have also been linked with decreased cellular methylation. It has been shown that EGCG inhibits dihydrofolate reductase (DHFR) disturbing the folate metabolism leading to decreased production of nucleotides and impeding the growth of colon cancer cells. Hypermethylation of tumor suppressors like p16 is observed in many cancers. EGCG also shifted the p16 methylation pattern to unmethylated form thus acting to rescue carcinogenesis (Navarro-Peran et al. 2007).

3.4.2.2   Effect on Apoptosis

Apoptosis or programmed cell death is body's way of getting rid of damaged or unwanted cells without alerting the immune system. Many cancers devise ways to bypass this process in order to survive and multiply indefinitely. EGCG has also been implicated to act of this process and control cancerous growth via inducing their apoptosis. EGCG was shown to cause mitochondrial transmembrane potentials loss and cytochrome c release, thus, leading to induction of apoptosis in myeloma cells. Intracellular reactive oxygen species (ROS) were elevated during EGCG induced apoptosis (Nakazato et al. 2005). In a study investigating the antiproliferative potential of EGCG on bladder cancer cells, it was found that treatment with EGCG modulated the Bcl-2 family proteins via inhibition of phosphoinositide 3-kinase/protein kinase b pathway and in turn leading to enhanced apoptosis (Qin et al. 2007). The induction of apoptosis and subsequent inhibition on growth of mouse viral mammary epithelial cancer cells has also been reported on EGCG treatment. Also, a significant reduction in tumor volume of breast cancer cells in a mouse model has been reported (Mineva et al. 2013; Rahmani et al. 2015).

3.4.2.3   Effect on Angiogenesis

Angiogenesis as discussed earlier is required by cancer cells to supply for the nutrients and oxygen to sustain their survival. Inhibition of angiogenesis can starve cancer cells resulting in their elimination. At least part of EGCG anticancer activity can be attributed to its effect on vascular endothelial growth factor (VEGF). A study investigating the effects of EGCG on VEGF induction and intracellular signaling in colon cancer cells, reported that EGCG treatment in addition to inhibiting extracellular signal regulated kinase activation also impeded the expression of VEGF. EGCG treatment of athymic Balb/c nude mice inoculated with human colon adenocarcinoma cells showed reduction in tumor growth, microvessel density and cell proliferation along with enhanced tumor cell and endothelial cell apoptosis (Jung et al. 2001). In another study by Gu JW tested the effect of EGCG on breast cancer cells. Significant reduction in tumor weight of those mice inoculated with breast cancer cells was observed that received EGCG. Significant inactivation of VEGF, hypoxia inducible factor-1 and nuclear factor kappa light chain enhancer of activated B cells (NF-kB) was also observed in cultured breast cancer cells. These finding supported the notion that EGCG can work in a multifactorial manner by targeting not only tumor cells but also tumor vasculature thus impeding growth, proliferation as well as angiogenesis of tumor cells (Gu et al. 2013).

3.4.2.4 Effect on Oncogenes

Activation and overexpression of oncogenes is the reason of a number cancers. EGCG is also reported to exert its effect by acting on various oncogenes. Alteration of EGFR signalling is observed in many cancers. In colon cancer cells, inactivation of epidermal growth factor receptor (EGFR) and human epidermal growth factor receptor 2 (HER2) has been reported as effects of EGCG. High level constitutive activation of human epidermal growth factor receptor 3 (HER3) on cultured colon cancer cells decreased within 6 h of EGCG treatment. Additionally, phosphorylated forms of extracellular signal regulated kinase and protein kinase B as downstream signaling proteins also decreased (Shimizu et al. 2005). In another study examined the effects of EGCG on the EGFR pathway in tamoxifen resistant breast cancer cells. EGFR mRNA and protein was down regulated in a dose dependent manner upon EGCG treatment. EGCG also decreased EGFR phosphorylation in tamoxifen resistant cells. It also showed reduction in proteins involved in invasion and metastasis like matrix metalloproteinase-2 (MMP-2) and MMP-9. Alternatively tissue inhibitor of metalloproteinases-1 (TIMP-1) and TIMP-2, which have role in down regulation of MMPs, increased. Thus, it was shown that EGCG could attenuate tamoxifen resistance in breast cancer cells and impede its growth and invasion (Farabegoli et al. 2011).

3.4.3 Resveratrol

Resveratrol is a polyphenol occurring in several plants like vine (climbing plants) produced in response to injury or pathogen attack. It is a hydroxyl derivative of stilbene belonging to phenylpropanoids family. It is reported to be effective against a wide variety of cancers like those of breast, colon, leukemia, ovarian, prostate, pancreatic and others. It is stipulated to act non-specifically on multiple targets as opposed to traditional chemotherapeutics (Pavan et al. 2016).

3.4.3.1 Effect on Transcription Factors

Overexpression of hypoxia inducible factor-1 alpha (HIF-1alpha) is a characteristic of many tumors and has an important role in promoting blood formation vessel that in turn feed the cancer cells enabling their survival. Zhang Q investigated the effect of resveratrol HIF-1alpha accumulation and vascular endothelial growth factor (VEGF) expression on carcinoma cells of tongue and liver. They observed a marked inhibition of HIF1alpha protein on treatment with resveratrol. Pretreatment with resveratrol was found to also reduce VEGF expression. Resveratrol mediated the enhancement of degradation by 26S proteasome system seemed to reduce the half-life of HIF-1alpha. It also impeded activation of extracellular-signal-regulated kinase and protein kinase B thereby decreasing HIF-1alpha accumulation and VEGF activation (Zhang et al. 2005). Many transformed cells show excessive aerobic glycolysis. A study on B cell lymphomas showed inhibition of phosphatidylinositol 3-kinase (PI-3K) along with glucose metabolism which corresponded with cell cycle arrest on treatment with resveratrol. Proteins like S6 ribosomal protein and protein kinase B were also inactivated (Faber et al. 2006).

Metastasis and cell clusters are characteristics of advanced stage ovarian cancer. A study investigating the importance of nuclear factor kappa light chain enhancer of activated B cells (NF-kB) in ovarian cancer 3D cell aggregates on resveratrol treatment found reduction in tumor growth as well as metabolism. The growth inhibition corresponded not only with reduced NF-kB levels but also diminished VEGF secretion. They suggested resveratrol mediated inhibition of ovarian cancer 3D cell aggregates via NF-kB controlled VEGF secretion (Tino and Chitcholtan 2016). Resveratrol mediated regulation of NF-kB has also been implicated in reduced expression of miR-221, which is an oncogenic microRNA, in melanoma cells (Wu and Cui 2017).

3.4.3.2 Effect on Tumor Suppressor

Normal functioning of the tumor suppressor and caretaker of genome, p53, is of utmost importance. Any alteration can leave the cell without any countermeasures to deal with any kind of stressed condition. A study by De Amicis F on breast cancer cells probed that effect of resveratrol on cell proliferation. The results showed the appraisal of resveratrol in adjuvant therapy against those breast cancers that had developed hormonal resistance. The cell proliferation inhibition was found to be particularly dependent on p53. The induction of p53 via p38 phosphorylation lead to inhibition of estrogen receptor alpha expression and these effects were overturned small interfering RNA targeting p53 (De Amicis et al. 2011). Aggregation of mutant p53 forming amyloid like structures has been indicated as a crucial factor in tumorigenesis. Defects in p53 resulting in its functional inactivation are linked to amyloid aggregation. A study to investigate if resveratrol had any effect on p53 aggregation found that resveratrol was able to interact with p53 core domain. Using breast cancer cell line, a reduction in protein aggregation of mutant p53 was observed on resveratrol treatment. Additionally, the proliferation and migration properties of the treated cells were significantly reduced (Yang-Hartwich et al. 2015; Ferraz da Costa et al. 2018).

3.4.3.3 Effect on Apoptosis

Apoptosis or programmed cell death occurs in a controlled manner to get rid of unwanted or infected cells. Tumor cells need to devise ways by which they can evade apoptosis ensuring their survival. Activating apoptosis against tumors can be a beneficial strategy to naturally get rid of cancerous cells. Resveratrol has been known to exhibit anti-inflammatory as well as tumor suppressor properties. A study probing the effects of resveratrol on renal cell carcinoma (RCC) concluded that it could suppress RCC progression. It was observed that resveratrol down regulated anti-apoptotic Bcl-2 and up regulated pro-apoptotic Bax while also promoting activation of adenosine monophosphate-activated protein kinase (AMPK) and p53 expression. Additionally, the autophagy genes were up regulated and mammalian target of rapamycin (mTOR) phosphorylation was inhibited. Thus, resveratrol could induce apoptosis and also lead to suppression of viability and migration in RCC (Liu et al. 2018b).

In another study investigating the anti-proliferative potential of resveratrol observed significant inhibition of cell proliferation, cell cycle arrest and subsequent apoptosis of hepatoma cells infected with hepatitis C virus. They also found an increase in expression of p21/Waf1 protein along with decrease in expression of cyclin-dependent kinase-2 (cdk-2) and cyclins A and E. Additionally, enhanced caspase activity, mitochondrial membrane depolarization, increased pro-apoptotic proteins and decreased anti-apoptotic proteins were observed (Liao et al. 2010). Resveratrol has also been investigated as a radiosensitizing agent in combination with radiation therapy (XRT) for localized prostate cancer (PCA) treatment. Synergistic enhancement in the induction of apoptosis as well as inhibition of cell proliferation were shown on using the resveratrol-XRT combination. These effects paralleled with an enhanced p21, p15 and mutant p53 expression with a reduction of cyclins D and B and cdk-2 expression. Additionally, enhanced apoptosis correlated well with enhanced Fas and TNF-related apoptosis-inducing ligand Receptor 1 (TRAILR 1) expression (Fang et al. 2012).

3.4.3.4 Effect on Angiogenesis

The malignant progression in hepatocellular carcinoma (HCC) is facilitated by hepatic stellate cells (HSCs). It has been reported that the induction of angiogenesis by HSCs in HCC was mediated by enhanced expression of Gli-1 which in turn stimulated the production of reactive oxygen species (ROS) and potentiated HCC cell invasiveness. Resveratrol diminished this induced angiogenesis along with suppression of ROS production. Additionally, it was also shown that Gli-1 expression down regulated along with C-X-C chemokine receptor type-4 receptor and interleukin-6 expression (Yan et al. 2017). A study investigating the effects of resveratrol on angiogenesis and inflammation in experimentally induced endometrial implants evaluated the endometriotic lesions, monocyte chemoattractant protein-1 (MCP-1) and vascular endothelial growth factor (VEGF) levels in blood and peritoneal fluid. Post treatment, the areas of implants and histopathological scores reduced. Also, the levels of MCP-1 and VEGF in serum and peritoneal fluid of resveratrol treated groups reduced. Thus, resveratrol showed promising potential as angiogenesis and inflammation inhibitor in endometriosis (Ozcan Cenksoy et al. 2015).

Co-administration of natural products with conventional therapeutics is an active area of research. Co-administration can be helpful in enhancing the efficacy and circumventing the problem of toxicity and resistance associated with conventional anticancer drugs. Many of the promising natural products showing anticancer properties have been investigated for combinatorial treatment. For example, when epigallocatechin gallate (EGCG) found in green tea and sulforaphane found in cruciferous vegetables, were administered along with cisplatin therapy for ovarian cancer cells, it was seen that cancer cell viability reduced which was time and dose dependent. Also, up regulation of G2/M arrest and apoptosis with increased p21 expression was noted (Chen et al. 2013). In another study, when esophageal squamous-cell carcinoma (ESCC) cells were pretreated with curcumin, the proapoptotic effects of irradiation therapy elevated. ESCC cells seem to have been sensitized to irradiation. The xenograft mice also survived longer upon co-administration (Liu et al. 2018a). Further, prostate cancer treatment by radiation therapy when combined with resveratrol lead to augmentation of apoptosis as well as impediment of cell proliferation which corresponded with elevated p21, p15 and p53 expression according to a study reported by Fang, DeMaro et al. (Fang et al. 2012).

Natural products as good as they seem, are still not devoid of limitations that have proven to be major stumbling blocks in their translational potential. Most of the natural products showing promising anticancer potential are of hydrophobic nature. This attribute has been a crucial obstacle in delivering these agents inside the system. Majority of these agents exhibit poor aqueous solubility and poor permeability leading to their low bioavailability in humans. Moreover, physiological processes like complex formation with other components and biotransformation may also contribute to low bioavailability (D'Ambrosio et al. 2011; Arora and Jaglan 2016).

3.5 Impact of Polymeric Nanocarriers on Natural Products as Alternative Chemotherapeutics

The emergence of nanotechnology has ignited growing interest into their potential applications to the field of medicine. It enables us to achieve various possibilities in the field of biology that could not have been thought of as achievable a few years before. Colloidal carrier systems of submicron sizes have long been investigated as drug carrier system. Their potential to alter the basic properties of drugs like solubility, stability and related toxicity to name a few have made them attractive area of research in the past few years. Due to their small size they possess a high surface area to volume ratio and may provide access to otherwise inaccessible sites within the body.

Its entry into the field of cancer chemotherapy promises to circumvent many limitations associated with conventional chemotherapeutics like low solubility, non-specific targeting, toxicity and low efficacy. Past few years have seen tremendous advancement in the development of chemotherapeutic drug loaded nanoparticles targeted against a wide variety of cancers. Nanoparticles with appropriate size, shape as well as surface properties have been designed to improve upon the properties of therapeutics that they carry. Nanoparticle loaded with therapeutics also have the capability to exploit the pathophysiological conditions of tumor microenvironment like enhanced permeability and retention. They can penetrate effectively into tumoral mass through passive targeting due to the leaky vasculature. Engineering nanoparticles decorated with suitable ligands like hyaluronic acid, folic acid, antibodies and peptides at their surface confer them with the ability to specifically deliver the payloads to cancerous cells via selective binding to overexpressed receptors. Additionally, more than one drug could be delivered via same carrier enabling combination chemotherapy. Thus, nanoparticles are emerging as the new generation delivery systems for cancer chemotherapy (Saneja et al. 2014; Sun et al. 2014; Zhong et al. 2014; Blanco et al. 2015) (Fig. 3.1).

Most of the natural products emerging are hydrophobic and face solubility problems. Using nanocarriers for their delivery can improve their solubility as well as stability. Controlled release, site specific delivery, lower dose and lower toxicity of bioactive compounds are some of the additional features that nanocarriers come with. Furthermore, the possibility of altering the physiochemical properties like size, shape and surface properties of the nanocarriers themselves serves as the

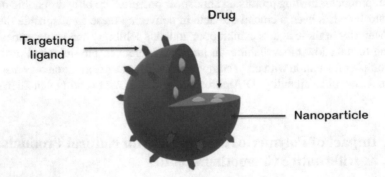

Fig. 3.1 Drug encapsulated nanoparticle. Several drugs can be encapsulated inside nanoparticles and delivered to otherwise inaccessible sites owing to the size of nanoparticles. They can also be decorated with ligands in order to deliver them to specific sites

cherry on the cake if one decide to use nanodelivery system for the intended drug (Din et al. 2017).

Though nanocarriers can be polymeric, lipid based or inorganic nanocarriers, here we will focus our discussion on polymeric nanocarriers as we believe they are the most promising option and can be engineered into various shapes and sizes and can be decorated with a variety of targeting ligands.

The polymers used to fabricate nanocarriers can be natural or synthetic polymers. Natural polymers include polymers like gelatin, collagen, chitosan, alginic acid and dextran. Synthetic polymers can be biodegradable like poly lactic-co-glycolic acid or non-biodegradable like poly (methyl methacrylate). However, natural polymers have some limitations like antigenicity, risk of infection and non-uniformity from batch to batch. Synthetic polymers seem to have an edge as they can be tuned to develop carriers with specific properties. A simple change in the building block or preparation method can be used to modulate the properties accordingly. Since their preparation methods are well established they also offer the advantage of being reproducible.

Amongst the synthetic polymers, biodegradable polymers like poly lactic-co-glycolic acid (PLGA) and polylactic acid (PLA) have particularly gained particular attention for biomedical applications (Coelho et al. 2010). Such polymers have a long history of use as suture materials and in many controlled release formulations (Frazza and Schmitt 1971; O'Hagan and Singh 2003). These polymers are biodegradable, biocompatible and Food and Drug Administration (FDA) approved. As polymers degrade naturally at in vivo conditions and their degradation products (lactic acid or glycolic acid) are normal physiological constituents they are non-toxic and non-immunogenic. Apart from safety, delivery using PLA/PLGA polymeric particulate systems offers other advantages like protection of therapeutic agents from cellular enzymes, better stability during storage and lower toxicity. They can also be tailored to provide sustained release of the bioactive compound thereby lowering the dose or the number of sittings required to achieve the desired effect. This can ultimately help cutting down on cost of treatment for the patients.

Many promising anticancer products have been delivered using polymeric nanocarriers to improve upon their potential. For instance, epigallocatechin gallate (EGCG), the green tea occurring polyphenol when delivered after loading into PLA nanoparticles coated with polyethylene glycol (PEG) showed significant elevation of pro-apoptotic factors along with lowering of anti-apoptotic factors. The most interesting fact in this study was that the dose required to achieve the same effects with free EGCG was 10 times higher. A remarkable reduction of dose required for angiogenesis inhibition was also demonstrated by in the same study (Siddiqui et al. 2009). A different study aiming at improvement of the oral bioavailability of curcumin entrapped curcumin in PLGA nanoparticles. The prepared particles manifested a 640-fold increase in water solubility. Oral delivery of these particles showed an increase of 5.6-fold relative bioavailability (Xie et al. 2011). In another study, curcumin loaded PLGA nanoparticles showed a tenfold increase in solubility as well as superior anti-migratory potential relative to free curcumin upon treatment of cancer cells. The particle treatment of breast and lung cancer cells lead to a notable

Table 3.3 Polylactic acid/poly lactic-co-glycolic acid based products approved for clinical use

Name	Active ingredient	Indication	Approval date
Lupron	Leuprolide Acetate	Prostate cancer	1989
Sandostatin	Octreotide acetate	Acromegaly	1998
Atridox	Doxycycline hyclate	Chronic adult periodontitis	1998
Trelstar	Triptorelin pamoate	Prostate cancer	2000
Eligard	Leuprolide acetate	Prostate cancer	2002
Bydureon	Exenatide synthetic	Type 2 diabetes	2012
Signifor	Pasireotide pamoate	Acromegaly	2014
Bydureon Bcise	Exenatide	Type 2 diabetes	2017
Triptodur Kit	Triptorelin pamoate	Central precocious puberty	2017
Sublocade	Buprenorphine	Moderate to severe opioid use disorder	2017

Adapted from Zhong et al. (2018)

reduction in invasive capacity. They concluded that the anticancer potential of curcumin increased three times upon incorporation into nanoparticles (Khan et al. 2018) (Table 3.3).

Now we discuss some of the natural products that have been successfully delivered using polymeric particles to improve upon their limitations and exert better anticancer action.

3.5.1 Curcumin

Curcumin, a polyphenolic compound described in ancient Ayurvedic medicine has shown promising anticancer potential in a multifaceted manner targeting many mechanisms at a time. Despite how promising curcumin may seem, its instability, rapid metabolism, insolubility in water, poor bioavailability and limited tissue distribution pose regulatory hurdles. Also since it reacts with several proteins like cytochrome P450 and glutathione S-transferase (GST), it poses toxicity problems if present appreciably in plasma. Regardless of its potential against various diseases, the associated inherent properties of curcumin as a molecule have been a stumbling block in its approval as a drug candidate (Jha et al. 2017).

The use of polymeric nanoparticles to improve upon the curcumin's inherent limitations have been described in many studies. Mukerjee and Vishwanatha successfully prepared poly lactic-co-glycolic acid (PLGA) nanoparticles with encapsulation efficiency as high as 90%. The particles were of range 35–100 nm aimed at improving the intracellular uptake in prostate cancer cells. They also reported a more prominent effect of curcumin entrapped PLGA nanoparticles compared to free curcumin through cell viability studies. They concluded curcumin entrapped nanoparticle to be of high potential as adjuvant therapy against prostate cancer (Mukerjee and Vishwanatha 2009). A study comparing two PLGA combinations with different lactide to glycolide ratios, one being 50:50 and the other being 75:25

found that curcumin entrapped nanoparticles prepared with PLGA 50:50 ratio showed higher encapsulation efficiency. Both the formulations showed enhanced cellular uptake evident from intracellular fluorescence studies. Nanoparticles with PLGA 50:50 showed better cytotoxicity and significantly better induction of apoptosis. Additionally, the superior efficacy of curcumin entrapped nanoparticles compared to free curcumin was shown by immunocytochemical analysis and electrophoretic mobility shift assay. Thus, it showed that curcumin entrapped in PLGA nanoparticles not only possessed enhanced aqueous solubility but also showed better anticancer properties (Nair et al. 2012).

Curcumin entrapped in poly lactic-co-glycolic acid (PLGA) nanoparticles were prepared to enhance the oral bioavailability. The prepared particles were of about 200 nm in size and exhibited a 640-fold water solubility compared to free curcumin. A relative bioavailability of 5.6-fold and a longer half-life was observed upon oral administration in rats. The results indicated that improved oral bioavailability of curcumin was associated with increased water solubility, higher release and also increased residence time in intestinal cavity (Xie et al. 2011). A study on breast and lung cancer cells focused at investigating the mechanism of action of PLGA nanoparticle entrapping curcumin under hypoxic conditions observed a reduction in of Hypoxia-Inducible Factor-1α and nuclear p65 levels which are usually elevated in metastatic cancers. They also showed that curcumin entrapped PLGA nanoparticles had superior anti-migratory capacity compared to free curcumin. A notable reduction of cell invasion was observed with both the cancer cells implying that curcumin entrapped in PLGA nanoparticle decreased the invasive capacity of cancer cells which is critical to the inhibition of metastasis. Overall, entrapping of curcumin in PLGA nanoparticle resulted in a tenfold enhancement of solubility with a threefold enhancement of anticancer activity compared to free curcumin (Khan et al. 2018). Another study reported that PLGA based curcumin nanoparticle led to inhibition of cell growth, induction of apoptosis and cell cycle arrest in cervical cancer cells. Nanoparticle formulation also modulated carcinogenesis associated microRNAs, transcription factors and proteins. Curcumin encapsulated PLGA nanoparticle mediated reduction of oncogenic microRNA-21, suppression of nuclear β-catenin and abrogation of expression of E6/E7 human papillomavirus oncoproteins led to a significant reduction in tumor burden (Zaman et al. 2016). The synergistic inhibitory effect of metformin and curcumin combination encapsulated in polyethylene glycol (PEG) coated PLGA nanoparticles has also been reported. It was found that cytotoxicity of nanoparticles against breast cancer cells was dose dependent with co-encapsulated metformin and curcumin showing an enhanced synergistic antiproliferative effect in addition to a notable arrest of cell growth. Inhibition expression of human telomerase reverse transcriptase (hTERT) gene was also observed with the combination nanoformulation exhibiting a further decline (Farajzadeh et al. 2018). A study aimed at overcoming the limitations of curcumin formulated curcumin entrapped PLGA nanoparticle surface coated with chitosan and PEG. The particles showed enhanced uptake and internalization by pancreatic cancer cells as validated by confocal microscopy and fluorescent spectrophotometry. They also showed higher cytotoxicity as compared to free curcumin. Further, migration and invasion assays revealed an effective inhibition of migratory and invasive properties in cells treated with nanoparticles. Further, enhanced expression of pro-apoptotic proteins like cleaved poly ADP-ribose polymerase, Bax and cleaved Caspase 3 along with decreased expression of anti-apoptotic proteins like Bcl2 and increased induction of apoptosis in nanoparticle treated cells was observed (Arya et al. 2018).

Curcumin encapsulated poly lactic-co-glycolic acid (PLGA) nanoparticles prepared by nanoprecipitation resulted in a highly dispersible particles. Optimization of nanoparticles showed two to sixfold enhancement of cellular uptake in ovarian as well as breast cancer cells. A better anticancer potential as revealed by cell proliferation as well as clonogenic assays was shown by the optimized PLGA encapsulated curcumin compared to free curcumin. This also correlated with increased induction of apoptosis by the PLGA encapsulated curcumin. The ability to conjugate antibodies was also shown demonstrating the potential for targeted delivery (Yallapu et al. 2010). A study also prepared curcumin loaded PLGA microspheres increasing its bioavailability. The microspheres were able to perturb mitochondrial membrane potential and induce apoptosis in breast cancer cells. The PLGA microspheres were also conjugated to folic acid to enable targeted delivery to cancer cells. Treatment with curcumin loaded and folic acid conjugated PLGA microspheres enhanced cleaved caspase-3 and reduced phosphorylated Protein kinase B leading to induction of apoptosis. They also led to increased tumor regression in Balb/c mice (Pal et al. 2019). In a study liquid-driven co-flow focusing (LDCF) process was used to prepare of curcumin loaded PLGA microspheres for application in ovarian cancer. The intraperitoneal delivery of microspheres prepared by LDCF showed slow systemic absorption and higher mean residence time compared to free curcumin as well as microspheres prepared by single emulsion method (Dwivedi et al. 2018). Another study investigating specific targeting of cancer cells prepared polyethylene glycol coated PLGA particles which were conjugated to epidermal growth factor receptor (EGFR) targeting peptides. These were then delivered into EGFR expressing cancer cells. Treatment of breast cancer cells with these targeted nanoparticles led to a suppression of phosphoinositide 3-kinase signalling, cell viability, drug clearance and tumor burden. More than 40-fold increment in half-life of curcumin in nanoparticle formulation was shown. A reduction in tumor burden by >50% was observed in a mouse xenograft tumor model (Jin et al. 2017).

Hormone therapies and therapies targeting human epidermal growth factor receptor 2 (HER-2) receptors are not able to control triple-negative breast cancers. A study evaluating the effects of curcumin on triple negative breast cancer used positively charged curcumin loaded nanoparticles prepared by nanoprecipitation using poly lactic-co-glycolic acid (PLGA) and detergent cetyl-tri-methyl ammonium bromide. Nanoparticles showed higher cellular uptake and cytotoxicity. The authors observed induction of reactive oxygen species that led to DNA damage and resulted in mitogen-activated protein kinase activation. This induction ultimately led to cell cycle arrest at G1/S phase and G2/M phase via increased expression of proteins such as p21, p16 and p53 along with decreased expression of proteins like

Table 3.4 Summary of poly lactic-co-glycolic acid nanoparticles for delivery of curcumin for cancer

Formulation	Cancer type	Size	Reference
PLGA	Prostate	35–100 nm	Mukerjee and Vishwanatha (2009)
PLGA	Cervical	100–200 nm	Nair et al. (2012)
PLGA	Breast, lung	200–600 nm	Khan et al. (2018)
PLGA	Cervical	76.2 ± 5.36 nm	Yallapu et al. (2010) and Zaman et al. (2016)
PLGA/PEG	Breast	202–257 nm	Farajzadeh et al. (2018)
PLGA-chitosan-PEG	Pancreatic	Mean 264 nm	Arya et al. (2018)
PLGA-FA	Breast	800–900 nm	Pal et al. (2019)
PLGA	Ovarian	Mean 20.26 ± 2.37 um	Dwivedi et al. (2018)
PLGA-PEG	Breast	210 ± 54 nm	Jin et al. (2017)
PLGA-CTAB	Breast	81.05 ± 3.85 nm	Meena et al. (2017)

PLGA poly lactic-co-glycolic acid, *PEG* polyethylene glycol, *FA* folic acid, *CTAB* cetyl tri methyl ammonium bromide

cyclin D1, cyclin E, cyclin-dependent kinase-2 (CDK-2) and CDK4. Additionally, reduction in expression of DNA repair genes resulting in persistent DNA damage leading to induction of apoptosis was also reported (Meena et al. 2017) (Table 3.4).

3.5.2 *Epigallocatechin-3-Gallate*

Epigallocatechin gallate (EGCG), the anticancer polyphenol occurring in green tea has been successfully entrapped in delivered using polymeric nanoparticles in many studies showing improved efficacy and reduced toxicity. In a study aimed at designing targeted nanoparticles used conjugated PLGA-polyethylene glycol (PLGA-PEG) di-block copolymer which was then conjugated to a prostate-specific membrane antigen (PSMA) targeting ligand. Encapsulation of EGCG in targeted PLGA-PEG nanoparticles enabled selective delivery of EGCG to prostate cancer cells. The functionalized nanoparticles showed selective efficacy against prostate cancer cells expressing PSMA. An increased anti-proliferative effect towards PSMA positive cancer cells with negligible effect on normal cell viability was observed. The enhanced potency of targeted nanoparticles could be attributed to increase in binding resulting in increased cellular uptake accumulation via receptor mediated endocytosis (Sanna et al. 2011). EGCG encapsulated into PLA-PEG nanoparticles conferred a remarkable dose advantage against prostate cancer cells. Enhance uptake and apoptosis were observed on treatment with nanoparticles. Additionally, a notable increase of pro-apoptotic protein Bax with a decrease of anti-apoptotic protein Bcl-2 was observed tipping the balance in favor of apoptosis. Also, enhanced expression of p21 and p27 protein levels with a remarkable dose advantage were also reported. They also showed a notable reduction of EGCG concentration required to inhibit angiogenesis when delivered in nanoparticle encapsulated form. It was also demonstrated that nanoparticle formation conferred a tenfold dose advantage for inhibition of tumor growth in mouse xenograft tumor model (Siddiqui et al. 2009).

In extension to their previous findings Sanna et al. has suggested that target specific delivery in the tumor cells can be achieved by functionalizing the nanoparticles with ligands that would specifically bind the over expressing prostate specific membrane antigen (PSMA), a transmembrane protein on prostate cancer epithelial cells. For their study they prepared three different types of nanoparticles for EGCG delivery including non-functionalized nanoparticle and functionalized nanoparticles. The size of these particles were analyzed by *dynamic light scattering* and found to be ranging from 130 nm to 250 nm in diameter. The anti-proliferative activity of the prepared formulations were tested on prostate cancer cell lines and they found a significant accumulation and cellular uptake of the functionalized nanoparticles and non-specific cellular uptake of non-functionalized nanoparticles but however the both of them provided similar level of cellular toxicity. However, upon treating cells with both types of nanoparticles led to an increase in Bax protein with the decreased levels of Bcl-2 protein, therefore resulting in apoptosis. Also, it has been observed that these formulations have resulted in cleaving poly ADP-ribose polymerase protein and inhibiting survivin protein. Along with this, all the formulations and the native EGCG resulted in an effective inhibition of cell cycle regulatory proteins, cyclins A, Cyclin-dependent kinase-2 (CDK2) and CDK- 6, cyclin D3 and cyclin B1 by activating the levels of cell cycle inhibitors. From in vivo studies, in athymic nude mice also, it has been reported that the functionalized nanoparticles had significantly higher effect on tumor growth inhibition while non-functionalized nanoparticles also showed a significant effect on tumor inhibition as compared to native-EGCG, due to passive targeting (Sanna et al. 2017).

Singh et al. reported EGCG loaded poly lactic-co-glycolic acid nanoparticles could offer a three to sevenfold dose reduction in half maximal inhibitory concentration (IC50) doses upon treatment of lung, cervical and acute monocytic leukemia cells. The nanoparticles enhanced the induction of caspase 3 and caspase 9 as well as shifted the Bax/Bcl2 ratio in favor of apoptosis. Cell cycle analysis showed that nanoparticles more efficiently sensitized lung cancer cells to cisplatin induced apoptosis. Further, the nanoparticle treatment was more effective in nuclear factor kappa light chain enhancer of activated B cells (NF-kB) reduction as well as down regulating matrix metalloproteinase-9, cyclin D1 and vascular endothelial growth factor expression thereby inhibiting invasion, proliferation and metastasis respectively. In addition, the mice bearing *Ehrlich*-Lettre *ascites* carcinoma cells upon treatment with a combination of nanoparticles and cisplatin showed a notable increase in mean survival time along with tumor volume regression (Singh et al. 2015) (Table 3.5).

Table 3.5 Summary of poly lactic-co-glycolic acid nanoparticles for delivery of EGCG for cancer

Formulation	Cancer	Size	Reference
PLGA-PEG	Prostate	80.53 +/− 15.0 nm	Sanna et al. (2011)
PLA-PEG	Prostate	260 nm	Siddiqui et al. (2009)
PLGA-PEG	Prostate	130–250 nm	Sanna et al. (2017)
PLGA	Lung, cervical, acute monocytic leukemia	239 ± 12 nm	Singh et al. (2015)

PLGA poly lactic-co-glycolic acid, *PEG* polyethylene glycol

3.5.3 Resveratrol

Resveratrol, a polyphenol occurring vine plants active against a wide variety of cancers also suffers from challenges such as low bioavailability, low aqueous solubility as well as rapid metabolism in the system and nanotechnology has been extensively investigated to circumvent these limitations.

Jung et.al prepared resveratrol entrapped polylactic acid nanoparticles coated with polyethylene glycol and reported the controlled delivery as well as its cytotoxic effect in vitro and in vivo on colon cancer cell lines. However, these cells have shown the significant reduction in cell numbers and its colony forming capacity. In recent studies, it has been found that apart from its anti-tumorous and antioxidant properties, it also suppresses tumor cells glucose metabolism due to down-regulation of reactive oxygen species (ROS) and with the activation of hypoxia inducible factor-1alpha. From in vivo studies, resveratrol has shown the efficient reduction in the tumor growth rate in tumor bearing mice by passive targeting of the non-functionalized nanoparticles in tumor cells by enhanced permeation and retention effect (EPR) (Jung et al. 2015).

In another study, Sanna et.al suggested the resveratrol encapsulation in polymeric nanoparticles for the treatment of prostate cancer. From the experiments, it has been reported that resveratrol mediates apoptosis in prostate cancer cells by mitogen-activated protein kinase (MAPK) and protein kinase C. However, it has been observed that it is also associated with reduction in oxidative stress and in the production of nitric oxide in and hence, reduces the growth, proliferation and spreading of cancer cells. From the anti-tumor studies, it has been found that, resveratrol entrapped in the nanoparticles were effective against prostate cancer cells in a dose dependent manner and significantly showed the cytotoxicity against these cells as compared to free resveratrol. Additionally, confocal fluorescence microscopy confirmed the efficient uptake of nanoparticles by prostate cancer cells (Sanna et al. 2013).

In a study, Nassir et.al designed resveratrol entrapped polymer nanoparticle by solvent displacement method and determined its cytotoxicity and apoptosis against prostate cancer cells. In vitro studies have also reported the high efficacy of resveratrol loaded nanoparticles as compared to resveratrol, which is due to selective internalization of polymeric nanoparticle along with the blockage in the efflux transport. From the experiments, it has been reported that upon resveratrol loaded nanoparticle treatment against prostate cancer cells, the mitochondrial membrane potential significantly dropped and hence induced apoptosis by mitochondrial dependent pathway. They also reported an increase in caspase-3 activity and enhancement of phosphatidylserine levels in the outer plasma membrane. Moreover, the alteration in the DNA material during apoptosis upon resveratrol loaded nanoparticle exposure was also observed (Nassir et al. 2018).

In a study by Xin, et. Al, resveratrol loaded poly lactic-co-glycolic acid (PLGA) nanoparticles decorated with folic acid (FA) and indocyanine green (ICG) were prepared by facile self-assembly-and-nanoprecipitation method and tested on glio-blastoma *cells*. Confocal fluorescence showed that nanoparticle formulation enhanced cellular uptake which could be attributed to FA receptor-mediated endo-cytosis. Also, the particle formulation showed a prolonged circulation time over-coming the short circulation time of free resveratrol. The decorated nanoparticle formulation showed a notably higher cell death of glioblastoma *cells* compared to the same dose of free resveratrol. This increased anticancer effect was attributed to the higher apoptosis of glioma cells as shown by Annexin V-fluorescein isothiocya-nate/propidium iodide staining (Xin et al. 2016).

Zhao et.al, suggested that the co-entrapment of resveratrol and doxorubicin in modified polymeric poly lactic-co-glycolic acid (PLGA) nanoparticles, i.e. by mod-ifying the PLGA surface with sigma receptor so as to bind the sigma receptor of over expressing cancerous cells could deliver the significant amount of resveratrol and doxorubicin into breast cancer cells with the aim of overcoming resistance caused due to doxorubicin. The co-delivery of resveratrol and doxorubicin enhanced their accumulation in the nucleus as observed by confocal laser scanning micros-copy (CSLM). It was also found that doxorubicin-resveratrol-loaded PLGA nanoparticle inhibited the expression of nuclear factor kappa light chain enhancer of activated B cells (NF-kB) and Bcl-2 proteins as well as increased the expression of Bax and caspase-3 proteins thereby inducing apoptosis. Furthermore, the expres-sion of drug resistance related proteins like *P-glycoprotein*, Multidrug resistance-associated protein 1 as well as Breast Cancer Resistance Protein was notably reduced in tumor bearing mice and a dose dependent decrease in tumor growth was observed (Zhao et al. 2016) (Table 3.6).

Table 3.6 Summary of poly lactic-co-glycolic acid nanoparticles for delivery of resveratrol for cancer

Formulation	Cancer	Size	Reference
PLGA	Breast	170 nm	Zhao et al. (2016)
FA-ICG -PLGA	Glioma	104.5–121.1 nm	Xin et al. (2016)
PEG-PLA	Colon	233.5+/− 14.0	Jung et al. (2015)
PCL:PLGA-PEG-COOH	Prostate	150 nm	Sanna et al. (2013)
PLGA	Prostate	202.8+/−2.64	Nassir et al. (2018)

PLGA poly lactic-co-glycolic acid, *FA* folic acid, *ICG* Indocyanine green, *PEG* polyethylene gly-col, *PLA* polylactic acid, *PCL* poly epsilon-caprolactone

3.5.4 Some Other Compounds

In a study targeting the treatment of liver cancer evaluated emodin encapsulated in poly lactic-co-glycolic acid – tocopherol polyethylene glycol succinate (PLGA-TPGS) nanoparticles. Emodin obtained from rhizome of rheum is a traditional Chinese medicine has a wide range of effects like anti-inflammatory, hepatoprotective, antioxidant and antitumor limited by its low solubility and stability. Using fluorescence inversion microscopy it was demonstrated that emodin loaded PLGA-TPGS nanoparticle showed an excellent liver targeting property. The particles ere internalized by liver cancer cells and apoptosis was enhanced. The particles showed a sustained release effect as revealed by a longer half-life and mean retention time (MRT). More in vivo antitumor activity was shown by emodin loaded PLGA-TPGS particles owing to sustained release and a higher cellular uptake (Liu et al. 2016).

In another study aimed at improving the efficacy of genistein for treatment of liver cancer used star-shaped block copolymer mannitol functionalized PLGA-TPGS (M-PLGA-TPGS) for making nanoparticles by a modified nanoprecipitation method. The prepared nanoparticles showed stability in terms of size and surface charge up to a period of 3 months. A higher cellular uptake efficiency was shown by the star shaped M-PLGA-TPGS nanoparticles compared to genistein solution. The mannitol functionalized nanoparticles also showed an increase in apoptosis which was dose dependent as revealed by Annexin V- fluorescein isothiocyanate/propidium iodide double staining flow cytometry. Studies on severe combined immunodeficiency mice bearing human liver carcinoma xenograft showed a significant inhibition of tumor growth compared to genistein solution (Wu et al. 2016).

Recently, statins have shown anticancer potential owing to their antiproliferative and pro-apoptotic effect (Campbell et al. 2006; Osmak 2012). A statin, named simvastatin which is insoluble but effective against breast cancer was used in a study. Nanocarriers with a cholic acid core (CA) and star shaped poly lactic-co-glycolic acid (PLGA) polymer was used for delivery of simvastatin. The star shaped nanoparticles showed internalization in breast cancer cells as observed via confocal microscopy. Star shaped nanoparticles loaded with simvastatin showed a higher cytotoxicity and a decreased expression of cyclin D1 compared with pristine simvastatin as well as simvastatin loaded linear PLGA nanoparticles. Moreover, a Balb/c nude mice xenograft of breast cancer cells revealed that the star shaped formulation had the greatest effect on inhibiting the tumor growth (Wu et al. 2015).

To overcome instability and low solubility associated with quercetin, quercetin loaded PLGA-TPGS particles were prepared in a different study investigating its effects on liver cancer. High performance liquid chromatography analysis revealed the particle uptake to be 50.87% and 61.09% on liver cancer cells. Annexin-Propidium iodide flow cytometry showed a dose dependent induction of apoptosis in cancer cells. Also, histological analysis was done to confirm that particles were indeed targeted to liver cells. A suppression of tumor growth by 59.07% was reported in the solid tumor bearing mice (Guan et al. 2016). A study aimed to overcome multidrug resistance in cancers prepared nanoparticles with quercetin in the

PLGA core and anthracycline drug on the surface. Adriamycin (ADR) and mitoxantrone (MTX) bound to bovine serum albumin (BSA) or histones (His) were coated over the PLGA loaded quercetin nanoparticles. Quercetin being an antioxidant helped reduce the damage to healthy cells affected by the action of anthracyclines (ADR and MTX). The drugs being co-delivered were expected to work synergistically and overcome multi-drug-resistance in cancer cells (Saha et al. 2016). In an effort to increase the low response rate of docetaxel (DTX) in the treatment of metastatic breast cancer, quercetin was used to improve its efficacy. For this, hyaluronic acid modified PLGA (HA-PLGA) nanoparticles were prepared using a polyethylenimine (PEI) linker arm. A modified solvent evaporation method was used to prepare dual loaded DTX/quercetin loaded HA-PLGA nanoparticles. Synergistic inhibition on cell motility was seen in wound healing assays. Reduction of cell migratory and invasion properties was bestowed by a decrease of phosphorylated protein kinase B and Matrix metallopeptidase-9 expression. Furthermore, an efficient uptake of particles by cancer cells was demonstrated. The particles also induced cytotoxicity, decreased colony formation and led to the promotion of apoptosis. An enhancement of drug accumulation in tumor as well as lungs demonstrated by biodistribution assays indicated that these particles could be effective for primary tumor as well as pulmonary metastasis (Li et al. 2017). A study by Halder, et al. functionalized PLGA nanoparticles entrapping quercetin were made and tested against triple negative breast cancer cells and Larynx epidermoid carcinoma. Once quercetin was loaded into PLGA nanoparticles, transferrin was decorated onto the nanoparticle surface as transferrin receptors are overexpressed in cancer cells. Fluorescence activated cell sorting analysis revealed a better uptake of the functionalized nanoparticles by receptor mediated mechanism. A marked anti proliferative and cytotoxic effect was observed in response to the functionalized nanoparticles compared to either quercetin loaded PLGA or free quercetin which was indicated to be a result of enhanced uptake. Flow cytometric analysis revealed a higher apoptosis in cells treated with the functionalized nanoparticles (Halder et al. 2018).

A study aiming towards improving the anti-cancerous activity of betulinic acid worked with methoxy poly ethylene glycol-PLGA (mPEG-PLGA) nanoparticles. Betulinic acid, a naturally occurring pentacyclic triterpenoid has been demonstrated to exert anticancer effects in a variety of cancers like pancreatic, breast, ovarian, cervix, prostate, lung and colorectal. Its clinical efficacy is however hampered by its low solubility and short half-life. To improve upon these limitations betulinic acid was loaded into mPEG surface modified PLGA by single emulsion solvent evaporation method. Flow cytometric analysis showed that mPEG-PLGA nanoparticles showed a reduced uptake by macrophage cells compared to PLGA nanoparticles indicating their long circulatory behavior. The mPEG-PLGA nanoparticle treatment led to a loss of mitochondrial membrane potential resulting in enhanced cellular apoptosis in pancreatic cancer cells. Furthermore, the reactive oxygen species generation and G1/G0 cell cycle arrest was enhanced in the particle treated group compared to free betulinic acid. A 7.21-fold enhancement of half-life was also revealed by the pharmacokinetic study. In vivo study on Ehrlich (solid) tumor bearing mice demonstrated a superior anti cancerous efficacy of betulinic acid loaded nanoparticles compared to native betulinic acid with no hematological, histological and biochemical toxicity (Saneja et al. 2017a). In another study, combinatorial PEG-PLGA nanoparticles were prepared using modified double emulsion solvent evaporation method co-encapsulating gemcitabine and betulinic acid. Annexin V-Propidium iodide staining followed by flow cytometric analysis revealed a higher apoptosis in cells treated with co-encapsulated formulation. Moreover, a significant enhancement of reactive oxygen species was also observed upon treatment with gemcitabine and betulinic acid co-encapsulated particles. Furthermore, studies on Ehrlich tumor bearing mice revealed that the combinatorial nanoparticles were more efficient in curbing the tumor growth showing improved antitumor efficacy (Saneja et al. 2019).

Venugopal et al. designed nanoparticles targeting the epidermal growth factor receptor (EGFR) protein expressed on triple negative breast cancer cells. The nanoparticles were prepared by nanoprecipitation method. Paclitaxel was entrapped in PEG-PLGA polymer which was then decorated with anti-EFGR protein via cross linking. The nanoparticles showed a notable reduction in cancer cell viability along with an excellent enhancement of paclitaxel accumulation in the tumor plasma. Surface decorated nanoparticles were able to reduce the expression of EGFR protein and specifically target the cancer cells. Studies on tumor induced athymic mice showed a prominent reduction of tumor volume with time (Venugopal et al. 2018). In another study, paclitaxel loaded rough PLGA microspheres were prepared with microporous surface and porous internal structure. The microspheres showed increased antitumor effects and could induce a higher apoptosis at a lower concentration compared to free paclitaxel. The microsphere treatment also lead to prolongation of microtubules along with increased expression of Bax and cyclin B1 proteins whereas expression of Bcl-2 protein decreased. Studies on mice bearing tumor revealed a higher tumor inhibition potency of the microspheres with potential to decrease frequency of drug administration. Additionally, the stable body weight of mice treated with microspheres showed that they have negligible side effects (Zhang et al. 2018). A study by Wang et al. proposed the combination therapy of the chemotherapeutic drugs in order to improve drug's therapeutic efficacy and co-encapsulated paclitaxel and etoposide PLGA nanoparticles were prepared. The co-loaded nanoparticles showed an enhanced cytotoxic effect and a considerable uptake in cancer cells. In apoptosis assay, assessment by Annexin V-fluorescein isothiocyanate/propidium iodide staining showed that treatment with the nanoparticles led to a significantly enhanced induction of apoptosis. Cell cycle studies showed that combination of drugs led to enhanced G2/M phase arrest and the maximum cell cycle arrest was seen upon treatment with the co-loaded nanoparticles (Wang et al. 2015) (Table 3.7).

Table 3.7 Some other compounds delivered using poly lactic-co-glycolic acid nanoparticles

Compound	Polymer	Cancer	Size	Reference
Emodin	PLGA-TPGS	Liver	100–200 nm	Liu et al. (2016)
Genistein	M-PLGA-TPGS	Liver	220–250 nm	Wu et al. (2016)
Simvastatin	CA-PLGA	Breast	100–150 nm	Wu et al. (2015)
Betulinic acid	mPEG-PLGA	Mammary adenocarcinoma	147 nm	Saneja et al. (2017a)
Betulinic acid	PLGA-PEG	Mammary adenocarcinoma	195.93 ± 6.83 nm	Saneja et al. (2019)
Quercetin	HA-PLGA	Breast	209.8 ± 10.8 nm	Li et al. (2017)
Quercetin	PLGA-TPGS	Liver	100–200 nm	Guan et al. (2016)
Quercetin	Tf-PLGA	Breast, larynx epidermoid carcinoma	150 nm	Halder et al. (2018)
Paclitaxel	PLGA-PEG	Breast	335.3+/−2.2 nm	Venugopal et al. (2018)
Paclitaxel	PLGA	Glioma	50 um	Zhang et al. (2018)
Paclitaxel	PLGA	Osteosarcoma	100 ± 3.68 nm	Wang et al. (2015)

PLGA poly lactic-co-glycolic acid, *TPGS* D -α-Tocopherol polyethylene glycol succinate, *M* mannitol, *CA* Cholic acid, *mPEG* methoxy poly ethylene glycol, *PEG* polyethylene glycol, *HA* Hyaluronic acid, *Tf* Transferrin

3.6   Conclusion

Biodegradable synthetic polymers have come a long way from being used as suture materials to now being applied in a wide variety of biomedical applications particularly for drug and biological delivery. Owing to their versatile nature they can be tailored in multiple ways to alter the particle properties with the goal to achieve the desired effect. Their use in the field of cancer chemotherapy promises new avenues for better treatment. The limitations like toxicity and repeated administrations associated with conventional chemotherapeutics have led to exploring new compounds for cancer treatment. Natural products though promising due to their non-toxic and pleiotropic nature suffer from similar limitations. Moreover, most of these face solubility issues owing to their hydrophobic nature. Delivering anticancer natural products via polymeric nanocarriers serves to offer the solution to many of the existential challenges posed by cancer chemotherapy field. Being safe and biodegradable, the byproducts of polymer hydrolysis are eliminated naturally from the system without causing any harm. The possibility to provide sustained release of the bioactive compound, they can help lower the dose or the number of sittings required, thereby lowering the cost. They can be tuned to different sizes and gain access to otherwise inaccessible sites within the body. Furthermore, the possibility of decorating them with specific ligands can deliver the bioactive agents selectively to tumor cells. Hence, delivering natural products loaded in polymeric nanocarriers seems as the most promising approach towards a safe, patient compliant and economical treatment of cancer.

References

Adams JM, Cory S (2007) The Bcl-2 apoptotic switch in cancer development and therapy. Oncogene 26:1324–1337. https://doi.org/10.1038/sj.onc.1210220

Amin ARMR, Karpowicz PA, Carey TE, Arbiser J, Nahta R, Chen ZG, Dong J-T, Kucuk O, Khan GN, Huang GS, Mi S, Lee H-Y, Reichrath J, Honoki K, Georgakilas AG, Amedei A, Amin A, Helferich B, Boosani CS, Ciriolo MR, Chen S, Mohammed SI, Azmi AS, Keith WN, Bhakta D, Halicka D, Niccolai E, Fujii H, Aquilano K, Ashraf SS, Nowsheen S, Yang X, Bilsland A, Shin DM (2015) Evasion of anti-growth signaling: a key step in tumorigenesis and potential target for treatment and prophylaxis by natural compounds. Semin Cancer Biol 35:S55–S77. https://doi.org/10.1016/j.semcancer.2015.02.005

Amit I, Citri A, Shay T, Lu Y, Katz M, Zhang F, Tarcic G, Siwak D, Lahad J, Jacob-Hirsch J, Amariglio N, Vaisman N, Segal E, Rechavi G, Alon U, Mills GB, Domany E, Yarden Y (2007) A module of negative feedback regulators defines growth factor signaling. Nat Genet 39:503–512. https://doi.org/10.1038/ng1987

Anderson JM, Shive MS (2012) Biodegradation and biocompatibility of PLA and PLGA microspheres. Adv Drug Deliv Rev 64:72–82. https://doi.org/10.1016/j.addr.2012.09.004

Arora D, Jaglan S (2016) Nanocarriers based delivery of nutraceuticals for cancer prevention and treatment: a review of recent research developments. Trends Food Sci Technol 54:114–126. https://doi.org/10.1016/j.tifs.2016.06.003

Arya G, Das M, Sahoo SK (2018) Evaluation of curcumin loaded chitosan/PEG blended PLGA nanoparticles for effective treatment of pancreatic cancer. Biomed Pharmacother 102:555–566. https://doi.org/10.1016/j.biopha.2018.03.101

Aye Y, Li M, Long MJ, Weiss RS (2015) Ribonucleotide reductase and cancer: biological mechanisms and targeted therapies. Oncogene 34:2011–2021. https://doi.org/10.1038/onc.2014.155

Balboni B, El Hassouni B, Honeywell RJ, Sarkisjan D, Giovannetti E, Poore J, Heaton C, Peterson C, Benaim E, Lee YB, Kim DJ, Peters GJ (2019) RX-3117 (fluorocyclopentenyl cytosine): a novel specific antimetabolite for selective cancer treatment. Expert Opin Investig Drugs 28:311–322. https://doi.org/10.1080/13543784.2019.1583742

Barr RD, Ferrari A, Ries L, Whelan J, Bleyer WA (2016) Cancer in adolescents and young adults: a narrative review of the current status and a view of the future. JAMA Pediatr 170:495–501. https://doi.org/10.1001/jamapediatrics.2015.4689

Berx G, van Roy F (2009) Involvement of members of the cadherin superfamily in cancer. Cold Spring Harb Perspect Biol 1:a003129. https://doi.org/10.1101/cshperspect.a003129

Bhowmick NA, Neilson EG, Moses HL (2004) Stromal fibroblasts in cancer initiation and progression. Nature 432:332–337. https://doi.org/10.1038/nature03096

Blanco E, Shen H, Ferrari M (2015) Principles of nanoparticle design for overcoming biological barriers to drug delivery. Nat Biotechnol 33:941–951. https://doi.org/10.1038/nbt.3330

Burkhart DL, Sage J (2008) Cellular mechanisms of tumour suppression by the retinoblastoma gene. Nat Rev Cancer 8:671–682. https://doi.org/10.1038/nrc2399

Calaf GM, Ponce-Cusi R, Carrion F (2018) Curcumin and paclitaxel induce cell death in breast cancer cell lines. Oncol Rep 40:2381–2388. https://doi.org/10.3892/or.2018.6603.

Campbell MJ, Esserman LJ, Zhou Y, Shoemaker M, Lobo M, Borman E, Baehner F, Kumar AS, Adduci K, Marx C, Petricoin EF, Liotta LA, Winters M, Benz S, Benz CC (2006) Breast cancer growth prevention by statins. Cancer Res 66:8707–8714. https://doi.org/10.1158/0008-5472.can-05-4061

Carmeliet P (2005) VEGF as a key mediator of angiogenesis in cancer. Oncology 69(Suppl 3):4–10. https://doi.org/10.1159/000088478.

Cavallaro U, Christofori G (2004) Cell adhesion and signalling by cadherins and Ig-CAMs in cancer. Nat Rev Cancer 4:118–132. https://doi.org/10.1038/nrc1276

Chen H, Landen CN, Li Y, Alvarez RD, Tollefsbol TO (2013) Enhancement of cisplatin-mediated apoptosis in ovarian cancer cells through potentiating G2/M arrest and p21 upregulation by combinatorial epigallocatechin gallate and sulforaphane. J Oncol 2013:872957. https://doi.org/10.1155/2013/872957

Cheng N, Chytil A, Shyr Y, Joly A, Moses HL (2008) Transforming growth factor-beta signaling-deficient fibroblasts enhance hepatocyte growth factor signaling in mammary carcinoma cells to promote scattering and invasion. Mol Cancer Res 6:1521–1533. https://doi.org/10.1158/1541-7786.mcr-07-2203

Chin K, de Solorzano CO, Knowles D, Jones A, Chou W, Rodriguez EG, Kuo WL, Ljung BM, Chew K, Myambo K, Miranda M, Krig S, Garbe J, Stampfer M, Yaswen P, Gray JW, Lockett SJ (2004) In situ analyses of genome instability in breast cancer. Nat Genet 36:984–988. https://doi.org/10.1038/ng1409

Coelho JF, Ferreira PC, Alves P, Cordeiro R, Fonseca AC, Góis JR, Gil MH (2010) Drug delivery systems: advanced technologies potentially applicable in personalized treatments. EPMA J 1:164–209. https://doi.org/10.1007/s13167-010-0001-x.

Conklin KA (2005) Coenzyme q10 for prevention of anthracycline-induced cardiotoxicity. Integr Cancer Ther 4:110–130. https://doi.org/10.1177/1534735405276191

Cortazar P, Justice R, Johnson J, Sridhara R, Keegan P, Pazdur R (2012) US Food and Drug Administration approval overview in metastatic breast cancer. J Clin Oncol Off J Am Soc Clin Oncol 30:1705–1711. https://doi.org/10.1200/JCO.2011.39.2613

Curto M, Cole BK, Lallemand D, Liu CH, McClatchey AI (2007) Contact-dependent inhibition of EGFR signaling by Nf2/Merlin. J Cell Biol 177:893–903. https://doi.org/10.1083/jcb.200703010

D'Ambrosio DN, Clugston RD, Blaner WS (2011) Vitamin A metabolism: an update. Nutrients 3:63–103. https://doi.org/10.3390/nu3010063

Danhier F, Ansorena E, Silva JM, Coco R, Le Breton A, Preat V (2012) PLGA-based nanoparticles: an overview of biomedical applications. J Control Release 161:505–522. https://doi.org/10.1016/j.jconrel.2012.01.043

Dasiram JD, Ganesan R, Kannan J, Kotteeswaran V, Sivalingam N (2017) Curcumin inhibits growth potential by G1 cell cycle arrest and induces apoptosis in p53-mutated COLO 320DM human colon adenocarcinoma cells. Biomed Pharmacother 86:373–380. https://doi.org/10.1016/j.biopha.2016.12.034

Davies MA, Samuels Y (2010) Analysis of the genome to personalize therapy for melanoma. Oncogene 29:5545–5555. https://doi.org/10.1038/onc.2010.323

De Amicis F, Giordano F, Vivacqua A, Pellegrino M, Panno ML, Tramontano D, Fuqua SA, Ando S (2011) Resveratrol, through NF-Y/p53/Sin3/HDAC1 complex phosphorylation, inhibits estrogen receptor alpha gene expression via p38MAPK/CK2 signaling in human breast cancer cells. FASEB J 25:3695–3707. https://doi.org/10.1096/fj.10-178871

Demain AL, Vaishnav P (2011) Natural products for cancer chemotherapy. Microb Biotechnol 4:687–699. https://doi.org/10.1111/j.1751-7915.2010.00221.x

Deshpande A, Sicinski P, Hinds PW (2005) Cyclins and cdks in development and cancer: a perspective. Oncogene 24:2909–2915. https://doi.org/10.1038/sj.onc.1208618

Din FU, Aman W, Ullah I, Qureshi OS, Mustapha O, Shafique S, Zeb A (2017) Effective use of nanocarriers as drug delivery systems for the treatment of selected tumors. Int J Nanomedicine 12:7291–7309. https://doi.org/10.2147/ijn.s146315

Du G-J, Zhang Z, Wen X-D, Yu C, Calway T, Yuan C-S, Wang C-Z (2012) Epigallocatechin Gallate (EGCG) is the most effective cancer chemopreventive polyphenol in green tea. Nutrients 4:1679–1691. https://doi.org/10.3390/nu4111679

Dwivedi P, Yuan S, Han S, Mangrio FA, Zhu Z, Lei F, Ming Z, Cheng L, Liu Z, Si T, Xu RX (2018) Core–shell microencapsulation of curcumin in PLGA microparticles: programmed for application in ovarian cancer therapy. Artif Cells Nanomed Biotechnol 46:S481–S491. https://doi.org/10.1080/21691401.2018.1499664

El-Kenawi AE, El-Remessy AB (2013) Angiogenesis inhibitors in cancer therapy: mechanistic perspective on classification and treatment rationales. Br J Pharmacol 170:712–729. https://doi.org/10.1111/bph.12344

Faber AC, Dufort FJ, Blair D, Wagner D, Roberts MF, Chiles TC (2006) Inhibition of phosphatidylinositol 3-kinase-mediated glucose metabolism coincides with resveratrol-induced cell cycle arrest in human diffuse large B-cell lymphomas. Biochem Pharmacol 72:1246–1256. https://doi.org/10.1016/j.bcp.2006.08.009

Fang Y, DeMarco VG, Nicholl MB (2012) Resveratrol enhances radiation sensitivity in prostate cancer by inhibiting cell proliferation and promoting cell senescence and apoptosis. Cancer Sci 103:1090–1098. https://doi.org/10.1111/j.1349-7006.2012.02272.x

Farabegoli F, Papi A, Orlandi M (2011) (−)-Epigallocatechin-3-gallate down-regulates EGFR, MMP-2, MMP-9 and EMMPRIN and inhibits the invasion of MCF-7 tamoxifen-resistant cells. Biosci Rep 31:99–108. https://doi.org/10.1042/bsr20090143

Farajzadeh R, Pilehvar-Soltanahmadi Y, Dadashpour M, Javidfar S, Lotfi-Attari J, Sadeghzadeh H, Shafiei-Irannejad V, Zarghami N (2018) Nano-encapsulated metformin-curcumin in PLGA/PEG inhibits synergistically growth and hTERT gene expression in human breast cancer cells. Artif Cells Nanomed Biotechnol 46:917–925. https://doi.org/10.1080/21691401.2017.1347879

Farhan M, Khan HY, Oves M, Al-Harrasi A, Rehmani N, Arif H, Hadi SM, Ahmad A (2016) Cancer therapy by catechins involves redox cycling of copper ions and generation of reactive oxygen species. Toxins 8:37–37. https://doi.org/10.3390/toxins8020037

Ferraz da Costa DC, Campos NPC, Santos RA, Guedes-da-Silva FH, Martins-Dinis M, Zanphorlin L, Ramos C, Rangel LP, Silva JL (2018) Resveratrol prevents p53 aggregation in vitro and in breast cancer cells. Oncotarget 9:29112–29122. https://doi.org/10.18632/oncotarget.25631.

Frazza EJ, Schmitt EE (1971) A new absorbable suture. J Biomed Mater Res 5:43–58. https://doi.org/10.1002/jbm.820050207

Fu H, Wang C, Yang D, Wei Z, Xu J, Hu Z, Zhang Y, Wang W, Yan R, Cai Q (2018) Curcumin regulates proliferation, autophagy, and apoptosis in gastric cancer cells by affecting PI3K and P53 signaling. J Cell Physiol 233:4634–4642. https://doi.org/10.1002/jcp.26190

Furlan V, Konc J (2018) Inverse molecular docking as a novel approach to study anticarcinogenic and anti-neuroinflammatory effects of curcumin. Molecules 23. https://doi.org/10.3390/molecules23123351

Galluzzi L, Kroemer G (2008) Necroptosis: a specialized pathway of programmed necrosis. Cell 135:1161–1163. https://doi.org/10.1016/j.cell.2008.12.004

Gan RY, Li HB, Sui ZQ, Corke H (2018) Absorption, metabolism, anti-cancer effect and molecular targets of epigallocatechin gallate (EGCG): an updated review. Crit Rev Food Sci Nutr 58:924–941. https://doi.org/10.1080/10408398.2016.1231168

Goswami KK, Ghosh T, Ghosh S, Sarkar M, Bose A, Baral R (2017) Tumor promoting role of anti-tumor macrophages in tumor microenvironment. Cell Immunol 316:1–10. https://doi.org/10.1016/j.cellimm.2017.04.005

Granja A, Frias I, Neves AR, Pinheiro M, Reis S (2017) Therapeutic potential of epigallocatechin gallate nanodelivery systems. Biomed Res Int 2017:5813793–5813793. https://doi.org/10.1155/2017/5813793

Greenberg RA (2005) Telomeres, crisis and cancer. Curr Mol Med 5:213–218

Grivennikov SI, Greten FR, Karin M (2010) Immunity, inflammation, and cancer. Cell 140:883–899. https://doi.org/10.1016/j.cell.2010.01.025

Gu JW, Makey KL, Tucker KB, Chinchar E, Mao X, Pei I, Thomas EY, Miele L (2013) EGCG, a major green tea catechin suppresses breast tumor angiogenesis and growth via inhibiting the activation of HIF-1alpha and NFkappaB, and VEGF expression. Vasc Cell 5:9. https://doi.org/10.1186/2045-824x-5-9

Guan X, Gao M, Xu H, Zhang C, Liu H, Lv L, Deng S, Gao D, Tian Y (2016) Quercetin-loaded poly (lactic-co-glycolic acid)-d-alpha-tocopheryl polyethylene glycol 1000 succinate nanoparticles for the targeted treatment of liver cancer. Drug Deliv 23:3307–3318. https://doi.org/10.1080/10717544.2016.1176087

Halder A, Mukherjee P, Ghosh S, Mandal S, Chatterji U, Mukherjee A (2018) Smart PLGA nanoparticles loaded with Quercetin: cellular uptake and in-vitro anticancer study. Mater Today: Proc 5:9698–9705. https://doi.org/10.1016/j.matpr.2017.10.156

Hanahan D, Weinberg RA (2011) Hallmarks of cancer: the next generation. Cell 144:646–674. https://doi.org/10.1016/j.cell.2011.02.013

He K, Ma Y, Yang B, Liang C, Chen X, Cai C (2017) The efficacy assessments of alkylating drugs induced by nano-Fe3O4/CA for curing breast and hepatic cancer. Spectrochim Acta A Mol Biomol Spectrosc 173:82–86. https://doi.org/10.1016/j.saa.2016.08.047

Hezel AF, Bardeesy N (2008) LKB1; linking cell structure and tumor suppression. Oncogene 27:6908–6919. https://doi.org/10.1038/onc.2008.342

Huang T, Sun L, Yuan X, Qiu H (2017) Thrombospondin-1 is a multifaceted player in tumor progression. Oncotarget 8:84546–84558. https://doi.org/10.18632/oncotarget.19165.

Ikushima H, Miyazono K (2010) TGFbeta signalling: a complex web in cancer progression. Nat Rev Cancer 10:415–424. https://doi.org/10.1038/nrc2853.

Jafri MA, Ansari SA, Alqahtani MH, Shay JW (2016) Roles of telomeres and telomerase in cancer, and advances in telomerase-targeted therapies. Genome Med 8:69–69. https://doi.org/10.1186/s13073-016-0324-x

Jha A, Mohapatra PP, AlHarbi SA, Jahan N (2017) Curcumin: not so spicy after all. Mini Rev Med Chem 17:1425–1434. https://doi.org/10.2174/1389557517666170228114234.

Jiang BH, Liu LZ (2009) PI3K/PTEN signaling in angiogenesis and tumorigenesis. Adv Cancer Res 102:19–65. https://doi.org/10.1016/s0065-230x(09)02002-8.

Jin H, Pi J, Zhao Y, Jiang J, Li T, Zeng X, Yang P, Evans CE, Cai J (2017) EGFR-targeting PLGA-PEG nanoparticles as a curcumin delivery system for breast cancer therapy. Nanoscale 9:16365–16374. https://doi.org/10.1039/c7nr06898k

Joerger AC, Fersht AR (2016) The p53 pathway: origins, inactivation in cancer, and emerging therapeutic approaches. Annu Rev Biochem 85:375–404. https://doi.org/10.1146/annurev-biochem-060815-014710

Jung YD, Kim MS, Shin BA, Chay KO, Ahn BW, Liu W, Bucana CD, Gallick GE, Ellis LM (2001) EGCG, a major component of green tea, inhibits tumour growth by inhibiting VEGF induction in human colon carcinoma cells. Br J Cancer 84:844–850. https://doi.org/10.1054/bjoc.2000.1691

Jung K-H, Lee JH, Park JW, Quach CHT, Moon S-H, Cho YS, Lee K-H (2015) Resveratrol-loaded polymeric nanoparticles suppress glucose metabolism and tumor growth in vitro and in vivo. Int J Pharm 478:251–257. https://doi.org/10.1016/j.ijpharm.2014.11.049

Karnoub AE, Dash AB, Vo AP, Sullivan A, Brooks MW, Bell GW, Richardson AL, Polyak K, Tubo R, Weinberg RA (2007) Mesenchymal stem cells within tumour stroma promote breast cancer metastasis. Nature 449:557–563. https://doi.org/10.1038/nature06188

Kessenbrock K, Plaks V, Werb Z (2010) Matrix metalloproteinases: regulators of the tumor microenvironment. Cell 141:52–67. https://doi.org/10.1016/j.cell.2010.03.015

Khan MN, Haggag YA, Lane ME, McCarron PA, Tambuwala MM (2018) Polymeric nano-encapsulation of curcumin enhances its anti-cancer activity in breast (MDA-MB231) and lung (A549) cancer cells through reduction in expression of HIF-1alpha and nuclear p65 (Rel A). Curr Drug Deliv 15:286–295. https://doi.org/10.2174/1567201814666171019104002

Klymkowsky MW, Savagner P (2009) Epithelial-mesenchymal transition: a cancer researcher's conceptual friend and foe. Am J Pathol 174:1588–1593. https://doi.org/10.2353/ajpath.2009.080545

Kuttan R, Sudheeran PC, Josph CD (1987) Turmeric and curcumin as topical agents in cancer therapy. Tumori 73:29–31

Lee CK, Ki SH, Choi JS (2011) Effects of oral curcumin on the pharmacokinetics of intravenous and oral etoposide in rats: possible role of intestinal CYP3A and P-gp inhibition by curcumin. Biopharm Drug Dispos 32:245–251. https://doi.org/10.1002/bdd.754

Lemmon MA, Schlessinger J (2010) Cell signaling by receptor tyrosine kinases. Cell 141:1117–1134. https://doi.org/10.1016/j.cell.2010.06.011

Levine B, Kroemer G (2008) Autophagy in the pathogenesis of disease. Cell 132:27–42. https://doi.org/10.1016/j.cell.2007.12.018

Li J, Zhang J, Wang Y, Liang X, Wusiman Z, Yin Y, Shen Q (2017) Synergistic inhibition of migration and invasion of breast cancer cells by dual docetaxel/quercetin-loaded nanoparticles via Akt/MMP-9 pathway. Int J Pharm 523:300–309. https://doi.org/10.1016/j.ijpharm.2017.03.040

Li X, Wang X, Xie C, Zhu J, Meng Y, Chen Y, Li Y, Jiang Y, Yang X, Wang S, Chen J, Zhang Q, Geng S, Wu J, Zhong C, Zhao Y (2018a) Sonic hedgehog and Wnt/beta-catenin pathways mediate curcumin inhibition of breast cancer stem cells. Anti-Cancer Drugs 29:208–215. https://doi.org/10.1097/cad.0000000000000584.

Li Y, Domina A, Lim G, Chang T, Zhang T (2018b) Evaluation of curcumin, a natural product in turmeric, on Burkitt lymphoma and acute myeloid leukemia cancer stem cell markers. Future Oncol 14:2353–2360. https://doi.org/10.2217/fon-2018-0202

Liao PC, Ng LT, Lin LT, Richardson CD, Wang GH, Lin CC (2010) Resveratrol arrests cell cycle and induces apoptosis in human hepatocellular carcinoma Huh-7 cells. J Med Food 13:1415–1423. https://doi.org/10.1089/jmf.2010.1126

Liu E, Wu J, Cao W, Zhang J, Liu W, Jiang X, Zhang X (2007) Curcumin induces G2/M cell cycle arrest in a p53-dependent manner and upregulates ING4 expression in human glioma. J Neuro-Oncol 85:263–270. https://doi.org/10.1007/s11060-007-9421-4

Liu H, Gao M, Xu H, Guan X, Lv L, Deng S, Zhang C, Tian Y (2016) A promising emodin-loaded poly (lactic co glycolic acid)-d-alpha-tocopheryl polyethylene glycol 1000 succinate nanoparticles for liver cancer therapy. Pharm Res 33:217–236. https://doi.org/10.1007/s11095-015-1781-4

Liu G, Wang Y, Li M (2018a) Curcumin sensitized the antitumour effects of irradiation in promoting apoptosis of oesophageal squamous-cell carcinoma through NF-kappaB signalling pathway. J Pharm Pharmacol 70:1340–1348. https://doi.org/10.1111/jphp.12981

Liu Q, Fang Q, Ji S, Han Z, Cheng W, Zhang H (2018b) Resveratrol-mediated apoptosis in renal cell carcinoma via the p53/AMPactivated protein kinase/mammalian target of rapamycin autophagy signaling pathway. Mol Med Rep 17:502–508. https://doi.org/10.3892/mmr.2017.7868.

Lowe SW, Cepero E, Evan G (2004) Intrinsic tumour suppression. Nature 432:307–315. https://doi.org/10.1038/nature03098

Mannargudi MB, Deb S (2017) Clinical pharmacology and clinical trials of ribonucleotide reductase inhibitors: is it a viable cancer therapy? J Cancer Res Clin Oncol 143:1499–1529. https://doi.org/10.1007/s00432-017-2457-8

Massague J (2008) TGFbeta in Cancer. Cell 134:215–230. https://doi.org/10.1016/j.cell.2008.07.001.

Meena R, Kumar S, Kumar R, Gaharwar US, Rajamani P (2017) PLGA-CTAB curcumin nanoparticles: fabrication, characterization and molecular basis of anticancer activity in triple negative breast cancer cell lines (MDA-MB-231 cells). Biomed Pharmacother 94:944–954. https://doi.org/10.1016/j.biopha.2017.07.151

Mehrmohamadi M, Jeong SH, Locasale JW (2017) Molecular features that predict the response to antimetabolite chemotherapies. Cancer Metab 5:8. https://doi.org/10.1186/s40170-017-0170-3

Mendonsa AM, Na T-Y, Gumbiner BM (2018) E-cadherin in contact inhibition and cancer. Oncogene 37:4769–4780. https://doi.org/10.1038/s41388-018-0304-2

Micalizzi DS, Farabaugh SM, Ford HL (2010) Epithelial-mesenchymal transition in cancer: parallels between normal development and tumor progression. J Mammary Gland Biol Neoplasia 15:117–134. https://doi.org/10.1007/s10911-010-9178-9

Mineva ND, Paulson KE, Naber SP, Yee AS, Sonenshein GE (2013) Epigallocatechin-3-gallate inhibits stem-like inflammatory breast cancer cells. PLoS One 8:e73464–e73464. https://doi.org/10.1371/journal.pone.0073464

Mordente A, Meucci E, Martorana GE, Tavian D, Silvestrini A (2017) Topoisomerases and anthracyclines: recent advances and perspectives in anticancer therapy and prevention of cardiotoxicity. Curr Med Chem 24:1607–1626. https://doi.org/10.2174/0929867323666161214120355.

Mukerjee A, Vishwanatha JK (2009) Formulation, characterization and evaluation of curcumin-loaded PLGA nanospheres for cancer therapy. Anticancer Res 29:3867–3875

Nair KL, Thulasidasan AK, Deepa G, Anto RJ, Kumar GS (2012) Purely aqueous PLGA nanoparticulate formulations of curcumin exhibit enhanced anticancer activity with dependence on the combination of the carrier. Int J Pharm 425:44–52. https://doi.org/10.1016/j.ijpharm.2012.01.003

Nakazato T, Ito K, Ikeda Y, Kizaki M (2005) Green tea component, catechin, induces apoptosis of human malignant B cells via production of reactive oxygen species. Clin Cancer Res 11:6040–6049. https://doi.org/10.1158/1078-0432.ccr-04-2273.

Nassir AM, Shahzad N, Ibrahim IAA, Ahmad I, Md S, Ain MR (2018) Resveratrol-loaded PLGA nanoparticles mediated programmed cell death in prostate cancer cells. Saudi Pharm J 26:876–885. https://doi.org/10.1016/j.jsps.2018.03.009

Navarro-Peran E, Cabezas-Herrera J, Campo LS, Rodriguez-Lopez JN (2007) Effects of folate cycle disruption by the green tea polyphenol epigallocatechin-3-gallate. Int J Biochem Cell Biol 39:2215–2225. https://doi.org/10.1016/j.biocel.2007.06.005

O'Hagan DT, Singh M (2003) Microparticles as vaccine adjuvants and delivery systems. Expert Rev Vaccines 2:269–283

Okada T, Lopez-Lago M, Giancotti FG (2005) Merlin/NF-2 mediates contact inhibition of growth by suppressing recruitment of Rac to the plasma membrane. J Cell Biol 171:361–371. https://doi.org/10.1083/jcb.200503165

Osmak M (2012) Statins and cancer: current and future prospects. Cancer Lett 324:1–12. https://doi.org/10.1016/j.canlet.2012.04.011

Ozcan Cenksoy P, Oktem M, Erdem O, Karakaya C, Cenksoy C, Erdem A, Guner H, Karabacak O (2015) A potential novel treatment strategy: inhibition of angiogenesis and inflammation by resveratrol for regression of endometriosis in an experimental rat model. Gynecol Endocrinol 31:219–224. https://doi.org/10.3109/09513590.2014.976197

Pal K, Laha D, Parida PK, Roy S, Bardhan S, Dutta A, Jana K, Karmakar P (2019) An in vivo study for targeted delivery of curcumin in human triple negative breast carcinoma cells using biocompatible PLGA microspheres conjugated with folic acid. J Nanosci Nanotechnol 19:3720–3733. https://doi.org/10.1166/jnn.2019.16292

Partanen JI, Nieminen AI, Klefstrom J (2009) 3D view to tumor suppression: Lkb1, polarity and the arrest of oncogenic c-Myc. Cell Cycle 8:716–724. https://doi.org/10.4161/cc.8.5.7786

Pavan AR, Silva GDB d, Jornada DH, Chiba DE, Fernandes GFDS, Man Chin C, Dos Santos JL (2016) Unraveling the anticancer effect of curcumin and resveratrol. Nutrients 8:628. https://doi.org/10.3390/nu8110628

Pento JT (2017) Monoclonal antibodies for the treatment of cancer. Anticancer Res 37:5935–5939. https://doi.org/10.21873/anticanres.12040.

Perez-Herrero E, Fernandez-Medarde A (2015) Advanced targeted therapies in cancer: drug nanocarriers, the future of chemotherapy. Eur J Pharm Biopharm 93:52–79. https://doi.org/10.1016/j.ejpb.2015.03.018

Peters GJ (2014) Novel developments in the use of antimetabolites. Nucleosides Nucleotides Nucleic Acids 33:358–374. https://doi.org/10.1080/15257770.2014.894197

Puyo S, Montaudon D, Pourquier P (2014) From old alkylating agents to new minor groove binders. Crit Rev Oncol Hematol 89:43–61. https://doi.org/10.1016/j.critrevonc.2013.07.006

Qian BZ, Pollard JW (2010) Macrophage diversity enhances tumor progression and metastasis. Cell 141:39–51. https://doi.org/10.1016/j.cell.2010.03.014

Qin J, Xie LP, Zheng XY, Wang YB, Bai Y, Shen HF, Li LC, Dahiya R (2007) A component of green tea, (−)-epigallocatechin-3-gallate, promotes apoptosis in T24 human bladder cancer cells via modulation of the PI3K/Akt pathway and Bcl-2 family proteins. Biochem Biophys Res Commun 354:852–857. https://doi.org/10.1016/j.bbrc.2007.01.003

Qin J, Chen HG, Yan Q, Deng M, Liu J, Doerge S, Ma W, Dong Z, Li DW (2008) Protein phosphatase-2A is a target of epigallocatechin-3-gallate and modulates p53-Bak apoptotic pathway. Cancer Res 68:4150–4162. https://doi.org/10.1158/0008-5472.can-08-0839

Rahmani AH, Al Shabrmi FM, Allemailem KS, Aly SM, Khan MA (2015) Implications of green tea and its constituents in the prevention of cancer via the modulation of cell signalling pathway. Biomed Res Int 2015:925640–925640. https://doi.org/10.1155/2015/925640

Rauf A, Imran M, Butt MS, Nadeem M, Peters DG, Mubarak MS (2018) Resveratrol as an anticancer agent: a review. Crit Rev Food Sci Nutr 58:1428–1447. https://doi.org/10.1080/10408398.2016.1263597

Rawluk J, Waller CF (2018) Gefitinib. Recent Results Cancer Res 211:235–246. https://doi.org/10.1007/978-3-319-91442-8_16.

Raynaud CM, Hernandez J, Llorca FP, Nuciforo P, Mathieu MC, Commo F, Delaloge S, Sabatier L, Andre F, Soria JC (2010) DNA damage repair and telomere length in normal breast, preneoplastic lesions, and invasive cancer. Am J Clin Oncol 33:341–345. https://doi.org/10.1097/COC.0b013e3181b0c4c2

Ronca R, Giacomini A, Rusnati M, Presta M (2015) The potential of fibroblast growth factor/fibroblast growth factor receptor signaling as a therapeutic target in tumor angiogenesis. Expert Opin Ther Targets 19:1361–1377. https://doi.org/10.1517/14728222.2015.1062475

Roskoski R Jr (2007) Sunitinib: a VEGF and PDGF receptor protein kinase and angiogenesis inhibitor. Biochem Biophys Res Commun 356:323–328. https://doi.org/10.1016/j.bbrc.2007.02.156

Saha C, Kaushik A, Das A, Pal S, Majumder D (2016) Anthracycline drugs on modified surface of Quercetin-loaded polymer nanoparticles: a dual drug delivery model for cancer treatment. PLoS One 11:e0155710. https://doi.org/10.1371/journal.pone.0155710

Saneja A, Dubey RD, Alam N, Khare V, Gupta PN (2014) Co-formulation of P-glycoprotein substrate and inhibitor in nanocarriers: an emerging strategy for cancer chemotherapy. Curr Cancer Drug Targets 14:419–433

Saneja A, Kumar R, Singh A, Dhar Dubey R, Mintoo MJ, Singh G, Mondhe DM, Panda AK, Gupta PN (2017a) Development and evaluation of long-circulating nanoparticles loaded with betulinic acid for improved anti-tumor efficacy. Int J Pharm 531:153–166. https://doi.org/10.1016/j.ijpharm.2017.08.076

Saneja A, Sharma L, Dubey RD, Mintoo MJ, Singh A, Kumar A, Sangwan PL, Tasaduq SA, Singh G, Mondhe DM, Gupta PN (2017b) Synthesis, characterization and augmented anticancer potential of PEG-betulinic acid conjugate. Mater Sci Eng C Mater Biol Appl 73:616–626. https://doi.org/10.1016/j.msec.2016.12.109

Saneja A, Kumar R, Mintoo MJ, Dubey RD, Sangwan PL, Mondhe DM, Panda AK, Gupta PN (2019) Gemcitabine and betulinic acid co-encapsulated PLGA−PEG polymer nanoparticles for improved efficacy of cancer chemotherapy. Mater Sci Eng C 98:764–771. https://doi.org/10.1016/j.msec.2019.01.026

Sanna V, Pintus G, Roggio AM, Punzoni S, Posadino AM, Arca A, Marceddu S, Bandiera P, Uzzau S, Sechi M (2011) Targeted biocompatible nanoparticles for the delivery of (−)-epigallocatechin 3-gallate to prostate cancer cells. J Med Chem 54:1321–1332. https://doi.org/10.1021/jm1013715

Sanna V, Siddiqui IA, Sechi M, Mukhtar H (2013) Resveratrol-loaded nanoparticles based on poly(epsilon-caprolactone) and poly(D,L-lactic-co-glycolic acid)-poly(ethylene glycol) blend for prostate cancer treatment. Mol Pharm 10:3871–3881. https://doi.org/10.1021/mp400342f

Sanna V, Singh CK, Jashari R, Adhami VM, Chamcheu JC, Rady I, Sechi M, Mukhtar H, Siddiqui IA (2017) Targeted nanoparticles encapsulating (−)-epigallocatechin-3-gallate for prostate cancer prevention and therapy. Sci Rep 7:41573. https://doi.org/10.1038/srep41573

Schoubben A, Ricci M, Giovagnoli S (2019) Meeting the unmet: from traditional to cutting-edge techniques for poly lactide and poly lactide-co-glycolide microparticle manufacturing. J Pharm Investig 49:381–404. https://doi.org/10.1007/s40005-019-00446-y

Shanmugam MK, Rane G, Kanchi MM, Arfuso F, Chinnathambi A, Zayed ME, Alharbi SA, Tan BKH, Kumar AP, Sethi G (2015) The multifaceted role of curcumin in cancer prevention and treatment. Molecules (Basel, Switzerland) 20:2728–2769. https://doi.org/10.3390/molecules20022728.

Shehzad A, Lee J, Huh TL, Lee YS (2013) Curcumin induces apoptosis in human colorectal carcinoma (HCT-15) cells by regulating expression of Prp4 and p53. Mol Cells 35:526–532. https://doi.org/10.1007/s10059-013-0038-5

Sherr CJ, McCormick F (2002) The RB and p53 pathways in cancer. Cancer Cell 2:103–112

Shimizu M, Deguchi A, Joe AK, McKoy JF, Moriwaki H, Weinstein IB (2005) EGCG inhibits activation of HER3 and expression of cyclooxygenase-2 in human colon cancer cells. J Exp Ther Oncol 5:69–78

Siddiqui IA, Sanna V (2016) Impact of nanotechnology on the delivery of natural products for cancer prevention and therapy. Mol Nutr Food Res 60:1330–1341. https://doi.org/10.1002/mnfr.201600035

Siddiqui IA, Adhami VM, Bharali DJ, Hafeez BB, Asim M, Khwaja SI, Ahmad N, Cui H, Mousa SA, Mukhtar H (2009) Introducing nanochemoprevention as a novel approach for cancer control: proof of principle with green tea polyphenol epigallocatechin-3-gallate. Cancer Res 69:1712–1716. https://doi.org/10.1158/0008-5472.can-08-3978

Singh M, Bhatnagar P, Mishra S, Kumar P, Shukla Y, Gupta KC (2015) PLGA-encapsulated tea polyphenols enhance the chemotherapeutic efficacy of cisplatin against human cancer cells and mice bearing Ehrlich ascites carcinoma. Int J Nanomedicine 10:6789–6809. https://doi.org/10.2147/ijn.s79489.

Sleeboom JJF, Eslami Amirabadi H, Nair P, Sahlgren CM, den Toonder JMJ (2018) Metastasis in context: modeling the tumor microenvironment with cancer-on-a-chip approaches. Dis Model Mech 11:dmm033100. https://doi.org/10.1242/dmm.033100

Sun T, Zhang YS, Pang B, Hyun DC, Yang M, Xia Y (2014) Engineered nanoparticles for drug delivery in cancer therapy. Angew Chem Int Ed Engl 53:12320–12364. https://doi.org/10.1002/anie.201403036.

Taube JH, Herschkowitz JI, Komurov K, Zhou AY, Gupta S, Yang J, Hartwell K, Onder TT, Gupta PB, Evans KW, Hollier BG, Ram PT, Lander ES, Rosen JM, Weinberg RA, Mani SA (2010) Core epithelial-to-mesenchymal transition interactome gene-expression signature is associated with claudin-low and metaplastic breast cancer subtypes. Proc Natl Acad Sci U S A 107:15449–15454. https://doi.org/10.1073/pnas.1004900107

Thakur VS, Gupta K, Gupta S (2012) Green tea polyphenols increase p53 transcriptional activity and acetylation by suppressing class I histone deacetylases. Int J Oncol 41:353–361. https://doi.org/10.3892/ijo.2012.1449.

Thiery JP, Acloque H, Huang RY, Nieto MA (2009) Epithelial-mesenchymal transitions in development and disease. Cell 139:871–890. https://doi.org/10.1016/j.cell.2009.11.007

Tino AB, Chitcholtan K (2016) Resveratrol and acetyl-resveratrol modulate activity of VEGF and IL-8 in ovarian cancer cell aggregates via attenuation of the NF-kappaB protein. J Ovarian Res 9:84

Uray IP, Dmitrovsky E, Brown PH (2016) Retinoids and rexinoids in cancer prevention: from laboratory to clinic. Semin Oncol 43:49–64. https://doi.org/10.1053/j.seminoncol.2015.09.002

van Vuuren RJ, Visagie MH, Theron AE, Joubert AM (2015) Antimitotic drugs in the treatment of cancer. Cancer Chemother Pharmacol 76:1101–1112. https://doi.org/10.1007/s00280-015-2903-8

Venugopal V, Krishnan S, Palanimuthu VR, Sankarankutty S, Kalaimani JK, Karupiah S, Kit NS (2018) Anti-EGFR anchored paclitaxel loaded PLGA nanoparticles for the treatment of triple negative breast cancer. In-vitro and in-vivo anticancer activities. PloS one 13:e0206109. https://doi.org/10.1371/journal.pone.0206109

Wang B, Yu X-C, Xu S-F, Xu M (2015) Paclitaxel and etoposide co-loaded polymeric nanoparticles for the effective combination therapy against human osteosarcoma. J Nanobiotechnol
13:22. https://doi.org/10.1186/s12951-015-0086-4

Watkins R, Wu L, Zhang C, Davis RM, Xu B (2015) Natural product-based nanomedicine: recent
advances and issues. Int J Nanomedicine 10:6055–6074. https://doi.org/10.2147/ijn.s92162.

White E, DiPaola RS (2009) The double-edged sword of autophagy modulation in cancer. Clin
Cancer Res 15:5308–5316. https://doi.org/10.1158/1078-0432.ccr-07-5023

Willis SN, Adams JM (2005) Life in the balance: how BH3-only proteins induce apoptosis. Curr
Opin Cell Biol 17:617–625. https://doi.org/10.1016/j.ceb.2005.10.001

World Health Organization (2014) World cancer report. http://publications.iarc.fr/Non-Series-
Publications/World-Cancer-Reports/World-Cancer-Report-2014 (on 23-07-19)

World Health Organization (n.d.) Retrieved from https://www.who.int/news-room/fact-sheets/
detail/cancer (on 23-07-19)

Wu F, Cui L (2017) Resveratrol suppresses melanoma by inhibiting NF-kappaB/miR-221
and inducing TFG expression. Arch Dermatol Res 309:823–831. https://doi.org/10.1007/
s00403-017-1784-6

Wu Y, Wang Z, Liu G, Zeng X, Wang X, Gao Y, Jiang L, Shi X, Tao W, Huang L, Mei L (2015)
Novel simvastatin-loaded nanoparticles based on cholic acid-core star-shaped PLGA for breast
cancer treatment. J Biomed Nanotechnol 11:1247–1260

Wu B, Liang Y, Tan Y, Xie C, Shen J, Zhang M, Liu X, Yang L, Zhang F, Liu L, Cai S, Huai DZ,
Zhang R, Zhang C, Chen K, Tang X, Sui X (2016) Genistein-loaded nanoparticles of star-
shaped diblock copolymer mannitol-core PLGA-TPGS for the treatment of liver cancer. Mater
Sci Eng C Mater Biol Appl 59:792–800. https://doi.org/10.1016/j.msec.2015.10.087

Xie X, Tao Q, Zou Y, Zhang F, Guo M, Wang Y, Wang H, Zhou Q, Yu S (2011) PLGA nanoparticles
improve the oral bioavailability of curcumin in rats: characterizations and mechanisms. J Agric
Food Chem 59:9280–9289. https://doi.org/10.1021/jf202135j

Xin Y, Liu T, Yang C (2016) Development of PLGA-lipid nanoparticles with covalently conjugated
indocyanine green as a versatile nanoplatform for tumor-targeted imaging and drug delivery.
Int J Nanomedicine 11:5807–5821. https://doi.org/10.2147/ijn.s119999

Yallapu MM, Gupta BK, Jaggi M, Chauhan SC (2010) Fabrication of curcumin encapsulated
PLGA nanoparticles for improved therapeutic effects in metastatic cancer cells. J Colloid
Interface Sci 351:19–29. https://doi.org/10.1016/j.jcis.2010.05.022

Yan Y, Zhou C, Li J, Chen K, Wang G, Wei G, Chen M, Li X (2017) Resveratrol inhibits hepatocellular carcinoma progression driven by hepatic stellate cells by targeting Gli-1. Mol Cell
Biochem 434:17–24. https://doi.org/10.1007/s11010-017-3031-z

Yang J, Wang C, Zhang Z, Chen X, Jia Y, Wang B, Kong T (2017a) Curcumin inhibits the survival and metastasis of prostate cancer cells via the Notch-1 signaling pathway. APMIS
125:134–140. https://doi.org/10.1111/apm.12650

Yang Z, Hackshaw A, Feng Q, Fu X, Zhang Y, Mao C, Tang J (2017b) Comparison of gefitinib, erlotinib and afatinib in non-small cell lung cancer: a meta-analysis. Int J Cancer 140:2805–2819.
https://doi.org/10.1002/ijc.30691

Yang-Hartwich Y, Bingham J, Garofalo F, Alvero AB, Mor G (2015) Detection of p53 protein
aggregation in cancer cell lines and tumor samples. Methods Mol Biol 1219:75–86. https://doi.
org/10.1007/978-1-4939-1661-0_7.

Yuan TL, Cantley LC (2008) PI3K pathway alterations in cancer: variations on a theme. Oncogene
27:5497–5510. https://doi.org/10.1038/onc.2008.245

Zaman MS, Chauhan N, Yallapu MM, Gara RK, Maher DM, Kumari S, Sikander M, Khan S, Zafar
N, Jaggi M, Chauhan SC (2016) Curcumin nanoformulation for cervical cancer treatment. Sci
Rep 6:20051. https://doi.org/10.1038/srep20051

Zhang Q, Tang X, Lu QY, Zhang ZF, Brown J, Le AD (2005) Resveratrol inhibits hypoxia-induced
accumulation of hypoxia-inducible factor-1alpha and VEGF expression in human tongue
squamous cell carcinoma and hepatoma cells. Mol Cancer Ther 4:1465–1474. https://doi.
org/10.1158/1535-7163.mct-05-0198.

Zhang Z, Wang X, Li B, Hou Y, Yang J, Yi L (2018) Development of a novel morphological paclitaxel-loaded PLGA microspheres for effective cancer therapy: in vitro and in vivo evaluations. Drug Deliv 25:166–177. https://doi.org/10.1080/10717544.2017.1422296

Zhang Y, Lin XY, Zhang JH, Xie ZL, Deng H, Huang YF, Huang XH (2019) Apoptosis of mouse myeloma cells induced by curcumin via the Notch3-p53 signaling axis. Oncol Lett 17:127–134. https://doi.org/10.3892/ol.2018.9591

Zhao Y, Huan ML, Liu M, Cheng Y, Sun Y, Cui H, Liu DZ, Mei QB, Zhou SY (2016) Doxorubicin and resveratrol co-delivery nanoparticle to overcome doxorubicin resistance. Sci Rep 6:35267. https://doi.org/10.1038/srep35267

Zhong Y, Meng F, Deng C, Zhong Z (2014) Ligand-directed active tumor-targeting polymeric nanoparticles for cancer chemotherapy. Biomacromolecules 15:1955–1969. https://doi.org/10.1021/bm5003009

Zhong H, Chan G, Hu Y, Hu H (2018) A comprehensive map of FDA-approved pharmaceutical products. Pharmaceutics 10. https://doi.org/10.3390/pharmaceutics10040263

Chapter 4
Chemistry, Pharmacology and Therapeutic Delivery of Major Tea Constituents

Ajay Rana and Sanjay Kumar

Abstract Tea (*Camellia sinensis*) is a renowned plant of theaceae family whose processed leaves are brewed with hot water and consumed as a refreshing and rejuvenating beverage. Usually, fresh tender shoots of this plant are processed for manufacturing of various types of teas (green, black, oolong and white) in different parts of the world. Tea is a rich source of various bioactive phytochemicals. These phytochemicals are not only responsible for colour, aroma and taste of tea, but also imparts numerous therapeutic and health beneficial properties.

In the recent past, numerous studies have reported evident role of tea bioactives in prevention and delay of various metabolic diseases like diabetes, obesity, cancer and cardiovascular disorders. Despite the numerous pharmacological properties, the clinical application of tea constituents is facing a challenging situation due to low bioavailability and poor efficacy. Hence, there is urgent need of novel and sustainable technological interventions through which therapeutic delivery of these highly valuable natural compounds is enhanced for extensive clinical applications.

Keywords Tea · Phenolics · Antioxidants · Pharmacology · Drug delivery

4.1 Introduction

Tea comes from the *Camellia sinensis* plant, which is native to Southeast Asia. Currently tea is grown in more than thirty countries around the world. It flourishes well in tropical and sub-tropical region with good water drainage and slightly acidic soils (pH from 4.5–5.5) are preferred for its cultivation. Green tender leaves of this plant are plucked manually or with the help of machines and then processed for manufacture of different types of teas. Numerous epidemiological studies conducted in the past have demonstrated plethora of health beneficial properties

A. Rana · S. Kumar (✉)
CSIR-Institute of Himalayan Bioresource Technology, Palampur, Himachal Pradesh, India
e-mail: sanjaykumar@ihbt.res.in

© The Editor(s) (if applicable) and The Author(s), under exclusive license to
Springer Nature Switzerland AG 2020
A. Saneja et al. (eds.), *Sustainable Agriculture Reviews 43*, Sustainable
Agriculture Reviews 43, https://doi.org/10.1007/978-3-030-41838-0_4

associated with tea consumption. Various pharmacological studies have reported role of tea constituents in prevention and delay of various metabolic ailments (obesity, diabetes, hypertension, atherosclerosis and cancer) (Hara 2001; Demeule et al. 2002; Stangl et al. 2007; Ikeda 2008; Shahidi et al. 2009; Sharma and Rao 2009; Chacko et al. 2010; Jankun et al. 2012). These pharmacological properties are directly linked to the occurrence of diverse range of bioactive molecules in the tea plant. The major constituents of the tea plant are flavonoids (catechin) and their dimeric derivatives (theaflavins), along with caffeine and theanine (Stodt and Engelhardt 2013). In tea, mainly two types of flavonoid compounds are present: the catechins which comprise of group of monomeric flavanols (EGC, EC, EGCG and ECG) and theaflavins that comprise of dimeric flavanols (theaflavin, theaflavin monogallates and theaflavin digallate) which are formed from catechins during tea processing in presence of endogenous enzymes (Sharma et al. 2009; Rana and Singh 2012).

Apart from flavanols, other major compounds present in tea are caffeine and theanine. Caffeine (1, 3, 7-trimethylxanthine) is an alkaloid molecule. It is one of the most widely consumed active food ingredient with various health beneficial properties. Caffeine helps in enhancing mood and alertness by acting mainly upon the central nervous system (Heckman et al. 2010). Theanine is another very important compound that occurs exclusively in tea and apart of tea it has been reported in a mushroom, *Xerocomus badius*. It is a non-protein amino acid, which imparts umami taste to tea. Theanine consumption has been associated with brain relaxation and improving learning ability (Vuong et al. 2011). These major constituents of tea being water soluble impart taste, colour, aroma to the tea infusion and are also responsible for various health promoting properties associated with tea consumption. However, despite of numerous pharmacological properties of tea constituents its clinical applications faces huge hurdles. This hindrance in clinical and therapeutic delivery of tea bioactives is due to their poor bioavailability and efficacy. Hence, it is very necessary to develop noble sustainable techniques for improving the therapeutic delivery of these invaluable natural compounds.

4.2 Different Types of Teas

Processing of fresh tea shoots followed by heat drying has been practiced since very long time. The processing methods vary from country to country. Based on the manufacturing process teas are classified as: green, black, oolong and white teas (Sajilata et al. 2008; Chen et al. 2011). All types of teas are made from tender shoots of the same plant *Camellia sinensis*, which comprise of an unopened bud (apical bud) and subtending two to three leaves. Due to variability in processing techniques each of these teas impart different colour, taste and aroma.

4.2.1   Green Tea

Green tea is the most popular type of tea. It is richer in antioxidant flavanoids called as catechins compared to other types of teas along with appropriate amount of caffeine and theanine. Green tea is manufactured in such a way so that majority of chemical constituents present in green leaf remain unchanged during processing. For manufacture of green tea, freshly plucked tender tea shoots are heat treated immediately after plucking in order to inactivate the endogenous enzymes (polyphenol oxidase and peroxidases). Different techniques are used for enzyme inactivation such as steaming, pan roasting, hot air and microwave treating the fresh tea shoots (Gulati et al. 2003). The heat-treated tea shoots are cooled, air dried partially to remove excess moisture and then rolled by applying gentle pressure using roller machines to give proper shape to the shoots. The rolled tea shoots were dried by using hot air to a final moisture level of 3–5%. This is very important for good shelf-life. Drying is followed by grading based on shape and size. The grading is the sorting process of made tea using different sieves.

4.2.2   Black Tea

Black tea is the highly consumed tea. It is generally of two main types, CTC black tea and orthodox black tea sometimes called as leaf tea. The CTC black tea is manufactured by using CTC (cut/crush, tear and curl) machine. CTC tea generally produces rich orange reddish coloured infusion when infused with boiling water. However, CTC is prepared mostly after excessively boiling with water and consumed after addition of small volume of milk. Usually CTC tea contains very low content of antioxidant catechins. It contains few theaflavins and higher content of thearubigins and related compounds. While orthodox black or leaf tea is manufactured, accordingly traditional tea manufacture process. The processing of orthodox tea involves withering of tea shoots for 12–16 h. Withering is the process of removing moisture (50–70%) from fresh leaves under ambient condition to reduce the turgidity of the leaf and increase the permeability of the cell wall. This helps in giving proper shape to the leaf without causing physical damage. Withering is followed by rolling of the shoots in orthodox tea rollers by applying appropriate pressure. This is very crucial step in tea manufacturing because during rolling, the cells get disrupted allowing various cellular components such as polyphenols, amino acids, sugars etc. to come in contact with each other as well as with various cellular enzymes (polyphenol oxidase and peroxidase) and proteins. This leads to initiation of numerous biochemical reactions, which initiates the process for formation of various aroma, flavor and coloured compounds. The rolled tea is allowed to further oxidize for 30–45 min duration at ambient temperature. The rolled leaves are finally dried with hot air to final moisture level to 3–5%. The drying is followed by grading. In comparison to CTC black tea, orthodox black tea is richer in aroma, flavour compounds as well as health beneficial tea constituents.

4.2.3 Oolong Tea

Oolong tea is the traditional form of Chinese tea (Chen et al. 2011). This tea processing lies between green and black tea. It is manufactured by partial enzyme inactivation of endogenous enzymes. The sun withered tea shoots are processed similar to orthodox tea manufacture. The withered tea shoots are processed by rolling with application of appropriate pressure. The rolled tea is dried by hot air followed by grading process. Oolong tea contains higher content of catechins as compared to black teas. Due to higher aroma and flavour compounds in oolong tea compared to green tea with equivalent health beneficial property, an immense rise in worldwide production of this high quality tea has been noticed in the recent years.

4.2.4 White Tea

It is a special type of tea made selectively from apical buds (unopened leaves). For manufacturing of white tea, the apical buds are collected by hand picking. These buds were then withered for few hours, rolled gently without affecting or destroying the shape and texture of the bud and finally dried with hot air. On drying, this tea gives white appearance because of the presence of dried bud pubescent hairs. Due to the high price, white tea has been confined to a very limited market.

4.2.5 Herbal Teas

Herbal teas are those teas, which are made from aromatic herbs, flowers, spices and medicinal plants along with and without tea (*C. sinensis*) leaves. The practice of consumption of herbal teas has been referenced in traditional Indian and Chinese medicinal system as medicated teas (Poswal et al. 2019). There is no standardized procedure for manufacture of herbal teas and processing varies from region to region based on composition of the aromatic and medicinal plant or part. Moreover, the limit of herbs and botanicals usage is undetermined. Now a day's various herbal teas are available in market, which are fortified with synthetic aroma and flavors compounds.

4.3 Chemistry of Major Tea Constituents

Diverse array of phytochemicals are known to be present in green tea leaves. Phenolics flavan-3-ols generally called as catechins occurs predominantly in the tea along with caffeine alkaloid and amino acid theanine (Fig. 4.1). Apart from these major constituents, various other compounds are also reported in tea like organic

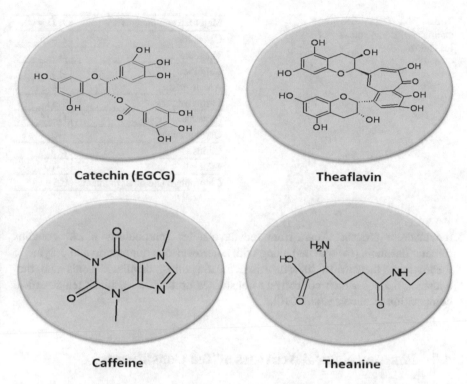

Fig. 4.1 Chemical structures of major tea constituents [catechin (epigallocatechin gallate), theaflavin, caffeine and theanine]

acids, carbohydrates, polysaccharides, minerals, methyl xanthines, proteins, lipids etc. (Table 4.1). Catechins are the predominant flavanols also called as 3-hydroxy derivatives of flavonoids as characterized by meta-5, 7-dihydroxy substituted A-ring with di- or trihydroxylic B-ring. They are also called as flavan-3-ols as the hydroxyl group is bound to 3rd position of the C-ring (Panche et al. 2016). They are derived from 2-phenylbenzopyran. Catechins constitute a group of monomeric flavanols comprising of epigallocatechin (EGC), epicatechin (EC), epigallocatechin gallate (EGCG) and epicatechin gallate (ECG) as naturally occurring phenolics in tea leaves. The catechins content in tea leaves decreases with maturity. They occurs maximum in fresh tender tea leaves which are generally used for tea manufacture as compared to coarse and mature tea leaves. Among catechins EGCG is the most frequently available and highly studied catechin molecule. The amount of EGCG among other catechins ranges up to 40–50%.

During the processing of tea selective catechins in presence of polyphenol oxidase enzyme forms reddish-orange colored dimeric theaflavins which have more antioxidant activity that catechins. Theaflavins are solely produced during tea processing as a consequence of biochemical reactions among different group of catechins. These compounds are not present in green tea leaves but are formed by enzymatic oxidation (earlier fermentation) and condensation of catechins during tea

Table 4.1 Data showing chemicals composition of tea shoots

Major tea components	% Dry wt.
Catechins	12–15
Theanine	2–3
Caffeine	2–3
Amino acids	3–4
Carbohydrates	8–11
Proteins	11–15
Organic acids	1–2
Lipids	2–3
Minerals	6–10
Chlorophyll and other pigments	0.5

manufacture process. Apart from theaflavins, tea (particularly black) contains dimeric flavanols (theaflavins) along with heterogeneous oligomeric and polymeric thearubigins theacitrins, theasinensins, theaflagallins, theaflavic acids and the-anaphtoquinones which collectively constitutes up to 70% of black tea chemical composition (Kuhnert et al. 2010).

4.4 Pharmacological Activities of Tea Constituents

Tea is widely recognized for plethora of health beneficial properties (Fig. 4.2). Tea consumption has been associated with prevention/delay of various diseases including lower risk of cardiovascular diseases (CVD) like stroke and artherosclerosis, prevention of cancer, and protection against dental caries and due to its beneficial effects on the gastrointestinal tract (Ruxton 2008; Sharma and Rao 2009). These pharmacological properties of tea have been associated with the presence of high contents of phenolic compounds. It has been evident from various *in vitro, in vivo* and animal model studies that tea phenolics act as powerful antioxidants and have direct effects on various biochemical processes that occurs via a range of complex mechanisms (Vuong 2014). Thus due to these assorted health properties, tea and it chemical constituents has attracted significant attention as novel therapeutics and nutraceuticals against metabolic disorders.

4.4.1 Antioxidants

Antioxidants are chemical compounds responsible for reducing/scavenging free radicals (ROS and RNS) that are generated in the body. They have strong ability to terminate oxidation chain reactions by removing free radical intermediates, and also inhibit other cellular oxidation reactions. They attain this by getting oxidized

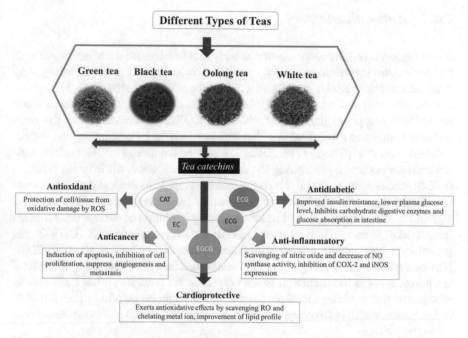

Fig. 4.2 Showing various types of teas (green, black, oolong and white) with major active constituents and their reported therapeutic potential as antioxidants, anticancer, cardio protective, anti-inflammatory and antidiabetic

themselves. The tea phenolics are well recognized antioxidant compounds. There are numerous reports that illustrate the antioxidant behaviour of tea phenolics (Wiseman et al. 1997; Rietveld and Wiseman 2003; Wang and Li 2006; Sharma and Rao 2009; Chu et al. 2017). The catechins and theaflavins are the major known polyphenolic antioxidant compounds that are present in tea. Polyphenols due to high conjugation in their structure, have the ability to donate proton or hydrogen which makes them potential antioxidant candidates. Numerous epidemiological studies have revealed that tea polyphenols help in protection of cell and tissue from oxidative damage by reactive oxygen species (ROS). Due to their free radical scavenging behavior, tea polyphenols can act as natural and reliable source of potent antioxidant compounds. Tea consumption has also been related to have a protective role during alcohol intoxication by preventing the reduction of liver glutathione peroxidase and reductase and catalase activity which would otherwise deplete cellular antioxidant defence system especially in liver (Luczaj and Skrzydlewska 2004). If we go for comparison between green and black tea the former has more antioxidant potential than latter which is the main reason for rising popularity of green tea.

4.4.2 Anti-inflammatory

Inflammation is cellular response of our body that involves influx of defensive cells like leukocytes (neutrophils, monocytes and eosinophils) to the site of injury. Any disturbance in this cellular mechanisms or prolonged inflammation leads to implication which could cause cancer and other diseased conditions. Generation of nitric oxide (NO) and prostaglandins by iNOS and COX-2 are considered as the most prominent molecular mechanisms that are involved in inflammation and cancer (Ahmed et al. 2002; Zhong et al. 2012). It was reported that green tea exhibits anti-inflammatory effects by inhibiting the expression of inducible nitric oxide synthase (iNOS) (Tedeschi et al. 2004). From the extensive epidemiological data it has been established that green and black tea polyphenols (catechins and theaflavins) play a significant role in suppression of inflammation by inhibition of various transcription factors like tumor necrosis factor-α (TNF-α), cyclooxygenase 2 (COX-2 and granulocyte-macrophage colony stimulating factor (GM-CSF) (Cao et al. 2007; Novilla et al. 2017). Although exact molecular mechanisms explicating the role of tea polyphenols in inflammation is not very clear till now. But it has been clearly recognized that drinking sufficient amount of tea might be helpful in prevention of inflammation which is also responsible for causing various other disease conditions.

4.4.3 Anticancer

Cancer is the disease condition wherein the cells of any specific part in the body starts uncontrolled growth and division. These cells are recognized as cancerous as they have the ability to invade and destroy surrounding healthy cells, tissues and organs as well. A diverse range of research articles are available in the literature reporting that tea and its constituents have positive effects in inhibition of tumorigenesis at various sites in our body; including skin, mammary gland, lung, oral cavity, esophagus, stomach, small intestine, colon, liver, prostrate and pancreas (Yang et al. 2002; Demeule et al. 2002; Boehm et al. 2010). Although, various clinical trials have been conducted in the past on both human and animal models for the investigation of possible anticancer properties of tea, but these studies lack evidences based consistency in few cases of cancer prevention (Demeule et al. 2002; Bettuzzi et al. 2006). But use of tea polyphenols as therapeutics has find lot of perspectives against prevention and delay of cancer which is amongst the severe killer of millions of people globally. Anticancer effect of green tea constituents through induction of detoxification enzyme i.e. glutathione S-transferase was also reported (Chow et al. 2007). Tea catechin, EGCG was reported to induce apoptosis and inhibit cell proliferation in different human cancer cell lines and also showed inhibitory effects on angiogenesis and metastasis (Singh et al. 2011). Regardless of large number of research data demonstrating significant role of tea and tea polyphenols against cancer progression, still there is long way to comprehend the unknown mechanisms behind cancer occurrence (Boehm et al. 2010; Yang and Wang 2011).

4.4.4   Antidiabetic

Diabetes is a disease state wherein blood glucose level of the patient is higher than normal level. This situation may usually arise due to the inability of body to pick up extra glucose from the blood stream or inability in glucose metabolism. Generally, diabetes has been characterized as two main types; type-I and type-II diabetes. Studies have found that diabetes has emerged as one of the leading cause of global deaths that occurs annually. In the last two decades various studies have found that tea especially green tea plays a significant role in prevention of diabetes (Miura et al. 2005; Islam and Choi 2007). Various investigators have reported diabetes preventive and curative properties exhibited by tea polyphenols (Chacko et al. 2010; Jankun et al. 2012). The documented research data on consumption of tea for diabetes prevention has revealed insulin-enhancing activity, lowering of plasma glucose level and inhibition of transcription factors by tea and its constituents. Tea catechins were reported to inhibit the enzymes involved in carbohydrate digestion and decreases glucose absorption in intestine (Sharma et al. 2019). As these factors are directly linked with diabetes occurrence, therefore tea consumption helps in prevention of diabetes. There is still lack of sufficient data in support of proper cellular and molecular mechanisms related to the prevention of diabetes in humans by tea.

4.4.5   Cardioprotective

Cardiovascular diseases are the diseases of heart and blood vessels and generally distinguished as coronary heart diseases, cerebro-vascular diseases, hypertension and heart failure. There are numerous causes that are directly associated with our diet and life style for elevated occurrence of cardiovascular diseases among populations. These major reasons of high prevalence of cardiovascular diseases are high blood pressure, high cholesterol, diabetes, extensive smoking and family history etc. Tea and its polyphenols have been recognized as potent and reliable source to fight against cardiovascular diseases. A large number of research data has been available supporting tea and its constituents and their role in prevention of these disease states. The tea polyphenols could act as possible cardioprotective candidates due to their potent antioxidant capacity. These constituents induce antioxidative effects by scavenging ROS and chelating metal ions (Velayutham et al. 2008). It has been found that free radicals play a key role in occurrence of cardiovascular disorders by causing oxidative stress (Aron and Kennedy 2008). Due to the repetitive induction of cellular oxidative stress there occurs induced cellular resistance to consequent exposure to reactive oxygen species. This could be inhibited by galloyl group in tea polyphenols (Stangl et al. 2007). The galloyl group may be interfering by modifying kinase activities in multiple pathways of signal transduction in cardiovascular relevant cells which could play a crucial role in the prevention and treatment of cardiovascular diseases (Stangl et al. 2007). Tea catechins were reported to improve blood lipid profile by inhibiting the main enzymes involved in biosynthesis of lipids (Velayutham et al. 2008). Thus tea polyphenols have huge perspective to be used as preventive therapeutics against cardiovascular disorders.

4.4.6 Antimicrobial

Antimicrobial behavior is the ability of tea polyphenols to act as toxic against harmful microorganisms such as bacteria, fungi and viruses. The role of tea polyphenols (catechins and theaflavins) in growth inhibition of various infectious microorganisms has been briefly discussed in numerous earlier reports (Liu et al. 2005; Friedman 2007). The tea polyphenols regulate the gene expression in *E. coli*, inhibits cellular division in some species of bacteria and also inhibit HIV-1 replication in human body by targeting at various steps in the life cycle of HIV-1 which is the major causative agent of AIDS (Liu et al. 2005). Although tea polyphenols could also act synergistically with selective antibiotics and have been found to be beneficial against multi drug resistance. Earlier Tiwari et al. (2005) have reported synergistic activity of tea extracts with chloramphenicol and other antibiotics like methicillin, nalidixic acid and gentamycin against various strains of enteropathogens. In another study, Betts et al. (2013) found synergistic anti-fungal effects of black tea theaflavin and green tea epicatechin against *Candida albicans* in combination. Thus, it is evident that tea and its constituents have good perspective to be used as antimicrobials.

4.5 Therapeutic Delivery of Tea Constituents

Numerous *in vitro* studies revealed the potential health benefits of tea catechins in various ailments. The majority of pharmacological properties of tea are associated with antioxidant activity of tea catechins, which is associated with phenolic groups. This polyphenolic nature of catechins is associated with efficacy and bioavailability of tea catechins. It effects the therapeutic application of tea catechins. To overcome these drawbacks considerable work is going on to enhance the bioavailability, cell permeability and inhibition of gastro intestinal degradation of tea catehins. Various matrix assisted encapsulations techniques has been used in recent past (Table 4.2). Generally, carbohydrate (chitosan, cellulose polymers, starch-based materials, gum arabic and sodium alginate), protein, lipids and gelatin derivatives are widely employed for encapsulation of tea bioactives (Cai et al. 2018). In a recent study (Ahmad et al. 2019), starch based nano particles were prepared for nano-encapsulation of catechin. It was showed that during simulated gastro-intestinal digestion, the biological properties of catechins were retained in encapsulated catechin compared to free catechin. In another study, chitosan was used for the nano-encapsulation of catechin that exhibited good surface topography and emulsion

Table 4.2 Different therapeutic delivery approaches to improve the bioefficacy of tea constituents

Sr. no.	Nanoparticle	Main matrix	Bioactive molecule/ Extract	Nature of study	Key findings	References
1	Casein micelles	Casein	EGCG	In vitro	EGCG-casein complex decrease the proliferation of HT-29 cells alike free EGCG	Haratifar et al. (2014)
2	Nanoemulsion	Soy protein	Catechins	In vitro	Nanoemulsions based on soy protein improved the stability, bioaccessibility and permeability of tea catechins.	Bhushani et al. (2016)
3	Protein isolate	Rice bran protein isolate	Catechins	In vitro	Binding catechins to rice bran protein isolate resulting in improved stability of tea catechins	Shi et al. (2017)
4	Emulsions	Sodium caseinate,	EGCG	in vitro	Sodium caseinate-stabilized emulsions can be employed for effective delivery of EGCG	Sabouri et al. (2017)
5	Liposomes	Soy lecithin	Green tea extract	in vitro	Improved bioavailability and stability of tea catechins	Dag and Oztop (2017)
6	Nanoparticles	Maltodextrin, gum arabic, egg-yolk l-alpha-phosphatidylcholine and stearylamine	EGCG	In vitro	Effective delivery of EGCG	Gomes et al. (2010)
7	Nanoparticles	Starch from horse chestnut (HSC), water chestnut (WSC) and lotus stem (LSC)	Catechins	In vitro	Helped in retaining biological properties of catechin during simulated gastro intestinal digestion	Ahmad et al. (2019)
8	Nanoparticles	Chitosan	Catechins	In vitro	Catechin-loaded folate-conjugated chitosan nanoparticles exhibited anti-proliferative effects in a concentration dependent manner for MCF-7 cells	Liu et al. (2017)
9	Nanoparticle	Gold	EGCG	In vitro	EGCG gold nanoparticles showed more apoptosis in cancer cells compared to EGCG	Chavva et al. (2019)

(continued)

Table 4.2 (continued)

Sr. no.	Nanoparticle	Main matrix	Bioactive molecule/ Extract	Nature of study	Key findings	References
10	Nanoparticle	Soya lecithin	EGCG	*In vitro*	EGCG loaded nanoparticles showed higher cytotoxicity against MDA-MB 231 and DU-145 cancer cells compared to free EGCG	Radhakrishnan et al. (2016)
11	Nanoparticles	β-chitosan	Catechins	*In vitro*	Improved the antibacterial activity against *E. coli* and *L. innocua*	Zhang et al. (2016)
12	Nanoparticles	Gold	EGCG	*In vivo*	EGCG-gold nanoparticles controlled growth of tumor cell by apoptosis	Hsieh et al. (2011)
13	Nanoparticles	Chitosan	EGCG	*In vivo*	Nano formulation of EGCG had greater activity against atherosclerosis.	Hong et al. (2014)

EGCG Epigallocatechin gallate, *HT 29* Human colorectal adenocarcinoma cell Line, *MCF 7* Breast cancer cell line, *DU 145* Human prostate cancer cell line, *MDA-MB 231* Epithelial breast cancer cell line, *E. coli – Escherichis coli*, *L. innocua – Listeria innocua*, *HSC* horse chestnut, *WSC* water chestnut, *LSC* lotus stem

stability (Kailaku et al. 2014). Encapsulation of green tea catechins in γ-cyclodextrin and coating with hydroxypropylmethyl cellulose phthalate were reported to increase their digestive stability upto 65% and 58%, respectively (Son et al. 2016). Bovine serum albumin, a globular protein was efficiently used as an effective carrier for encapsulation of EGC. It was reported that the stability of EGCG was enhanced when bound to bovine serum albumin (BSA) compared to free one (Yi et al. 2017). Moreover, chitosan coated BSA–EGCG nanoparticles improved absorption of EGCG. It was showed that chitosan coated BSA–EGCG nanoparticles showed higher permeability compared to free form of EGCG (Li et al. 2014).

4.6   Future Perspectives

Numerous epidemiological research evidences in recent years have clearly indicated that regular consumption of tea and its constituents could help in prevention and delay of various dreadful diseases such as diabetes, obesity, hypertension, cardiovascular disorders and cancer. Since tea is one of the most inexpensive and sustainable source to obtain these highly valuable natural antioxidants. Therefore, there is an immense need to intensify the research efforts to cope with the issues of poor stability, efficacy and bioavailability of tea constituents. The outcome of these research studies will open new horizons for exploration and valorization of tea constituents for the benefit of society.

4.7   Conclusion

Tea is undoubtedly a health-promoting beverage, which constitutes diverse array of bioactive molecules. The tea shoots are processed for preparation of various types of teas (green, black, oolong and white) each differs on the basis of manufacturing practices. These teas itself constitutes wide array of volatile and non-volatile chemical compounds which imparts unique aroma, flavor, colour and taste properties. Various research studies have shown the role of different tea and its chemical constituents in prevention and delay of metabolic ailments. The research efforts have made globally to understand the role of tea constituents against these dreadful diseases at cellular and molecular level. There is strong need to develop sustainable technologies to resolve the issue of poor stability and bioefficacy of tea constituents. Hence, it is clear that wide clinical applicability of tea constituents awaits use of novel and sustainable technological interventions.

References

Ahmad M, Mudgil M, Gani A, Hamed F, Masood FA, Maqsood S (2019) Nano encapsulation of catechin in starch nanoparticles: characterization; release behavior and bioactivity retention during in-vitro digestion. Food Chem 270:95–104. https://doi.org/10.1016/j.foodchem.2018.07.024

Ahmed S, Rehman A, Hasnain A, Lalonde M, Goldberg VM, Haqqi TM (2002) Green tea polyphenol epigallocatechin-3-gallate inhibits the IL-1β-induced activity and expression of cyclooxygenase-2 and nitric oxide synthase-2 in human chondrocytes. Free Radic Biol Med 33:1097–1105. https://doi.org/10.1016/s0891-5849(02)01004-3

Aron PM, Kennedy JA (2008) Flavan-3-ols: nature, occurrence and biological activity. Mol Nutr Food Res 52:79–104. https://doi.org/10.1002/mnfr.200700137

Betts JW, Wareham DW, Haswell SJ, Kelly SM (2013) Antifungal synergy of theaflavin and epicatechin combinations against *Candida albicans*. J Microbiol Biotechnol 23:1–5. https://doi.org/10.4014/jmb.1303.03010

Bettuzzi S, Brausi M, Rizzi F, Castagnetti G, Peracchia G, Corti A (2006) Chemoprevention of human prostate cancer by oral administration of green tea catechins in volunteers with high-grade prostate intraepithelial neoplasia: a preliminary report from a one-year proof-of-principle study. Cancer Res 66:1234–1240. https://doi.org/10.1158/0008-5472.CAN-05-1145

Bhushani JA, Karthik P, Anandharamakrishnan C (2016) Nanoemulsion based delivery system for improved bioaccessibility and Caco-2cell monolayer permeability of green tea catechins. Food Hydrocoll 56:372–382. https://doi.org/10.1016/j.foodhyd.2015.12.035

Boehm K, Borrelli F, Ernst E, Habacher G, Hung SK, Milazzo S, Horneber M (2010) Green tea (*Camellia sinensis*) for the prevention of cancer. Cochrane Database Syst Rev:1–60. https://doi.org/10.1002/14651858.CD005004.pub2.

Cai Z-Y, Li X-M, Liang J-P, Xiang L-P, Wang K-R, Shi Y-L, Yang R, Shi M, Ye J-H, Lu J-L, Zheng X-Q, Liang Y-R (2018) Bioavailability of tea catechins and its improvement. Molecules 23:2346. https://doi.org/10.3390/molecules23092346

Cao H, Kelly MA, Kari F, Dawson HD, Urban JF Jr, Coves S, Roussel AM, Anderson RA (2007) Green tea increases anti-inflammatory tristetraprolin and decreases pro-inflammatory tumor necrosis factor mRNA levels in rats. J Inflamm 4:1–12. https://doi.org/10.1186/1476-9255-4-1

Chacko SM, Thambi PT, Kuttan R, Nishigaki I (2010) Beneficial effects of green tea: a literature review. Chin Med 5:1–9. https://doi.org/10.1186/1749-8546-5-13

Chavva SR, Deshmukh SK, Kanchanapally R, Tyagi N, Coym JW, Singh AP, Singh S (2019) Epigallocatechin gallate-gold nanoparticles exhibit superior antitumor activity compared to conventional gold nanoparticles: potential synergistic interactions. Nano 9:396. https://doi.org/10.3390/nano9030396

Chen YL, Duan J, Jiang YM, Shi J, Peng L, Xue S, Kakuda Y (2011) Production, quality, and biological effects of oolong tea (Camellia sinensis). Food Rev Int 27:1–15. https://doi.org/10.1080/87559129.2010.518294

Chow H-H, Hakim IA, Vining DR, Crowell JA, Tome ME, Ranger-Moore J, Cordova CA, Mikhael DM, Briehl MM, Alberts DS (2007) Modulation of human glutathione S-transferases by polyphenon E intervention. Cancer Epidemiol Biomark Prev 16(8):1662–1666. https://doi.org/10.1158/1055-9965.EPI-06-0830

Chu C, Deng J, Man Y, Qu Y (2017) Green tea extracts epigallocatechin-3-gallate for different treatments. BioMed Res Int, Article ID 5615647, 9 pages

Dag D, Oztop MH (2017) Formation and characterization of green tea extract loaded liposomes. J Food Sci 82:463–470. https://doi.org/10.1111/1750-3841.13615

Demeule M, Annabi B, Gingras DB et al (2002) Green tea catechins as novel antitumor and antiangiogenic compounds. Curr Med Chem Anticancer Agents 2:441–463. https://doi.org/10.2174/1568011023353930

Friedman M (2007) Overview of antibacterial, antitoxin, antiviral, and antifungal activities of tea flavonoids and teas. Mol Nutr Food Res 51:116–134. https://doi.org/10.1002/mnfr.200600173

Gomes JFPS, Rochaa S, Pereiraa MDC, Peres I, Moreno S, Toca-Herrera J, Coelho MAN (2010) Lipid/particle assemblies based on maltodextrin–gum arabic core as bio-carriers. Colloids Surf B: Biointerfaces 76:449–455. https://doi.org/10.1016/j.colsurfb.2009.12.004

Gulati A, Rawat R, Singh B, Ravindranath SD (2003) Application of microwave energy in the manufacture of enhanced-quality green tea. J Agric Food Chem 51:4764–4768. https://doi.org/10.1186/s40529-016-0143-9

Hara Y (2001) Green tea health benefits and applications. Food Science and Technology, Marcel Dekker, Inc. Madison Avenue, New York

Haratifar S, Meckling KA, Corredig M (2014) Antiproliferative activity of tea catechins associated with casein micelles, using HT29 colon cancer cells. J Dairy Sci 97:672–678. https://doi.org/10.3168/jds.2013-7263

Heckman MA, Weil J, Demejia EG (2010) Caffeine (1, 3, 7-trimethylxanthine) in foods: a comprehensive review on consumption, functionality, safety, and regulatory matters. J Food Sci 75:R77–R87. https://doi.org/10.1111/j.1750-3841.2010.01561.x

Hong Z, Xu Y, Yin J-F, Jin J, Jiang Y, Du Q (2014) Improving the Effectiveness of (−)-epigallocatechin gallate (EGCG) against rabbit atherosclerosis by EGCG-loaded nanoparticles prepared from chitosan and polyaspartic Acid. J Agric Food Chem 62:12603–12609. https://doi.org/10.1021/jf504603n

Hsieh DS, Wang H, Tan SW, Huang YH, Tsai CY, Yeh MK, Wu CJ (2011) The treatment of bladder cancer in a mouse model by epigallocatechin-3-gallate-gold nanoparticles. Biomaterials 32:7633–7640. https://doi.org/10.1016/j.biomaterials.2011.06.073

Ikeda I (2008) Multifunctional effects of green tea catechins on prevention of the metabolic syndrome. Asia Pac J Clin Nutr 17:273–274. https://doi.org/10.6133/apjcn.2008.17.s1.65

Islam MS, Choi H (2007) Green tea anti-diabetic or diabetogenic: a dose response study. Biofactors 29:45–53. https://doi.org/10.1002/biof.5520290105

Jankun J, Al-Senaidy A, Skrzypczak-Jankun E (2012) Can drinking black tea fight diabetes: literature review and theoretical indication. Cent Eur J Immunol 37:167–172

Kailaku SI, Mulyawanti I, Alamsyah AN (2014) Formulation of nanoencapsulated catechin with chitosan as encapsulation material. Proc Chem 9:235–241. https://doi.org/10.1016/j.proche.2014.05.028

Kuhnert N, Drynan JW, Obuchowicz J, Clifford MN, Witt M (2010) Mass spectrometric characterization of black tea thearubigins leading to an oxidative cascade hypothesis for thearubigin formation. Rapid Commun Mass Spectrom 24(23):3387–3404. https://doi.org/10.1002/rcm.4778

Li Z, Ha J, Zou T, Gu L (2014) Fabrication of coated bovine serum albumin (BSA)-epigallocatechin gallate (EGCG) nanoparticles and their transport across monolayers of human intestinal epithelial Caco 2 cells. Food Funct 5:1278–1285. https://doi.org/10.1039/c3fo60500k

Liu S, Lu H, Zhao Q, He Y, Niu J, Debnath AK, Wu S, Jiang S (2005) Theaflavin derivatives in black tea and catechin derivatives in green tea inhibit HIV-1 entry by targeting gp41. Biochim Biophys Acta 1723:270–281. https://doi.org/10.1016/j.bbagen.2005.02.012

Liu B, Wang Y, Yu Q, Li D, Li F (2017) Synthesis, characterization of catechin-loaded folate-conjugated chitosan nanoparticles and their anti-proliferative effect. CyTA J Food 16:868–876. https://doi.org/10.1080/19476337.2018.1491625

Luczaj W, Skrzydlewska E (2004) Antioxidant properties of black tea in alcohol intoxication. Food Chem Toxicol 42:2045–2051. https://doi.org/10.1016/j.fct.2004.08.009

Miura T, Koike T, Ishida T (2005) Antidiabetic activity of green tea (Thea sinensis L.) in genetically type 2 diabetic mice. J Health Sci 51:708–710. https://doi.org/10.1248/jhs.51.708

Novilla A, Djamhuri DS, Nurhayati B, Rihibiha DD, Afifah E, Widowati W (2017) Anti-inflammatory properties of oolong tea (Camellia sinensis) ethanol extract and epigallocatechin gallate in LPS-induced RAW 264.7 cells. Asian Pac J Trop Biomed 7:1005–1009. https://doi.org/10.1016/j.apjtb.2017.10.002

Panche AN, Diwan AD, Chandra SR (2016) Flavonoids: an overview. J Nutr Sci 5:e47. https://doi.org/10.1017/jns.2016.41

Poswal FS, Russell G, Mackonochie M, MacLennan E, Adukwu EC, Rolfe V (2019) Herbal teas and their health benefits: a scoping review. Plant Foods Hum Nutr 74:266. https://doi.org/10.1007/s11130-019-00750-w

Radhakrishnan R, Kulhari H, Pooja D, Gudem S, Bhargava S, Shukla R, Sistla R (2016) Encapsulation of biophenolic phytochemical EGCG within lipid nanoparticles enhances its stability and cytotoxicity against cancer. Chem Phys Lipids 198:51–60. https://doi.org/10.1016/j.chemphyslip.2016.05.006

Rana A, Singh HP (2012) A Rapid HPLC-DAD method for analysis of theaflavins using C_{12} as stationary phase. J Liq Chromatogr Relat Technol 35:2272–2279. https://doi.org/10.1080/10826076.2011.631257

Rietveld A, Wiseman S (2003) Antioxidant effects of tea: evidence from human clinical trials. J Nutr 133:3285S–3295S. https://doi.org/10.1093/jn/133.10.3285S

Ruxton CHS (2008) Black tea and health. Br Nutr Found Nutr Bull 33:91–101

Sabouri S, Wright AJ, Corredig M (2017) In vitro, digestion of sodium caseinate emulsions loaded with epigallocatechin gallate. Food Hydrocoll 69:350–358. https://doi.org/10.1016/j.foodhyd.2017.02.008

Sajilata MG, Bajaj PR, Singhal RS (2008) Tea polyphenols as nutraceuticlas. Compr Rev Food Sci Food Saf 7:229–254. https://doi.org/10.1111/j.1541-4337.2008.00043.x

Shahidi F, Lin JK, Ho CT (2009) Tea and tea products: chemistry and health-promoting properties. CRC Press, Taylor & Francis Group

Sharma V, Rao LJM (2009) A thought on the biological activities of black tea. Crit Rev Food Sci Nutr 49:379–404. https://doi.org/10.1080/10408390802068066

Sharma K, Bari SS, Singh HP (2009) Biotransformation of tea catechins into theaflavins with immobilized polyphenol oxidase. J Mol Catal B: Enzym 56:253–258. https://doi.org/10.1016/j.molcatb.2008.05.016

Sharma V, Gupta AK, Walia A (2019) Effect of green tea on diabetes mellitus. Acta Sci Nutr 3:27–31

Shi M, Huang LY, Nie N, Ye JH, Zheng XQ, Lu JL, Liang YR (2017) Binding of tea catechins to rice bran protein isolate: interaction and protective effect during in vitro digestion. Food Res Int 93:1–7. https://doi.org/10.1016/j.foodres.2017.01.006

Singh BN, Shankar S, Srivastava RK (2011) Green tea catechin, epigallocatechin-3-gallate (EGCG): mechanisms, perspectives and clinical applications. Biochem Pharmacol 15(82):1807–1821. https://doi.org/10.1016/j.bcp.2011.07.093

Son Y-R, Chung J-H, Ko S, Shim S-M (2016) Combinational enhancing effects of formulation and encapsulation on digestive stability and intestinal transport of green tea catechins. J Microencapsul 33:183–190. https://doi.org/10.3109/02652048.2016

Stangl V, Dreger H, Stangl K, Lorenz M (2007) Molecular targets of tea polyphenols in the cardiovascular system. Cardiovasc Res 73:348–358. https://doi.org/10.1016/j.cardiores.2006.08.022

Stodt U, Engelhardt UH (2013) Progress in the analysis of selected tea constituents over the past 20 years. Food Res Int 53:236–248. https://doi.org/10.1016/j.foodres.2012.12.052

Tedeschi E, Menegazzi M, Yao Y, Suzuki H, Forstermann U, Kleinert H (2004) Green tea inhibits human inducible nitric-oxide synthase expression by down-regulating signal transducer and activator of transcription-1alpha activation. Mol Pharmacol 65:111–120. https://doi.org/10.1124/mol.65.1.111

Tiwari RP, Bharti SK, Kaur HD, Dikshit RP, Hoondal GS (2005) Synergistic antimicrobial activity of tea & antibiotics. Indian J Med Res 122:80–84

Velayutham P, Babu A, Liu D (2008) Green tea catechins and cardiovascular health: an update. Curr Med Chem 15(18):1840–1850. https://doi.org/10.2174/092986708785132979

Vuong QV (2014) Epidemiological Evidence Linking Tea Consumption to Human Health: A Review. Crit Rev Food Sci Nutr 54(4):523–536. https://doi.org/10.1080/10408398.2011.594184

Vuong QV, Bowyer MC, Roach PD (2011) L-Theanine: properties, synthesis and isolation from tea. J Sci Food Agric 91:1931–1939. https://doi.org/10.1002/jsfa.4373

Wang C, Li Y (2006) Research progress on property and application of Theaflavins. Afr J Biotechnol 5:213–218. https://doi.org/10.5897/AJB05.376.

Wiseman SA, Balentine DA, Frei B (1997) Antioxidants in tea. Crit Rev Food Sci Nutr 37:705–718. https://doi.org/10.1080/10408399709527798

Yang CS, Wang H (2011) Mechanistic issues concerning cancer prevention by tea catechins. Mol Nutr Food Res 55:819–831. https://doi.org/10.1002/mnfr.201100036

Yang CS, Maliakal P, Meng X (2002) Inhibition of carcinogenesis by tea. Annu Rev Pharmacol Toxicol 42:25–54. https://doi.org/10.1146/annurev.pharmtox.42.082101.154309

Yi W, Hao C, Chen Y, Chen L, Fang Z, Liang L (2017) Formation of a multiligand complex of bovine serum albumin with retinol, resveratrol, and (−)-epigallocatechin-3-gallate for the protection of bioactive components. J Agric Food Chem 65:3019–3030. https://doi.org/10.1021/acs.jafc.7b00326

Zhang H, Jung J, Zhao Y (2016) Preparation, characterization and evaluation of antibacterial activity of catechins and catechins–Zn complex loaded β-chitosan nanoparticles of different particle sizes. Carbohydr Polym 137:82–91. https://doi.org/10.1016/j.carbpol.2015.10.036

Zhong Z, Dong Z, Yang L, Chen X, Gong Z (2012) Inhibition of proliferation of human lung cancer cells by green tea catechins is mediated by upregulation of let-7. Exp Ther Med 4:267–272. https://doi.org/10.3892/etm.2012.580

Chapter 5
Galantamine Delivery for Alzheimer's Disease

Shweta Sharma

Abstract Alzheimer's disease (AD) is an irreversible neurodegenerative disorder characterized by progressive degeneration of different regions of brain and is the most common cause of dementia. The hallmark pathological feature of Alzheimer is the enhanced production and accumulation of amyloid-β peptide (Aβ) in form of senile plaques, which results in progressive neuro-degeneration. This leads to cognitive and memory impairment, anxiety, mood swings and language difficulties. Even though the correct pathogenesis is not clear, decline in cholinergic neurotransmission remains the universal neurochemical fact in the pharmacology of Alzheimer's. Therefore, cholinesterase inhibitors e.g. rivastigmine, donepezil, galantamine etc. remains the primary choice for treatment as they elevate the acetylcholine quantity in the synaptic cleft, which further leads to upsurge in cholinergic transmission in brain. Among these drugs, galantamine is the only naturally occurring acetylcholinesterase inhibitor and possess an additional activity of allosterically modulating nicotinic acetylcholine receptors.

This chapter provides the detailed insights about the galantamine which includes its natural source, chemistry, pharmacokinetics, pharmacodynamic and approved dosage forms. Further, in addition to conventional oral formulations, other drug delivery approaches such as transdermal, intranasal, long acting scaffolds and nanotechnology-based are being reviewed. In conclusion, this chapter supports that how galantamine therapy could be an effective for improving cognition and function in Alzheimer's patient comparable to other acetylcholinesterase inhibitors.

Keywords Donepezil · Rivastigmine · Memantine · Transdermal · Intranasal · Nanoparticles · Amyloid β · Cholinergic · Memogain · Razadyne

S. Sharma (✉)
Purdue University, Biomedical Engineering and Pharmaceutics, West Lafayette, IN, USA
e-mail: shwetainpharma@gmail.com

© The Editor(s) (if applicable) and The Author(s), under exclusive license to
Springer Nature Switzerland AG 2020
A. Saneja et al. (eds.), *Sustainable Agriculture Reviews 43*, Sustainable
Agriculture Reviews 43, https://doi.org/10.1007/978-3-030-41838-0_5

5.1 Introduction

Alzheimer's disease is an irreparable neurodegenerative brain disorder, affecting 4–8% of the elderly population world-wide. It is characterized by relatively slow, but progressive neurodegeneration (Godyń et al. 2016). The pathogenesis of Alzheimer is complex involving number of process such as metal and calcium dyshomeostasis, oxidative stress, excitotoxicity, neuroinflammation and mitochondrial damage and these processes are considered as promising targets for effective therapy of Alzheimer's (Castellani and Perry 2014). Though there is great hope of developing new therapies in near future, acetylcholinesterase inhibitors (AChEIs), such as donepezil, galantamine, rivastigmine and N-methyl-d-aspartate (NMDA) receptor antagonist memantine, remain the only US-FDA approved drugs until today for the treatment of Alzheimer's disease (Peters et al. 2015). These drugs do no inhibit the progression of disease rather temporary slowdown the process of loss of cognitive function by decreasing cholinesterase activity which results in higher acetylcholine levels and therefore improved brain function (Godyń et al. 2016). The data published from open-label extension studies of two currently available acetylcholinesterase inhibitors, Rivastigmine and donepezil, indicate that cognitive abilities evaluated using the cognitive subscale of the Alzheimer's Disease Assessment Scale (ADAS-cog) fell below baseline after 40 and 38 weeks, respectively (Farlow et al. 2000) (Rogers and Friedhoff 1998).

Because of the potential limitations of convention acetylcholinesterase inhibitors, there is a paradigm shift in drug development strategy for Alzheimer's with focusing on drugs having additional targets with broader and more sustained efficacy.

Galantamine is one of those drugs which has unique dual mode of action. It allosterically modulates nicotinic acetylcholine receptors (nAChRs) in addition to inhibiting acetylcholinesterase (AChE). A clinical trial program in patients with Alzheimer's showed that galantamine has broad as well as sustained benefits for at least 52 weeks in functional and cognitive abilities (Lilienfeld 2002). Galantamine also delays the onset of behavioral symptoms. The benefits were found to be constant across the different studies. Alzheimer treatment guidelines in USA and UK agree that only Galantamine has established benefits on cognitive and behavioral outcomes. Various studies are still ongoing, and the drug is also being assessed for other forms of dementia which may get benefit from this combination of AChE inhibition and the modulation of nAChRs (Knopman et al. 2007).

The goal of this chapter is to provide reader the detailed insight about the potential of galantamine as anti-Alzheimer's drug. The first part of the chapter discusses briefly about the Alzheimer's, its pathophysiology, treatment strategies and current approved treatment options. The rest part of the chapter provides more detailed description about galantamine, including its marketed formulations, chemistry, pharmacokinetics, pharmacodynamics and natural source. The latter part of the chapter covers the treatment focusing particularly on drug delivery strategies depending on the route of administration and in particular, role of nanotechnology-based systems for the delivery of galantamine.

5.2 Alzheimer's and Its Pathophysiology

Alzheimer's is a complex disease and still many events of its pathophysiology is not clearly understood and therefore several competing theories are there in the literature underlying the biology of the neuro-degeneration that have guided research into interventions to alter development of the disease and its clinical manifestations.

5.2.1 Cholinergic Hypothesis

Among the given theories, the cholinergic hypothesis was the earliest one formulated by Davies Maloney in 1976 and termed as Cholinergic deficit hypothesis (Davies and Maloney 1976). This was one of those hypotheses which revolutionized the field of Alzheimer's and received gripping justification when cholinesterase inhibitor treatment was shown to induce substantial symptomatic improvement in patients with Alzheimer's disease. As per this hypothesis, Alzheimer is caused by a primary neuro-degenerative process which selectively damage a set of cholinergic neurons in the amygadala, medial septum, frontal cortex, hippocampus, medial septum, regions and structure that serve important functional role in attention, memory, learning and conscious awareness. This selective degeneration of neurons causes the downregulation of cholinergic markers such as acetylcholinesterase and acetyl-transfcrase which leads to onset of memory impairment. The main variations in cholinergic pathway considered in this hypothesis are: impaired acetylcholine release, choline uptake, deficits in the expression of nicotinic and muscarinic receptors, deficiency in axonal transport and dysfunctional neurotrophin support (Wenk 2006). Further because glutamatergic system significantly interact with cholinergic system during neurotransmission, any alterations in the cholinergic signaling also leads glutamatergic disruptions found in Alzheimer disease, an aspect that enhances cholinergic hypothesis (Dong et al. 2009). This hypothesis has served the basis for most of current approved treatment strategies and drug development approaches (cholinergic precursors, acetylcholinesterase inhibitors, allosteric cholinergic receptor potentiators, cholinergic receptor agonists, N-methyl-D-aspartate receptor blockers) for Alzheimer's. Current research community is in agreement that there is a relationship between decreased cholinergic transmission and cognitive impairment in the brain which plays an important role in Alzheimer but itself does not establish ultimate cause of the disease (Doggrell and Evans 2003).

5.2.2 Amyloid Hypothesis

Though several theories have been proposed but from the last two decades Aβ theory seems to be prevailing. Amyloid β (Aβ) is a short chain peptide that is proteolytic byproduct of the transmembrane amyloid precursor protein (APP). Though its role is not clear but thought to be involved in neuronal development. As per amyloid cascade hypothesis, the disease occurs because of series of abnormalities in the process and secretion of the amyloid precursor protein (APP), where an imbalance between production and clearance of Aβ is the activating event (Salomone et al. 2012). This imbalance causes vivid alteration in conformation of Aβ monomers and leads to formation of β sheet-rich tertiary structure that assembles to develop amyloid fibrils. These amyloid fibrils get deposits on surface of the neurons in compact form called as *neuritic or senile plaques* whereas in less compact form they accumulate inside the walls of blood vessels as *diffusive plaques* (Perl 2010). Another hallmark of Alzheimer pathology is unusual accumulation of the hyperphosphorylated tau protein within the nerve cell bodies as neurofibrillary tangles (NFT) and as dystrophic neurites associated with Aβ plaques. Still the exact mechanism that facilitates the accumulation and development of neurofibrillary tangles and senile plaques remains unclear, but they are responsible to cause the injury and death of neurons and as a result memory loss and cognitive impairment appear (Ahmad et al. 2017). Current investigation has also emphasized the role of Aβ oligomers in synaptic impairment, signifying that these are primarily the only one among several other signals that are responsible to massive loss of synapses, dendrites and eventually neurons(Fig. 5.1) (Roth et al. 2005).

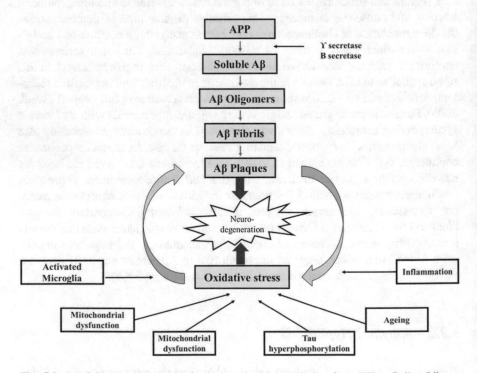

Fig. 5.1 Amyloid cascade hypothesis. (Adapted with permission from Wiley Online Library Alghamdi 2018)

The understanding of the pathogenesis of Alzheimer's is constantly changing; for instance, the neurofibrillary tangles, a pathological trademark of Alzheimer's, were earlier thought to be one of the core reasons for the disease but now they rather seem to reflect the damage to neurons that have occurred over a long time. In-fact, the notion that phosphorylated tau and Aβ peptide are pathologic hallmark is slowly changing, but now it seems that they might represent an adaptive strategy to oxidative stress. Apart from these hypothesis, various other deranged mechanisms such as chronic mitochondrial dysfunction, oxidative stress, calcium mishandling, hormone imbalance, mitotic dysfunction and genetic components may also play the role in the disease process (Anand et al. 2014).

5.3   Therapeutic Strategies for Alzheimer's Disease

Although Alzheimer is known for about a century now, only four cholinesterase inhibitors and memantine have been approved by US Food and drug administration for its treatment. These drugs provide symptomatic treatment, but they do not alter the course of the disease. Hence, continuous research is being done to find the modern therapeutic options that can target the disease modification. Table 5.1 summarize the existing therapeutic strategies till date with correlation to the pathophysiologic mechanisms for the disease(Anand et al. 2014)

5.4   U.S. Food and Drug Administration (US-FDA) Approved Treatments for Alzheimer's

There are five drugs currently approved by the U.S. Food and Drug Administration (FDA) to treat the Alzheimer symptoms. The first class of drugs are "Cholinesterase inhibitors (ChEIs)" — donepezil, Galantamine and Rivastigmine, which are licensed for mild to moderate Alzheimer's and the second class is glutamate-antagonist Memantine which is licensed for moderate to severe stage. A cholinesterase inhibitor works by blocking the activity of acetylcholinesterase and thereby slowing the breakdown of acetylcholine and in this way, it helps by maintaining acetylcholine levels. Though these drugs do not cure the root cause of the disease, but still they remain the only choice of drugs available in market for treatment (Table 5.2) (dos Santos Alves et al. 2014). The fourth drug is Memantine, it regulates the activity of altogether different chemical messenger Glutamate in the brain, which is also important for learning and memory. The fifth medication is a combination of one of the cholinesterase inhibitors (Donepezil) with Memantine.

Table 5.1 Therapeutic strategies in Alzheimer disease

Target	Details	References
Amyloid based therapies	Accumulation of Aβ fiber towards toxic level is one of the pathological symbol of Alzheimer's disease. Therefore, the amyloid based therapy targets various aspects of this Aβ fibrils-oligomers-monomers equilibrium. These approaches include, secretase enzymes modulation, preventing amyloid aggregation, amyloid transport amyloid based immunotherapy and promoting amyloid clearance	Anand et al. (2014) Atamna et al. (2008), Schenk et al. (2012)
Tau based therapies	Hyperphosphorylation of tau protein, pathological symbols of Alzheimer's disease, alters its property of microtubule stabilization and leads to neuronal degeneration. Hence the tau-based therapies such as tau-phosphorylation inhibition, blocking tau oligomerization, microtubule stabilization, enhancing tau degradation and tau based immunotherapy are the approaches which has been extensively investigated in research	Anand et al. (2014) Garcia and Cleveland (2001), Matsuoka et al. (2007)
Modulating neurotransmission	Dysfunction in cholinergic group of neurons is one of the main indications of Alzheimer's. Dysfunction in other neurotransmitter histamine, g-aminobutyric acid (GABA), and serotonin also lead to Alzheimer's. This approach involves the use of drugs that modulates this neurotransmission and is considered the best approach to providing symptomatic improvement in patients. This includes cholinesterase inhibitors, GABAergic modulation, N-methyl D-aspartate receptor antagonism, Histaminergic modulation, serotonin receptor modulation and adenosine receptor modulation	Anand et al. (2014) Revett et al. (2013), Marcade et al. (2008)
Oxidative stress reduction	Oxidative stress has also been reported as one of the crucial factors in the pathogenesis of Alzheimer's hence strategies are being designed to overcome this. This may either involves exogenous antioxidant supplementation or augmenting endogenous defense. For example, flavonoids and carotenoids which are known to have antioxidant property have shown neuroprotective effect in experimental setups	Anand et al. (2014) Zhu et al. (2007)
Mitochondria targeted therapy	Though traditional anti-oxidant therapy has shown encouraging results in Alzheimer's therapy, but they usually work by tackling the produced reactive oxygen species. Mitochondrial targeted therapy deals with altering the production of reactive oxygen species (ROS). Example CoQ10 supplementation has potential of neuroprotective effects which includes stabilization of mitochondrial function, suppression of ROS production and minimized ROS injury	Anand et al. (2014) Wang et al. (2009)

(continued)

Table 5.1   (continued)

Target	Details	References
Modulating intracellular signaling cascaded	Accumulation of amyloid β (Aβ) fibers activate various intracellular pathways, so drugs that effects these altered pathways have also been investigated for Alzheimer's. Example, Rolipram is a selective Phosphosdiesterases-4 inhibitor and is the first compound that effectively return cognitive deficits in animal models of Alzheimer's	Anand et al. (2014), Cheng et al. (2010)
Modulators of cellular calcium homeostasis	Alteration of Ca^{2+} homeostasis has also been found in Alzheimer's, therefore several drugs that target different Ca^{2+} signaling pathways are being investigated. Example, one of the mechanisms by which Memantine work in Alzheimer's is by producing moderate decreases in Ca^{2+} influx thus reducing excitotoxicity	Anand et al. (2014), Lipton (2006)
Ant—inflammatory therapy	Neuroinflammation in Alzheimer's has prompted the researchers to use non-steroidal anti-inflammatory drugs (NSAIDS) in Alzheimer's. They may work by variety of mechanisms such as cyclooxygenase inhibition like targeting γ-secretase, maintaining Ca^{2+} homeostasis etc. Though they have shown good results in experimental set-up, but clinical trials results have been disappointing	Anand et al. (2014), Choi et al. (2013)
Others	Gonadotropin supplementation, caspase inhibitors, lipid modifiers statins, metal chelation, growth factor supplementation, epigenetic modifiers, nitric oxide synthase modulation, nucleic acid drugs multi-target directed ligands are the other potential targets for the treatment of Alzheimer's	Anand et al. (2014)

5.5   Galantamine

Among the available AChEi, Galantamine is the only naturally occurring substance. It is a tertiary alkaloid which is isolated from *Amaryllidaceae plants* and bulbs of the Caucasian snowdrops, *Galanthus woronowi*. Apart from the natural sources, synthetic process has now been developed for the production of Galantamine (Sramek et al. 2000). Galantamine has been available in Eastern Europe for more than 40 years and is mainly used for the treatment of patient suffering from disorders associated with deficiency of acetylcholine, such as: myasthenia gravis and glaucoma.

5.5.1   Chemistry

Chemically Galantamine is (4aS, 6R, 8aS)-4a,5,9,10,11,12-hexahydro-3-methoxy-11-methyl-6H-benzofuro[3a, 3, 2-ef][2]benzazepin-6-ol hydrobromide. Galantamine possess three chiral centers; in both form such as base compound and its hydrobromide salt (Fig. 5.2).

Table 5.2 FDA approved treatment for Alzheimer's at a glance

Drugs	Class and indication	Mechanism of action	Proprietary name (date approved)	Formulations
Cholinesterase inhibitors				
Donepezil	Mild to moderate and moderate to severe Alzheimer's	Prevents the breakdown of Acetylcholine	Aricept® (1996)	Tablets, disintegrating tablets
Galantamine	Mild to moderate Alzheimer's	Prevents the breakdown of acetylcholine and stimulates nicotinic receptors to release more acetylcholine in brain	Razadyne ® (2001)	Immediate-release tablets, Oral solution, Extended release capsule
Rivastigmine	Mild to moderate Alzheimer's	Prevents the breakdown of acetylcholine as well as butyrylcholine in the brain	Exelon ® (2000)	Tablets, Oral solution, Transdermal patch
N-Methyl-D-Aspartate receptor antagonist				
Memantine	Mild to moderate Alzheimer's	Blocks the toxic effects associated with excessive glutamate in brain	Namenda ® (2003)	Tablets, Oral solution
Combination drugs				
Donepezil + Memantine	Moderate to severe Alzheimer's	Combine the action of two drugs	Namzaric ® (2014)	Extended release capsule

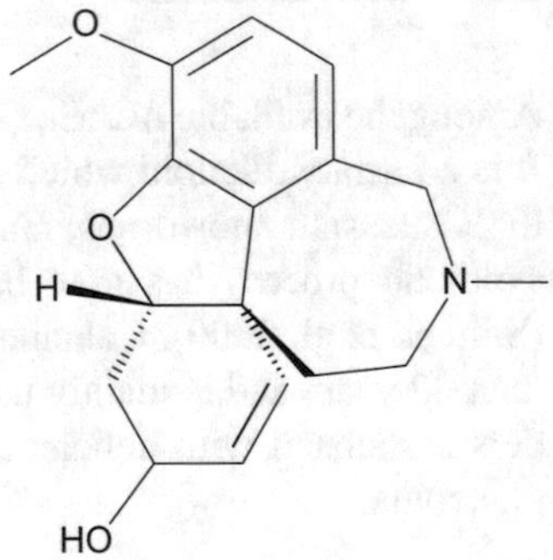

Fig. 5.2 Chemical structure of Galantamine

5.5.2 Pharmacodynamic Profile

Galantamine has a dual mode of action, along with inhibiting acetylcholinesterase, it also performs the positive allosteric modulation of Nicotinic acetylcholine receptors (nAChRs). It is the only marketed drug for Alzheimer's that demonstrates both modes of action (Fig. 5.3) (Lilienfeld 2002). The affinity to inhibits acetylcholinesterase (AChE) is 53-fold higher rather than butyrlcholinesterase (BuChE) in human

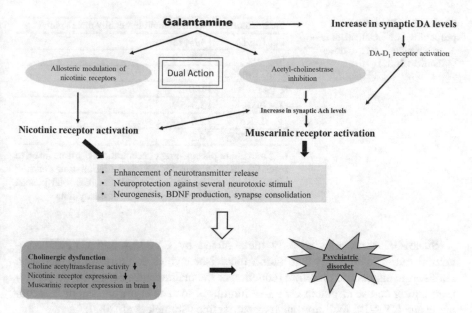

Fig. 5.3 Galantamine mechanism of action. (Adapted with permission from Elsevier Ago et al. 2011)

plasma and erythrocytes. On oral administration of Galantamine 5 mg three times daily for 2–3 month to Alzheimer patients, erythrocyte AChE activity get inhibited by 21–41% in 2 h after the morning dose whereas recovery of enzyme activity occurs within approximately 30 h of the final dose of the drug. In addition to this, Galantamine interacts directly with nicotinic acetylcholine receptors (nAChRs) and potentiates their action. Molecularly it allosterically binds to the alpha subunit of nAChRs (Scott and Goa 2000).

5.5.3 Pharmacokinetic Profile

Galantamine is a drug with good oral bioavailability (88.5%), low clearance, moderate volume distribution and low plasma protein binding (18%). Its terminal half-life ($t_{1/2}$) in human is approximately 8 h. The mean maximum plasma drug concentrations (C_{max}) after a single oral dose of 10 mg in healthy volunteers ranged from 49.2 to 1150 µg/l and reached at a mean of 0.88 or 2 h in two different studies (Table 5.3). At steady state, C_{max} and trough plasma concentrations (C_{min}) following 12 or 60 mg twice daily doses, were between 42–137 µg/l and 29–97 µg/l, respectively. Galantamine absorption does not get effected by food thus it has no interaction with food It is majorly metabolize via cytochrome P450 isoenzymes (mainly CYP2D6 and CYP3A4) in the liver. It predominantly excretes in urine and only negligible amounts of the metabolites galantaminone and epiGalantamine are produced and neither of have any clinically relevant AChE activity in-vivo.

Table 5.3 Pharmacokinetic properties of Galantamine after single oral doses (Lilienfeld 2002)

Parameters	Measurement in healthy male volunteers
C_{max}(μg/L)	49–1150
t_{max} (h)	0.9–2.0
AUC (mg/L.h)	4.77
F (%)	100
$T_{1/2}$ (h)	5.3–5.7
V_{ss} (L/Kg)	2.6
CL (L/h/Kg)	0.3

C_{max} maximum plasma drug concentration, t_{max} time taken to reach C_{max}, *AUC* area under the concentration–time curve, *F* bioavailablility, $T_{1/2}$ elimination half-life, *CL* total plasma clearance, V_{SS} volume of distribution at steady state

Since Galantamine is majorly metabolized by CYP3A4 and CYP2D6, co-administration with drugs that inhibit these isoenzymes could result in increased cholinergic effects and therefore reduction of the maintenance dose must be considered during co-use of potent CYP3A4 inhibitors such as erythromycin or ketoconazole and CYP2D6 inhibitors such as paroxetine (Sramek et al. 2000).

5.5.4 Sources of Galantamine

Galantamine is naturally occurring alkaloid and marketed as hydrobromide salt (Galantamine HBr) for the treatment of Alzheimer's disease, poliomyelitis and other neurological diseases. Although chemical synthesis of Galantamine has been done successfully still the plants remain the main the main source of production. Natural Galantamine is basically isolated from four different species: Leucojum aestivum, Ungernia victoris, Narcissus ssp. and Lycoris radiata. (Berkov et al. 2009)

5.5.4.1 Leucojum aestivum

Leucojum aestivum is one of the plant species used for the extraction of Galantamine, mainly under license in Bulgaria. This plant usually found to be distributed in the Eastern Europe and Mediterranean region. The bulbs are 2–4 cm in diameter whereas leaves are 30–60 cm long. It is commercially cultivated to supply the industry with leaf biomass in Bulgaria, which is used for Galantamine extraction. The Galantamine content in the leaves was found to vary 0.1–0.3% Dry weight (DW), depending on the geographical location. Amon the different location species, Southeast Bulgarian populations were found to possess highest content of Galantamine (Parolo et al. 2011). Apart from Galantamine, other alkaloids with a higher AChE inhibitory activity than Galantamine, namely *N*-allylnorgalantamine and *N*-(14-methylallyl) nor-galantamine, sanguinine, have also been isolated from snowflake.

The amount of Galantamine has been found to vary from 4% to 99% of all compounds in the alkaloid mixtures of dormant bulbs from Bulgarian- populations (Georgieva et al. 2007).

5.5.4.2 Ungernia Victoris

Ungernia victoris is another species used for isolation of Galantamine. It is an endemic plant which is mainly found on the Gissar Ridge and its southern spurs (Tadzhikistan and Uzbekistan). It is a perennial species with leaves up to 20–25 cm long and bulbs 4–7 cm in diameter. This plant is being cultivated near its natural habitat locations since 1970. Both the leaves and bulbs have been used for Galantamine extraction. The total alkaloids content in the leaves and bulbs were found to be 1.18–1.65% and 0.27–0.71% of DW, respectively and the proportion of Galantamine in this alkaloid mixtures was found to be around 56–57% and 47–48% in the foliage and bulbs, respectively.

5.5.4.3 Lycoris radiata

Lycoris radiata is a plant species dispersed in Korea, Japan and China. The leaves are 50 cm-long and bulbs are 1–3 cm in diameter. This plant is usually found in two different varieties, one is: var. *pumila* which has a diploid genome, and the other is var. *radiata*, with a triploid genome. This plant is cultivated in China for Galantamine extraction since 2002 (Hayashi et al. 2005).

5.5.4.4 *Narcissus* ssp.

Bulbs of *Narcissus* plants are also widely used for extraction of Galantamine. Cultivators has screened different species of *Narcissus* for this drug and more than 33 species have shown 0.1% (referred to DW) of Galantamine in their bulbs. The main commercial source of Galantamine is *N. pseudonarcissus* with bulb size (4–5 cm in diameter) and high Galantamine content, which is about 0.10–0.13%(Heinrich 2002). It is commercially cultivated in Netherlands and United Kingdom by various bulb growers. Kreh et al. reported 58% of Galantamine in alkaloid mixture obtained from bulb of *N. pseudonarcissus* cv (Kreh et al. 1995). Another promising species for cultivation and isolation of Galantamine is *N. confuses*. It is an endemic plant which occurs only in Spain and has been found to contain around 0.6% DW of Galantamine. The most interesting thing about this species is that high level of Galantamine is present in the aerial parts of the plant such as leaves, stems and flowers and at the end of the ontogenic cycle, it can reach up to 2.5% DW (Bastida et al. 2006).

Full Synthesis vs Plant Extraction

To overcome the supply bottleneck, in 1990 two different approaches were developed, first is commercial extraction by daffodil bulbs and the most important second one is the chemical synthesis of the drug. The full synthesis was optimized and upscaled which allowed the development at larger level and finally allowing the production batch size of 100 kg to be manufactured under GMP conditions. Clinical trials were conducted and in year 2000 and finally Galantamine was launched in USA and Europe for the treatment of Alzheimer's disease, originally as Reminyl®, later changed to Razadyne® in the USA (Mucke 2015).

5.6 US-FD Approved Galantamine Formulation for Alzheimer Treatment

Galantamine is one of the three major cholinesterase inhibitors (ChEIs) to be approved by the US-FDA. It has been formulated as both immediate release and extended release dosage form for clinical practice. The immediate release circular biconvex film-coated tablets are available in three strength of Galantamine (free base) i.e. 4 mg which is off-white, 8 mg which is pink and 12 mg which is orange brown. Galantamine is also available in oral solution form at concentration of 4 mg/ml. Galantamine appears to have almost similar efficacy as of other first line drugs such as rivastigmine and donepezil in patients with Alzheimer's (Galasko et al. 2004).

Another marketed formulation of Galantamine which got approved for use in 2004 in United States (US) is opaque hard gelatin extended release capsules (Razadyne ER). The capsules are available in three strengths of Galantamine base i.e. 8 mg which is white, 16 mg which is pink and 24 mg which is caramel. The capsule are designed in such a way that 25% of the dose is in immediate- release form whereas 75% are available in controlled release form (Robinson and Plosker 2006). Galantamine capsules, which are designed for extended release, given at dose of 24 mg once daily under fasting conditions are bioequivalent to Galantamine immediate-release tablets administered twice daily at dose of 12 mg in reference to C_{min} and AUC_{24h}. The Cmax is reached at 4 h for Extended Release capsules and in 1 h for immediate release tablets (Zhao et al. 2005). Both the dosage forms were superior to placebo as far as cognitive benefits are concerned, but none of the improved the global functioning (Brodaty et al. 2005). Though both the formulations showed adverse events but subjects reporting nausea, the mean percentage of days with nausea and the subsequent antiemetic use was lower with Galantamine ER than with Galantamine immediate release (Table 5.4).

Table 5.4 US-FDA approved dosage forms of Galantamine

Formulation type	Route	Description	Excipients
Galantamine tablet (Razadyne®)	Oral	4 mg circular biconvex, off-white tablet imprinted with "JANSSEN" and "G 4" on the opposite sides	Diethyl phthalate, ethylcellulose, gelatin, hypromellose, polyethylene glycol, sugar spheres (sucrose and starch), and titanium dioxide. The 16 mg capsule also contains red ferric oxide. The 24 mg capsule also contains red ferric oxide and yellow ferric oxide
		8 mg circular biconvex, pink tablet imprinted with "JANSSEN" on one side and "G 8" on the other side	
		12 mg circular biconvex, orange-brown tablet imprinted with "JANSSEN" and "G 12" on the opposite sides	
Galantamine solution (Razadyne®)	Oral	Razadyne® is clear colorless 4 mg/mL oral supplied in 100 mL bottles with a calibrated pipette. The maximum calibrated volume is 4 mL while the minimum calibrated volume is 0.5 mL	Methylparaben, Propylparaben, purified water, Sodium hydroxide, and Saccharin sodium
Galantamine capsule, extended release, (Razadyne ER®)	Oral	Capsule contain white to off-white pellets and are available in the following strengths. 8 mg: It is white opaque hard gelatin capsule having size 4 with the inscription "GAL 8"	Colloidal silicon dioxide, crospovidone, hypromellose, lactose monohydrate, magnesium stearate, microcrystalline cellulose, propylene glycol, talc, and titanium dioxide. The 4 mg tablets contain yellow ferric oxide. The 8 mg tablets contain red ferric oxide whereas the 12 mg tablets contains red ferric oxide as well as FD&C yellow #6 aluminum lake
		16 mg: It is pink opaque, size 2 hard gelatin capsule with the inscription "GAL 16"	
		24 mg: It is caramel opaque, size 1 hard gelatin capsule with the inscription "GAL 24"	

5.7   Combination of Galantamine and Memantine

Cholinergic and glutamatergic systems, N-methyl-D-aspartate (NMDA) receptor and alpha-7 nicotinic acetylcholine receptor (α-7nAChR) are strongly affected with cognitive impairments in Alzheimer's disease. Donepezil, a primary drug in the treatment of Alzheimer's, is only an acetylcholinesterase inhibitor (AChEI), whereas

Galantamine has dual mode of action. It is not only an AChEI but also positive allosteric modulator of the α-7 nicotinic receptors and α4β2 receptors. Memantine, on the other hand is an NMDA receptor antagonist. Other than individual AChEIs, donepezil-memantine in only combination (Namzaric) which is US Food and Drug Administration (FDA)–approved for the treatment of Alzheimer's. Several randomized controlled trials (RCTs) have shown that the donepezil-memantine combination is better than either drug alone to improve cognition in Alzheimer dementia. Based on this data and the unique properties of Galantamine, several studies have been conducted evaluating the efficacy of Galantamine-memantine combination in Alzheimer's disease. In a 1-year Randomized Controlled Trial with 232 participants with mild-to-moderate AD, the combination of Galantamine and memantine showed significantly better cognitive improvements compared to Galantamine alone in prodrome Alzheimer's and Donepezil-Memantine combination in Alzheimer's (Koola and Parsaik 2018). Memantine is a non-competitive antagonist for NMDA receptors and inhibits it in a voltage-dependent manner to reduces the excitotoxicity caused by excess glutamate release (Rogawski and Wenk 2003). On the other hand, Galantamine increases glutamate release. Thus, even though the two drugs appear to act in an opposing manner, reports reveal that these medications work synergistically to provide a more normal neurological response and improve cognitive impairments in Alzheimer's disease patients. When given in combination, Galantamine increases synaptic activities and long-term potential whereas memantine prevents cell damage due to electrophysiological noise (Grossberg et al. 2006). The use of memantine and Galantamine in combination is also supported by pharmacodynamic and pharmacokinetic studies. Therefore, combined treatment of these two medications do not affect each other metabolism. Galantamine is metabolized by cytochrome CYP3A4 and P450 (CYP) 2D6, which are not affected by memantine. This evidence support that combined modulation of NMDA and nicotinic receptors by memantine and Galantamine may provide an optimal combination therapy for the treatment of Alzheimer's (Koola et al. 2018). But even though both are FDA approved for the treatment of Alzheimer's, the combination is still underutilized in clinical practice.

5.8 Novel Delivery Strategies for Galantamine

5.8.1 Transdermal

The transdermal drug delivery system (TDDS) is an approach where an adhesive patch is placed over the skin to deliver specific dose of a specific drug in the blood stream through the skin. This adhesive patch is basically composed of three things: polymers, drugs and penetration enhancers. The multi-layered patch is made up of a release liner, a drug reservoir and an impermeable backing (Saravanakumar et al. 2015). This route of drug delivery offers several advantages compared to the conventional oral and parenteral systems, especially to the elderly patients and the patients suffering from chronic neurological disorder. These patients often show unwillingness to swallow table, capsule or solution, so the TDDs are particularly useful because it allows to circumvent the patient's unwillingness and/or incapability to swallow the oral dosage form such as tablet or capsule. The other advantages of transdermal drug delivery route are

- Extended duration action, so reduced dosing frequency
- Better patient compliance as they can be self-administered and are non-invasive
- Improve the bioavailability, by bypassing the firs-pass metabolism
- More uniform plasma levels and therefore lesser side effects
- Generally inexpensive

Alzheimer's is one of those diseases where the application of transdermal system can provide various benefits especially for the delivery of cholinesterase inhibitors. Cholinesterase inhibitors exhibit dose-response relationship which means that higher plasma levels can lead to higher inhibition of acetylcholinesterase (Bickel et al. 1991). However, increasing the dose by oral route also increase the gastrointestinal side effects such as nausea,vomiting and therefore patient withdrawal also increase in clinical practice. These side effects are thought to be the result of rapid absorption and attainment of high peak plasma concentrations (C_{max}) of the drug in a short time t_{max}, and the magnitude and frequency of the resulting fluctuations in drug plasma level (Imbimbo 2001). A TDDS can provide both smooth and continuous delivery of the drug thereby reducing the C_{max} and prolonging the t_{max} while maintaining drug exposure. This shows that a transdermal system has the capability of altering the pharmacokinetic profile of the drug and therefore can reduce the incidence of side effects by allowing patients to optimum therapeutic doses (Cummings et al. 2007). Also, this way of administration includes a simplified treatment regimen which is convenient and easy to use by patient.

Wei-Ze and co-worker reported the use of microneedles having length of 70–80 µm for pre skin pretreatment to improve the permeation of Galantamine across the stratum corneum and live epidermis. To understand the mechanism, histological examination of the skin was performed, and it revealed that microneedles were forming skin conduits by pressing and swaying against the backing layer at constant pressure. These conduits or channels were acting as pathway for transport of Galantamine across the skin. *In vitro* diffusion studies indicated that this kind of piercing of skin can significantly increase the Galantamine permeability. Moreover, this permeability also increased on increasing the retention time, microneedle number and insertion force in a certain range. They also proved that this kind of piercing is safe as skin was not damaged in such a way that they could allow the permeation of pathogens into the body through these channels across the skin. This microneedle system was shown to be a useful drug delivery device for the transdermal administration of Galantamine, but further more studies are required to translate these findings into clinical practice (Wei-Ze et al. 2010).

In 2012, Chun-Woong Park and co-workers, first time reported the development of transdermal patch of Galantamine to overcome the problems associated with oral administration. They develop a Drug-in-adhesive patch using various acrylate and acrylate-vinylacetate based pressure sensitive adhesive (PSA) and different enhancers such as oleic acid, benzyl alcohol, isopropyl myristate and transcutol. *In vitro* skin permeation studies showed that oleic acid was the most promising enhancer and DT-2510 was the most appropriate pressure-sensitive-adhesive for Galantamine patch. The optimized patch was found to physiochemically stable for 28 days at the conditions of 40°C/75% RH. *In vivo* studies of patch showed the bioavailability of around 80% and plasma levels were sustained for 24 h. The plasma concentration profiles of Galantamine after administration of 5 mg/kg by different route is present in Fig. 5.4. It clearly shows that Galantamine could not be detected more that 4–6 h after oral and parenteral administration whereas the plasma levels of Galantamine from patches were stable from 14 to 112 ng/mL and decreased until the patch was removed (at 24 h). The results suggested that transdermal drug-in-adhesive patch might be the better dosage form for Galantamine to prevent the gastrointestinal side effects during the treatment of Alzheimer disease (Park et al. 2012)

In 2015, Woo fong yen and co-workers., reported the development of gel based transdermal patch of Galantamine and Galantamine hydrobromide (HBr). In this technique, first they developed gel-based drug reservoir using carbopol polymer, galantamine HBr, propylene glycol, triethanolamine and deionized water and then

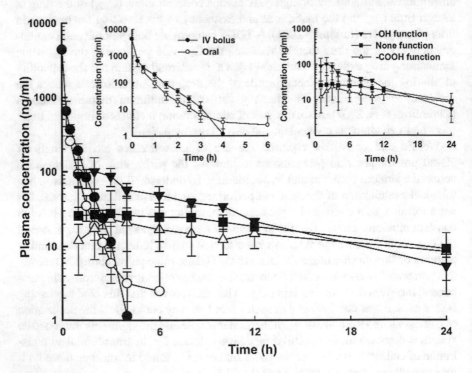

Fig. 5.4 Plasma concentration–time profiles of Galantamine observed in rabbits after intravenous (i.v), oral and transdermal administration at a dose 5 mg/kg (as Galantamine base) (mean ± S.D., n = 3). (Reproduced with permission from Elsevier Park et al. 2012)

this drug reservoir was sandwiched between the release line of Opside flexigrid and backing layer. The formulated patch system was found to have high drug content, suitable pH for skin and controlled drug-release pattern. They proposed this could have much great potential to be used for the therapy of Alzheimer's (Fong Yen et al. 2015; Woo et al. 2015).

Worranan Rangsimawong and co-workers investigated the effects of Limonene containing pegylated liposomes and sonophoresis in improving the trans-permeation of Galantamine HBr across the skin. They perform two different sets of experiment, in one experiment Galantamine loaded pegylated liposomes with different amount of limonene as penetration enhancer were developed and studied for the penetration across the skin. In another experiment they studied the permeation of Galantamine HBr solution in the presence of sonophoresis. Both the techniques were clearly able to enhance the formulation penetration across the skin; however, the effect was stronger with pegylated liposomes. They proposed the possible mechanism for this and as per them because of the small particle size and ultra-deformable structure, liposomes can enter the subcutaneous layer carrying the entrapped drug along with them into the skin through intercellular regions of the subcutaneous layer. Second, the limonene released from some of the vesicles during their penetration enhanced the fluidity of lipid lamellae in the skin. Whereas, with regard to Sonophoresis, the enhancing effect of might be due to removing some fractions of the subcutaneous barrier and creating aqueous channels at discrete regions, which leads the delivery of hydrophilic molecules through an intracellular pathway (Rangsimawong et al. 2018)

Very recently in 2019 Dina Ameen and Co-worker also reported the development of matrix-type transdermal patch for Galantamine. They investigated four pressure sensitive adhesives, ten penetration enhancers and four different drug loadings to obtain the optimized patch and then tested their permeation efficacy through human cadaver skin using vertical Franz diffusion cell. The optimized patch was composed of 10% w/w Galantamine, 5% w/w Oleic acid as crystallization inhibitor, 5%w/w Limonene as penetration enhancer, GELVA GMS 788 as pressure-sensitive-adhesive and was casted on ScotchpakTM1022 release liner and laminated with Scotchpak 9723 as a backing film. Additives showed a synergistic enhancement of Galantamine permeation while inhibiting the drug crystallization. Ex-vivo permeation studies using human cadaver skin indicated that optimized patch can achieve therapeutic plasma level with patch size of about 20 cm². Further pharmacokinetic are required to confirm the results (Ameen and Michniak-Kohn 2019).

5.8.2 Intranasal Delivery

Just like the transdermal route, Intranasal (IN) delivery is another attractive alternative option for oral delivery and invasive delivery methods. It is gaining lot of attraction for brain delivery because its various advantages over other system (Y. Wang et al. 1998) (Chow et al. 1999). This include

- Higher surface area leads to rapid absorption and rapid onset of action
- Rapid and higher levels can be achieved in brain for brain targeting
- Gastrointestinal mediated drug degradation is avoided
- Avoid gastrointestinal related adverse effects
- Hepatic first pass metabolism is absent
- Noninvasive and therefore better patient comfort and compliance.
- Lesser systemic exposure of drugs and therefore fewer side effects

The possible pathway of absorption of drugs and transport to brain via nasal route could be the combination of axonal transport from the olfactory epithelium to the olfactory bulb and of extracellular transport involving diffusion and bulk flow within the peri-neuronal channels and the lymphatic channels directly connected to the cerebrospinal fluid (Sakane et al. 1991) (Thorne et al. 2004). In case of Galantamine, intranasal route is one of preferred choice is because gastrointestinal adverse effects is one of the major limitation of Galantamine after oral delivery and often leads to the discontinuation of the treatment (Costantino et al. 2008).

In 2005, Alexis Kays Leonard et al., developed the high concentration formulation for Intranasal delivery. This was a conventional solution-based formulation developed for intranasal delivery. Since, the marketed hydrobromide salts is not suitable for intranasal delivery because of its limited solubility properties therefor this group developed a novel Galantamine lactate formulation by salt exchange method for the intranasal delivery. The typical dose of Galantamine is 8 mg per day and therefore 80 mg/ml solution is required for intranasal delivery considering minimum spray volume of 100 μl where conventional HBr has solubility of only 35 mg/ml. Therefore, to overcome this limitation, they developed a lactate-based formulation which showed solubility of 400 mg/ml (>12 fold). The new salt was found to possess same stability and safety to cells in line with conventional salt. Also, this novel lactate salt permeation across the epithelial tissue was of higher rate than the conventional HBr salt at the concentration of 35 mg/mL (Fig. 5.5). The in-vivo pharmacokinetic data in rat revealed that lactate formulation was either superior or

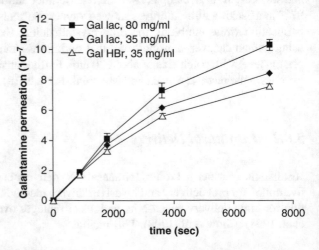

Fig. 5.5 Permeation of Galantamine lactate (Gal lac) and Galantamine HBr (Gal HBr) across epithelial tissue *in vitro*) at different time points. (Reproduced with permission from Elsevier publishing house Leonard et al. 2005)

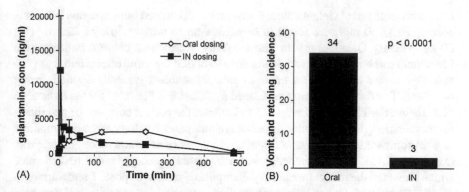

Fig. 5.6 *In vivo* Comparison of intranasally administered Galantamine lactate (200 mg/ml) with orally administered RAZADYNE® (20 mg/ml). Ferrets were dosed at 20 mg/kg Galantamine free base (n = 8). (**a**) Pharmacokinetics. (**b**) Emetic response. (Reproduced with permission from Elsevier Leonard et al. 2007)

comparable to the conventional HBr salt. For example, when given at dose of 1.75 mg/kg, the obtained AUC_{last} was 18,170 min.ng/ml with lactate salt as compared to 11,031 min.ng/ml with HBr salt. This was the first report of feasible intranasal formulation of Galantamine (Leonard et al. 2005). To further improve the bioavailability after intranasal administration, this group formulated Galantamine lactate at a concentration of 35–80 mg/ml along with excipients such as didecanoyl-L-α-phosphatidylcholine (1.7 mg/ml), methyl-β-cyclodextrin (30 mg/ml), and edetate disodium dihydrate (2 mg/ml). These permeation enhancers resulted in a four-fold increase in free Galantamine permeation across the epithelial tissue with low cytotoxicity and high cell viability. Pharmacokinetic kinetic data showed that intranasal Galantamine had shorter T_{max} in comparison to oral formulation (5 min vs 240 min, respectively) and nearly four-fold higher C_{max} (12,100 ± 8000 ng/ml versus 3200 ± 200 ng/ml). Finally, the hypothesis of reduced Gastrointestinal related side effects was also proved by testing in ferret model where fewer side effects was observed with intranasal administered Galantamine compared to oral route (Fig. 5.6) (Leonard et al. 2007).

5.8.2.1 Memogain

Memogain is an in-active prodrug of Galantamine given through nasal route. It is more hydrophobic and has potential of 15-fold higher bioavailability than free Galantamine at the same dose. It gets cleaved enzymatically by non-endogenous and liberates Galantamine as the active compound. Much higher levels of Galantamine were found to achieve in brain, and corresponding much lower levels were observed in blood when Memogain and Galantamine administered at same dose. The objective of developing Memogain was to develop a formulation that possess lower gastrointestinal side effect as compared to Galantamine. As reported by Maelicke et al.,2010 the observed gastrointestinal side effects were much lower as compared to standard Galantamine. Based on the calculated human equivalent dose, Memogain at 20 mg/were found to be equivalent to a daily dose of 226 mg (for 60-kg patient) Galantamine in ferrets. This correlation clearly indicates the Memogain can be given at much lower dose to obtain the same effects and thus possess high safety margin when given to patients at doses typically applied during treatment. The drug is presently developed by "Galantos Pharma" for the Indication of Alzheimer's. This drug is supposes to increase the patient compliance and therefor may also help to fully explore the advantageous pharmacology of the Galantamine as a neuroprotective treatment and sensitizing agent of neuronal nicotinic receptors (Maelicke et al. 2010). Memogain was also tested on scopolamine treated mice which indicated the model for memory disruption. For this purpose, T maze arrangement was used with task of spontaneous alternations of mice over a series of successive runs. Memogain cause the reversion of 80% scopolamine-induced amnesia at dose of 1 mg/kg as compared to Galantamine where the same effect was observed at 5 mg/kg. The calculated ED50 was and 0.34 mg/kg for Memogain and 1.07 mg/kg for Galantamine. Moreover, Memogain was able to fully reverse scopolamine activity. (Bhattacharya and Montag 2015)

5.8.3 Nanotechnology Based Systems

Nanotechnology is one of the well-established areas in the field of pharmaceutical science however it mostly has been explored for the delivery of anticancer agents and not yet has been investigated much for the treatment of neuro-degenerative disorders. Therefore, a great improvement to the current therapy for neuro-diseases is anticipated with the application of nanotechnology. Localized and targeted delivery are significant challenges in Alzheimer's treatment because of severe side effects of the approved drugs which occur due to the non-specific distribution to the healthy tissues and this can be overcome by application of nanoparticulate systems in Alzheimer's. Among the several available systems, polymeric, liposomes, or lipidic micro/nano-particles, polymeric micelles, dendrimers seem to be the most preferred choice because they can effectively interact with biological systems. The advantage of nanoparticulate systems are their small and tunable size and also surface tailorability which allows the conjugation of moieties that has affinity to the blood brain barrier (BBB) and have the promise of overpowering the limitations imposed by the BBB (McNeil 2005). Though Galantamine itself can cross the BBB from conventional formulations, but much higher dose is needed to give to achieve the levels required for the treatment and this higher dose often leads to the adverse effects cause by non-specific cholinesterase inhibition. However, there is also concern regarding the biodegradability and biocompatibility of nanotechnology based drug delivery systems as both biodegradable or non-biodegradable nanotechnology devices can interact with the neighboring tissue and trigger allergic-type immune reactions (Fournier et al. 2003). Though there are concerns but several polymers and lipids if properly sterilized have been proven to be safe for administration. Nanotechnology in clinical neurology can specifically target the drugs past the BBB and can develop potential regenerative therapies but despite that none of the nanotechnology-based is approved for use in Alzheimer's patients.

Liposomes are one of the delivery systems that has been approved by US-FDA for clinical use. Maluta S. Mufamadi and co-worker developed the neuronal cells targeted liposomes for the Galantamine delivery. They coupled the peptide (Lys-Val-Leu-Phe-Leu-Ser) onto the exterior of liposomes to interact with SEC-R receptors overexpressed on hepatoma cells and neuronal cells (PC 12 cells). These nanoliposomes with particle size from 127 to 165 nm, Zeta potential (ZP) values of -18 mV to -36 mV showed the higher accumulation in PC12 neuronal cells as compared to Galantamine solution and untargeted liposomes. Cytotoxicity assay revealed the low toxicity on cells when exposed to plain nanoliposomes and the ligand-conjugated nanoliposomes. Though further in-vivo studies are required but their data clearly established that nanoliposomes could be a potential delivery vehicle for the targeted delivery of Galantamine (Mufamadi et al. 2013). Just like liposomes, PLGA is another polymer which is biocompatible as well as biodegradable and is approved by US FDA for clinical purpose. To explore the potential of PLGA in Alzheimer's. C. Fornaguera and co-workers developed the Galantamine loaded polysorbate 80 coated PLGA nanoparticles for its parenteral delivery. They proposed the hypothesis that nano-emulsification can be useful approach for the entrapment of hydrophilic molecules like Galantamine in PLGA nanoparticles. Polysorbate 80 was selected because of its BBB targeting property. The particles were characterized and investigated for their acetylcholinesterase inhibiting potential. The data revealed that nanoparticles were effectively able to retain the enzyme inhibition activity of the drug. The in-vitro compatibility data showed that particles were nano-toxic and safe to administer by parenteral route. Though they did not perform the biodistribution studies, based on their previous results they claimed that this kind of system can surely improve the pharmacological therapy for Alzheimer's in the near future (Fornaguera et al. 2015). Though the conventional monotherapy is very common to treat Alzheimer's, this method is generally less effective and more newer approaches are required that do not rely on the activity of single drug Combination therapy is one such approach where two different drugs are given together for the better treatment of the disease. Sanaa M.R.Wahba and coworkers developed hydroxyapatite nanoparticles for the co-delivery of nanoceria and Galantamine. Nanoceria is recently gaining lot of attention for its potential antioxidant property and excellent capability to scavenge most of the reactive oxygen species produced from several intracellular processes. Since, oxidative stress is also one of the hallmarks of Alzheimer and therefore they propose the hypothesis that concurrent delivery of nanoceria and Galantamine can have better therapeutic efficacy than of the individual treatment. Two different nanoceria-Galantamine-hydroxyapatite composites were formed, one coated with carboxymethyl chitosan (GAL-Ce@HAp/CMC) and one uncoated (GAL-Ce@HAp). The in-vivo activity performed in ovariectomized Alzheimer's albino-rats indicated that these rod like nanoparticles could be a promising candidate for Alzheimer's because all the stress markers which were upregulated, successfully recovered and AB plaques vanished on i.p treatment of these nanocomposites more (Wahba et al. 2016). Amira S. Hanafy et al., in 2016 developed nanoparticles made up of complex of Galantamine and chitosan (CX-NP2) for the management of Alzheimer's. These nanoparticles administered to rats at dose of 3 mg/kg for 12 consecutive days, remarkable decrease in AChE protein level was observed compared to conventional nasal and oral solution-based formulations. Also, the intranasal nanoparticle formulations treated rats did not show any sign of toxicity or histopathological manifestation. Their results proved that this simple formulation can be a promising strategy for nasal delivery of Alzheimer drugs to brain (Hanafy et al. 2015). Aditi Poddar in 2017, used a systematic Design of experiment (DOE) approach to develop Galantamine loaded Bovine Serum Albumin (BSA) nanoparticles The group reported that the development of nanoparticles was simple and effective, however they did not perform any in-vitro or in-vivo studies (Poddar and Sawant 2017).

Another group Kavita R. Gajbhiye1 and coworkers developed the ascorbic acid modified nanoparticles for the brain targeted delivery of Galantamine. Ascorbic acid was conjugated to PLGA-PEG to promote SVCT2 transportation of Galantamine into the brain. Cellular uptake data presented in Fig. 5.7a showed that targeted nanoparticles (PLGA-b-PEG-Asc) accumulation is higher than that of free Galantamine and non-targeted nanoparticles at all time points. The *in vivo* pharmacodynamic study evaluated in scopolamine induced amnetic rats established significantly higher therapeutic levels of Galantamine and also sustained action by ascorbic acid tethered PLGA-b-PEG NPs than free Galantamine, Plain PLGA nanoparticles as well as PLGA-b-mPEG nanoparticles (Fig. 5.7b, c). Bio-distribution data revealed the higher distribution of targeted nanoparticles in brain compared to the non-targeted nanoparticles. The research suggested ascorbic acid successfully

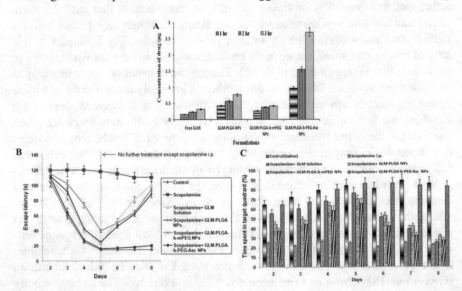

Fig. 5.7 (a) Cellular uptake of Galantamine from different formulations at different time points as estimated by High performance Liquid Chromatography [mean ± SD (n = 3)]. (b) Effect of different Galantamine formulations on the escape latency determined in the Morris water maze test and (c) Effect of various Galantamine formulations on time spent (%) in the target quadrant attained in the Morris water maze test. (Reproduced with permission from Nature Publishing house Gajbhiye et al. 2017)

target the nanoparticles to brain and thereby can deliver high concentration of drug to brain than other part of the body. The system holds a strong potential for the therapy of Alzheimer's disease (Gajbhiye et al. 2017).

5.8.4 Long-Acting Scaffolds for Localized Delivery

Polymer based scaffolds is another approach that has proven potential in Alzheimer's diseases as they can deliver the drug for prolong period and therefore reduce the dosing intervals. In very recent study in 2019, Maluta S. Mufamadi and co-workers developed liposome based polymeric scaffolds for prolonger delivery of Galantamine (LEPS). This LEPS were composed of Galantamine loaded liposomes embedded in highly porous chitosan, eudragit® RSPO and polyvinyl alcohol-based scaffold. The system was characterized using various studies and in-vitro release studies revealed that they could deliver the drug for prolonged period. Further cell proliferation of PC12 cells indicated that scaffold was suitable for their cultivation and proliferation. Overall finding concluded that this kind of system is suitable for prolonged release and precise delivery of Galantamine in nanoparticles. Though further work is required to confirm the potential of the delivery system for the intracranial localized delivery for the treatment of Alzheimer (Mufamadi et al. 2019).

5.9 Conclusion

Galantamine is a well-tolerated and effective treatment for the treatment of Alzheimer's. It provides sustained, broad, consistent and clinically relevant benefits in crucial areas of Alzheimer's It is the exceptional dual mode of action of Galantamine which could explicate its efficacy profile, and the potential additional benefits in comparison to other conventional cholinesterase inhibitors). It is one of the drugs which has a good safety profile, and do not possess any serious safety concerns on administration. Just like other cholinergic drugs, the only limitation is the occurrence of gastrointestinal side effects such as vomiting, nausea and diarrhea on oral administration. It's combination with memantine has also been shown better in some context in comparison to donepezil (most preferred choice of drug for Alzheimer) and memantine.

Research is continuously being done to overcome of these side effects by developing formulations that can be administered through altered route. Further the current approved formulation for Galantamine is oral which must be taken per day. Considering the condition of Alzheimer's patient and to improve its compliance, current research is focused on targeted and long acting formulation. Though obtained preliminary results have shown potential, but most of the results are preclinical based (Table 5.5). Therefore, more attention and more thorough investigation is required to prove its benefits and then only it can be assured that the benefits can be extended to patients or not.

Table 5.5 Summary of novel delivery systems of Galantamine for the therapy of Alzheimer's

Drug	Dosage form type	Details	References
Transdermal			
Galantamine	Microneedle pretreated	Microneedles having length of 70–80 μm were used for pre skin pretreatment to improve the permeation of Galantamine across the stratum corneum and live epidermis. Pre-piercing of skin significantly improved the Galantamine permeability	(Wei-Ze et al. 2010)
Galantamine	Drug-in-adhesive	Dug-in-adhesive patch using various acrylate and acrylate-vinylacetate based pressure sensitive adhesive (PSA) and different enhancers such as oleic acid, Benzyl alcohol, Isopropyl myristate and transcutol were prepared. Optimized Galantamine patch showed the bioavailability of around 80% and sustained plasma levels for 24 h	(Park et al. 2012)
Galantamine	Gel based	Gel-based drug reservoir using carbopol polymer, Galantamine HBr, propylene glycol, triethanolamine and deionized water. The proposed system controlled the release for longer period of time	(Fong Yen et al. 2015; Woo et al. 2015)
Galantamine	Liposomes	Effects of Limonene containing pegylated liposomes and sonophoresis on permeation of Galantamine was studied. Both the techniques were clearly able to enhance the formulation; however, the effect was stronger with pegylated liposomes	Rangsimawong et al. (2018)
Galantamine	Matrix based	Matrix-type transdermal patch for Galantamine was developed using 10% w/w Galantamine, 5% w/w Oleic acid (crystallization inhibitor), 5%w/w Limonene (penetration enhancer), GELVA GMS 788 as pressure-sensitive-adhesive. Permeation studies on human cadaver skin indicated that optimized patch can achieve therapeutic plasma level with patch size of ~20 cm²	(Ameen and Michniak-Kohn 2019)
Intranasal			
Galantamine	Memogain Prodrug	In-active hydrophobic prodrug of Galantamine given through nasal route and has potential of 15-fold higher bioavailability than Galantamine at the same dose	(Maelicke et al. 2010)
	Galantamine lactate salt	Galantamine lactate salt with solubility 400 mg/ml was prepared to be given through nasal route	(Leonard et al. 2005)

(continued)

Table 5.5 (continued)

Drug	Dosage form type	Details	References
	Galanatmaine lactate co-formulated with penetration enhancers	Galantamine lactate formulated with several penetration enhancers for intranasal delivery. These permeation enhancers resulted in four-fold enhancement of Galantamine permeation across the epithelial barrier with low cytotoxicity and high cell viability.	(Leonard et al. 2007)

Nanotechnology based systems

Drug	Dosage form type	Details	References
Galantamine	Galantamine and chitosan complex	Remarkable decrease in AChE protein level was observed compared to conventional nasal and oral solution-based formulations	(Hanafy et al. 2015).
Galantamine	Bovine Serum Albumin nanoparticles	The group reported that the development of nanoparticles was simple and effective, however they did not perform any in-vitro or in-vivo studies	(Poddar and Sawant 2017)
Galantamine	Liposomes	Lys-Val-Leu-Phe-Leu-Ser coupled liposomes were developed for Galantamine delivery to neuronal cells. Higher accumulation in PC12 neuronal cells as compared to non-targeted liposomes and plain Galantamine solution	(Mufamadi et al. 2013)
Galantamine	PLGA nanoparticles	Galantamine loaded polysorbate 80 coated PLGA nanoparticles for its parenteral delivery. Polysorbate 80 was used for blood brain barrier targeting	(Fornaguera et al. 2015)
Galantamine	Hydroxyapatite nanoparticles	Nanoparticles were developed to co-deliver nanoceria and Galantamine. Nanoceria was used to reduce the oxidative stress. In-vivo efficacy data showed that stress markers which were upregulated, successfully recovered and AB plaques vanished on i.p treatment of these nanoparticles	(Wahba et al. 2016)
Galantamine	Ascorbic acid tethered PLGA-PEG nanoparticles	Ascorbic acid was conjugated to PLGA-PEG to promote SVCT2 transportation of Galantamine into the brain. The *in vivo* study evaluated in scopolamine induced amnetic rats established higher therapeutic and sustained action by ascorbic acid tethered PLGA-b-PEG NPs than free drug, PLGA plain particles as well as PLGA-b-mPEG NPs	(Gajbhiye et al. 2017)

Scaffolds

Drug	Dosage form type	Details	References
Galantamine	Liposomes embedded in polymeric scaffolds (LEPS)	LEPS were composed of Galantamine loaded liposomes embedded in highly porous chitosan, eudragit® RSPO and polyvinyl alcohol-based scaffold	(Mufamadi et al. 2019)

References

Ago Y, Koda K, Takuma K, Matsuda T (2011) Pharmacological aspects of the acetylcholinesterase inhibitor galantamine. J Pharmacol Sci. https://doi.org/10.1254/jphs.11R01CR

Ahmad J, Akhter S, Rizwanullah M, Khan MA, Pigeon L, Addo RT et al (2017) Nanotechnology based theranostic approaches in Alzheimer's disease management: current status and future perspective. Curr Alzheimer Res. https://doi.org/10.2174/1567205014666170508121031

Alghamdi BS (2018) The neuroprotective role of melatonin in neurological disorders. J Neurosci Res. https://doi.org/10.1002/jnr.24220

Ameen D, Michniak-Kohn B (2019) Development and in vitro evaluation of pressure sensitive adhesive patch for the transdermal delivery of galantamine: effect of penetration enhancers and crystallization inhibition. Eur J Pharm Biopharm. https://doi.org/10.1016/j.ejpb.2019.04.008

Anand R, Gill KD, Mahdi AA (2014) Therapeutics of Alzheimer's disease: past, present and future. Neuropharmacology. https://doi.org/10.1016/j.neuropharm.2013.07.004

Atamna H, Nguyen A, Schultz C, Boyle K, Newberry J, Kato H, Ames BN (2008) Methylene blue delays cellular senescence and enhances key mitochondrial biochemical pathways. FASEB J. https://doi.org/10.1096/fj.07-9610com

Bastida J, Lavilla R, Viladomat F (2006) Chapter 3 chemical and biological aspects of Narcissus alkaloids. Alkaloids: Chem Biol. https://doi.org/10.1016/S1099-4831(06)63003-4

Berkov S, Georgieva L, Kondakova V, Atanassov A, Viladomat F, Bastida J, Codina C (2009) Plant sources of galanthamine: phytochemical and biotechnological aspects. Biotechnol Biotechnol Equip. https://doi.org/10.1080/13102818.2009.10817633

Bhattacharya S, Montag D (2015) Acetylcholinesterase inhibitor modifications: a promising strategy to delay the progression of Alzheimer's disease. Neural Regen Res. https://doi.org/10.4103/1673-5374.150648

Bickel U, Thomsen T, Weber W, Fischer JP, Bachus R, Nitz M, Kewitz H (1991) Pharmacokinetics of galanthamine in humans and corresponding cholinesterase inhibition. Clin Pharmacol Ther. https://doi.org/10.1038/clpt.1991.159

Brodaty H, Corey-Bloom J, Potocnik FCV, Truyen L, Gold M, Damaraju CRV (2005) Galantamine prolonged-release formulation in the treatment of mild to moderate Alzheimer's disease. Dement Geriatr Cogn Disord. https://doi.org/10.1159/000086613

Castellani RJ, Perry G (2014) The complexities of the pathology-pathogenesis relationship in Alzheimer disease. Biochem Pharmacol. https://doi.org/10.1016/j.bcp.2014.01.009

Cheng YF, Wang C, Lin HB, Li YF, Huang Y, Xu JP, Zhang HT (2010) Inhibition of phosphodiesterase-4 reverses memory deficits produced by Aβ25-35 or Aβ1-40 peptide in rats. Psychopharmacology. https://doi.org/10.1007/s00213-010-1943-3

Choi SH, Aid S, Caracciolo L, Sakura Minami S, Niikura T, Matsuoka Y et al (2013) Cyclooxygenase-1 inhibition reduces amyloid pathology and improves memory deficits in a mouse model of Alzheimer's disease. J Neurochem. https://doi.org/10.1111/jnc.12059

Chow HHS, Chen Z, Matsuura GT (1999) Direct transport of cocaine from the nasal cavity to the brain following intranasal cocaine administration in rats. J Pharm Sci. https://doi.org/10.1021/js9900295

Costantino HR, Leonard AK, Brandt G, Johnson PH, Quay SC (2008) Intranasal administration of acetylcholinesterase inhibitors. BMC Neurosci. https://doi.org/10.1186/1471-2202-9-S2-S6

Cummings J, Lefèvre G, Small G, Appel-Dingemanse S (2007) Pharmacokinetic rationale for the rivastigmine patch. Neurology. https://doi.org/10.1212/01.wnl.0000281846.40390.50

Davies P, Maloney AJF (1976) Selective loss of central cholinergic neurons in Alzheimer's disease. Lancet. https://doi.org/10.1016/S0140-6736(76)91936-X

Doggrell SA, Evans S (2003) Treatment of dementia with neurotransmission modulation. Expert Opin Investig Drugs. https://doi.org/10.1517/13543784.12.10.1633

Dong XX, Wang Y, Qin ZH (2009) Molecular mechanisms of excitotoxicity and their relevance to pathogenesis of neurodegenerative diseases. Acta Pharmacol Sin. https://doi.org/10.1038/aps.2009.24

dos Santos Alves R, de Souza AS et al (2014) FDA- approved treatments for Alzheimer's. Igarss 2014. https://doi.org/10.1007/s13398-014-0173-7.2

Farlow M, Anand R, Messina J, Hartman R, Veach J (2000) A 52-week study of the efficacy of rivastigmine in patients with mild to moderately severe Alzheimer's disease. Eur Neurol. https://doi.org/10.1159/000008243

Fong Yen W, Basri M, Ahmad M, Ismail M (2015) Formulation and evaluation of galantamine gel as drug reservoir in transdermal patch delivery system. Sci World J. https://doi.org/10.1155/2015/495271

Fornaguera C, Feiner-Gracia N, Calderó G, García-Celma MJ, Solans C (2015) Galantamine-loaded PLGA nanoparticles, from nano-emulsion templating, as novel advanced drug delivery systems to treat neurodegenerative diseases. Nanoscale. https://doi.org/10.1039/c5nr03474d

Fournier E, Passirani C, Montero-Menei CN, Benoit JP (2003) Biocompatibility of implantable synthetic polymeric drug carriers: focus on brain biocompatibility. Biomaterials. https://doi.org/10.1016/S0142-9612(03)00161-3

Gajbhiye KR, Gajbhiye V, Siddiqui IA, Pilla S, Soni V (2017) Ascorbic acid tethered polymeric nanoparticles enable efficient brain delivery of galantamine: an in vitro-in vivo study. Sci Rep. https://doi.org/10.1038/s41598-017-11611-4

Galasko D, Kershaw PR, Schneider L, Zhu Y, Tariot PN (2004) Galantamine maintains ability to perform activities of daily living in patients with Alzheimer's disease. J Am Geriatr Soc. https://doi.org/10.1111/j.1532-5415.2004.52303.x

Garcia ML, Cleveland DW (2001) Going new places using an old MAP: tau, microtubules and human neurodegenerative disease. Curr Opin Cell Biol. https://doi.org/10.1016/S0955-0674(00)00172-1

Georgieva L, Berkov S, Kondakova V, Bastida J, Viladomat F, Atanassov A, Codina C (2007) Alkaloid variability in Leucojum aestivum from wild populations. Z Naturforsch C J Biosci 62(9-10):627–635

Godyń J, Jończyk J, Panek D, Malawska B (2016) Therapeutic strategies for Alzheimer's disease in clinical trials. Pharmacol Rep. https://doi.org/10.1016/j.pharep.2015.07.006

Grossberg GT, Edwards KR, Zhao Q (2006) Rationale for combination therapy with galantamine and memantine in Alzheimer's disease. J Clin Pharmacol. https://doi.org/10.1177/0091270006288735

Hanafy AS, Farid RM, Elgamal SS (2015) Complexation as an approach to entrap cationic drugs into cationic nanoparticles administered intranasally for Alzheimer's disease management: preparation and detection in rat brain. Drug Dev Ind Pharm. https://doi.org/10.3109/03639045.2015.1062897

Hayashi A, Saito T, Mukai Y, Kurita S, Hori TA (2005) Genetic variations in Lycoris radiata var. radiata in Japan. Genes Genet Syst. https://doi.org/10.1266/ggs.80.199

Heinrich M (2002) Narcissus and Daffodil—the genus Narcissus. J Ethnopharmacol. https://doi.org/10.1016/s0378-8741(02)00207-6

Imbimbo BP (2001) Pharmacodynamic-tolerability relationships of cholinesterase inhibitors for Alzheimer's disease. CNS Drugs. https://doi.org/10.2165/00023210-200115050-00004

Knopman DS, DeKosky ST, Cummings JL, Chui H, Corey-Bloom J, Relkin N, Stevens JC (2007) Appendix B: practice parameter: diagnosis of dementia (an evidence-based review): report of the quality standards subcommittee of the American academy neurology. Continuum Lifelong Learn Neurol. 13: 210–221. https://doi.org/10.1212/01.CON.0000267232.84626.41

Koola MM, Parsaik AK (2018) Galantamine-memantine combination effective in dementia: translate to dementia praecox? Schizophr Res Cogn. https://doi.org/10.1016/j.scog.2017.11.001

Koola MM, Nikiforuk A, Pillai A, Parsaik AK (2018) Galantamine-memantine combination superior to donepezil-memantine combination in Alzheimer's disease: critical dissection with an emphasis on kynurenic acid and mismatch negativity. J Geriatr Care Res 5(2):57–67

Kreh M, Matusch R, Witte L (1995) Capillary gas chromatography-mass spectrometry of Amaryllidaceae alkaloids. Phytochemistry. https://doi.org/10.1016/0031-9422(94)00725-9

Leonard AK, Sileno AP, Macevilly C, Foerder CA, Quay SC, Costantino HR (2005) Development of a novel high-concentration galantamine formulation suitable for intranasal delivery. J Pharm Sci. https://doi.org/10.1002/jps.20389

Leonard AK, Sileno AP, Brandt GC, Foerder CA, Quay SC, Costantino HR (2007) In vitro formulation optimization of intranasal galantamine leading to enhanced bioavailability and reduced emetic response in vivo. Int J Pharm. https://doi.org/10.1016/j.ijpharm.2006.11.013

Lilienfeld S (2002) Galantamine – a novel cholinergic drug with a unique dual mode of action for the treatment of patients with Alzheimer's disease. CNS Drug Rev

Lipton SA (2006) Paradigm shift in neuroprotection by NMDA receptor blockade: memantine and beyond. Nat Rev Drug Discov. https://doi.org/10.1038/nrd1958

Maelicke A, Hoeffle-Maas A, Ludwig J, Maus A, Samochocki M, Jordis U, Koepke AKE (2010) Memogain is a galantamine pro-drug having dramatically reduced adverse effects and enhanced efficacy. J Mol Neurosci. https://doi.org/10.1007/s12031-009-9269-5

Marcade M, Bourdin J, Loiseau N, Peillon H, Rayer A, Drouin D et al (2008) Etazolate, a neuroprotective drug linking GABAA receptor pharmacology to amyloid precursor protein processing. J Neurochem. https://doi.org/10.1111/j.1471-4159.2008.05396.x

Matsuoka Y, Gray AJ, Hirata-Fukae C, Minami SS, Waterhouse EG, Mattson MP et al (2007) Intranasal NAP administration reduces accumulation of amyloid peptide and tau hyperphosphorylation in a transgenic mouse model of Alzheimer's disease at early pathological stage. J Mol Neurosci. https://doi.org/10.1007/s12031-007-0016-5

McNeil SE (2005) Nanotechnology for the biologist. J Leukoc Biol. https://doi.org/10.1189/jlb.0205074

Mucke HA (2015) The case of galantamine: repurposing and late blooming of a cholinergic drug. Futur Sci OA. https://doi.org/10.4155/fso.15.73

Mufamadi MS, Choonara YE, Kumar P, Modi G, Naidoo D, Van Vuuren S et al (2013) Ligand-functionalized nanoliposomes for targeted delivery of galantamine. Int J Pharm. https://doi.org/10.1016/j.ijpharm.2013.03.037

Mufamadi MS, Kumar P, du Toit LC, Choonara YE, Obulapuram PK, Modi G et al (2019) Liposome-embedded, polymeric scaffold for extended delivery of galantamine. J Drug Deliv Sci Technol. https://doi.org/10.1016/j.jddst.2019.02.001

Park CW, Son DD, Kim JY, Oh TO, Ha JM, Rhee YS, Park ES (2012) Investigation of formulation factors affecting in vitro and in vivo characteristics of a galantamine transdermal system. Int J Pharm. https://doi.org/10.1016/j.ijpharm.2012.06.057

Parolo G, Abeli T, Rossi G, Dowgiallo G, Matthies D (2011) Biological flora of Central Europe: Leucojum aestivum L. Perspect Plant Ecol Evol Syst. https://doi.org/10.1016/j.ppees.2011.05.004

Perl DP (2010) Neuropathology of Alzheimer's disease. Mt Sinai J Med. https://doi.org/10.1002/msj.20157

Peters O, Fuentes M, Joachim LK, Jessen F, Luckhaus C, Kornhuber J, Pantel J, Hüll M, Schmidtke K, Rüther E, Möller HJ, Kurz A, Wiltfang J, Maier W, Wiese B, Frölich L, Heuser I (2015) Combined treatment with memantine and galantamine-CR compared with galantamine-CR only in antidementia drug naïve patients with mild-to-moderate Alzheimer's disease. Alzheimer's and Dementia: Translational Research and Clinical Interventions. https://doi.org/10.1016/j.trci.2015.10.001

Poddar A, Sawant KK (2017) Optimization of Galantamine loaded bovine serum albumin nanoparticles by quality by design and its preliminary characterizations. J Nanomed Nanotechnol. https://doi.org/10.4172/2157-7439.1000459

Rangsimawong W, Obata Y, Opanasopit P, Ngawhirunpat T, Takayama K (2018) Enhancement of Galantamine HBr skin permeation using Sonophoresis and Limonene-containing PEGylated liposomes. AAPS PharmSciTech. https://doi.org/10.1208/s12249-017-0921-z

Revett TJ, Baker GB, Jhamandas J, Kar S (2013) Glutamate system, amyloid β peptides and tau protein: functional interrelationships and relevance to Alzheimer disease pathology. J Psychiatry Neurosci. https://doi.org/10.1503/jpn.110190

Robinson DM, Plosker GL (2006) Galantamine extended release. CNS Drugs. https://doi.org/10.2165/00023210-200620080-00006

Rogawski MA, Wenk GL (2003) The neuropharmacological basis for the use of memantine in the treatment of Alzheimer's disease. CNS Drug Rev

Rogers SL, Friedhoff LT (1998) Long-term efficacy and safety of donepezil in the treatment of Alzheimer's disease: An interim analysis of the results of a US multicentre open label extension study. Eur Neuropsychopharmacol. https://doi.org/10.1016/S0924-977X(97)00079-5

Roth AD, Ramírez G, Alarcón R, Von Bernhardi R (2005) Oligodendrocytes damage in Alzheimer's disease: beta amyloid toxicity and inflammation. Biol Res. https://doi.org/10.4067/S0716-97602005000400011

Sakane T, Akizuki M, Yoshida M, Yamashita S, Nadai T, Hashida M, Sezaki H (1991) Transport of cephalexin to the cerebrospinal fluid directly from the nasal cavity. J Pharm Pharmacol. https://doi.org/10.1111/j.2042-7158.1991.tb03510.x

Salomone S, Caraci F, Leggio GM, Fedotova J, Drago F (2012) New pharmacological strategies for treatment of Alzheimer's disease: focus on disease modifying drugs. Br J Clin Pharmacol. https://doi.org/10.1111/j.1365-2125.2011.04134.x

Saravanakumar K, Swapna P, Nagaveni P, Vani P, Pujitha K (2015) Transdermal drug delivery system: a review. J Glob Trends Pharm Sci 1, 70:–81

Schenk D, Basi GS, Pangalos MN (2012) Treatment strategies targeting amyloid β-protein. Cold Spring Harb Perspect Med. https://doi.org/10.1101/cshperspect.a006387

Scott LJ, Goa KL (2000) Galantamine: a review of its use in Alzheimer's disease. Drugs. https://doi.org/10.2165/00003495-200060050-00008

Sramek JJ, Frackiewicz EJ, Cutler NR (2000) Review of the acetylcholinesterase inhibitor galanthamine. Expert Opin Investig Drugs. https://doi.org/10.1517/13543784.9.10.2393

Thorne RG, Pronk GJ, Padmanabhan V, Frey WH (2004) Delivery of insulin-like growth factor-I to the rat brain and spinal cord along olfactory and trigeminal pathways following intranasal administration. Neuroscience. https://doi.org/10.1016/j.neuroscience.2004.05.029

Wahba SMR, Darwish AS, Kamal SM (2016) Ceria-containing uncoated and coated hydroxyapatite-based galantamine nanocomposites for formidable treatment of Alzheimer's disease in ovariectomized albino-rat model. Mater Sci Eng C. https://doi.org/10.1016/j.msec.2016.04.041

Wang Y, Aun R, Tse FLS (1998) Brain uptake of dihydroergotamine after intravenous and nasal administration in the rat. Biopharm Drug Dispos. https://doi.org/10.1002/(SICI)1099-081X(199812)19:9<571::AID-BDD142>3.0.CO;2-O

Wang X, Su B, Lee HG, Li X, Perry G, Smith MA, Zhu X (2009) Impaired balance of mitochondrial fission and fusion in Alzheimer's disease. J Neurosci. https://doi.org/10.1523/JNEUROSCI.1357-09.2009

Wei-Ze L, Mei-Rong H, Jian-Ping Z, Yong-Qiang Z, Bao-Hua H, Ting L, Yong Z (2010) Supershort solid silicon microneedles for transdermal drug delivery applications. Int J Pharm. https://doi.org/10.1016/j.ijpharm.2010.01.024

Wenk GL (2006) Neuropathologic changes in Alzheimer's disease: potential targets for treatment. J Clin Psychiatry

Woo FY, Basri M, Masoumi HRF, Ahmad MB, Ismail M (2015) Formulation optimization of galantamine hydrobromide loaded gel drug reservoirs in transdermal patch for Alzheimer's disease. Int J Nanomedicine. https://doi.org/10.2147/IJN.S80253

Zhao Q, Janssens L, Verhaeghe T, Brashear HR, Truyen L (2005) Pharmacokinetics of extended-release and immediate-release formulations of galantamine at steady state in healthy volunteers. Curr Med Res Opin. https://doi.org/10.1185/030079905X61965

Zhu X, Lee H g, Perry G, Smith MA (2007) Alzheimer disease, the two-hit hypothesis: an update. Biochim Biophys Acta Mol basis Dis. https://doi.org/10.1016/j.bbadis.2006.10.014

Chapter 6
An Overview of Paclitaxel Delivery Systems

Prabakaran A, Sourav Kar, K. Vignesh, and Ujwal D. Kolhe

Abstract As per the World Health Organization, cancer is the second leading cause of death globally and is responsible for an estimated 9.6 million deaths in 2018. More than 2 million cases of lung and breast cancer have been reported in 2018. Paclitaxel is a natural product based anti-mitotic agent useful for the treatment of lung cancer, ovarian cancer and breast cancer. Paclitaxel exhibits low oral bioavailability due to poor aqueous solubility and poor permeability. Paclitaxel is the first blockbuster anticancer drug with annual sales of more than US $1 billion in 1997. Taxol®, Abraxane® and Genexol PM® are commercial injectable preparations of paclitaxel with a drug loading of not more than 17% w/w. The toxicity issues of Cremophor® EL used in Taxol® led to the development of different delivery systems sans Cremophor® EL. The maximum tolerated dose of Taxol® is 135 mg/m^2. Taxol® exhibited hypersensitivity reactions and required use of special IVEX-2 filter to avoid leaching of plasticizer into the product. Genexol PM® was physically unstable and was ineffective against multidrug-resistant tumor treatment. Abraxane® exhibited limited tumor exposure, tumor uptake, tumor regression and higher half-maximal inhibitory concentration.

In this chapter, we reviewed the literature of paclitaxel delivery systems. Micelles, liposomes, nanoparticles, lipid systems, microparticles, emulsions, solid dispersions, cyclodextrin complexes, implants, prodrugs and hybrid systems have been reported for paclitaxel delivery. The major points of our analysis of the literature are (1) Efficiency of solubility enhancement of paclitaxel was found in the decreasing order for prodrugs, mixed micelles, cyclodextrin complexes and solid dispersions, (2) The oral bioavailability enhancement for nanoparticles, micelles and emulsions of paclitaxel was found to be 7-fold, 5-fold and 4-fold respectively, (3) Decreasing order of reduction in tumor growth was found in emulsion, liposome, prodrug and nanoparticulate delivery system of paclitaxel as compared to that with Taxol®, (4) Genexol PM® and Abraxane® do not require special filters to avoid leaching of plasticizer into the product, (5) Genexol PM® and Abraxane® being free from Cremophor® EL do not exhibit hypersensitivity reaction, (6) Paclitaxel mixed

Prabakaran A · S. Kar · K. Vignesh · U. D. Kolhe (✉)
Department of Pharmaceutics, National Institute of Pharmaceutical Education and Research, Raebareli (NIPER-R), Lucknow, Uttar Pradesh, India

© The Editor(s) (if applicable) and The Author(s), under exclusive license to Springer Nature Switzerland AG 2020
A. Saneja et al. (eds.), *Sustainable Agriculture Reviews 43*, Sustainable Agriculture Reviews 43, https://doi.org/10.1007/978-3-030-41838-0_6

micelles composed of poly (ethylene glycol-co-lactic acid) and D-α-tocopheryl polyethylene glycol 1000 succinate were effective against multidrug-resistant tumor cells in contrast to Genexol PM®, (7) Recombinant chimeric polypeptide conjugated paclitaxel nanoparticles exhibited 2-fold systemic tumor exposure, tumor uptake as compared to Abraxane® and almost complete tumor regression, (8) Paclitaxel loaded polycaprolactone-co-D-α-tocopheryl polyethylene glycol 1000 succinate nanoparticles exhibited 8-fold lower half maximal inhibitory concentration as compared to Abraxane®, (9) Paclitaxel delivery systems such as implants, nanoparticles, solid dispersions, lipid nanoparticles and micelles have been reported with more than 20% drug loading, (10) The maximum tolerated dose of Genexol PM® and Abraxane® could be increased from 135 mg/m^2 to 300 mg/m^2, and (11) Orally effective paclitaxel formulations such as DHP107 and Oraxol® are under phase II clinical trials.

Keywords Paclitaxel · Delivery system · Stimuli sensitive delivery · Targeted delivery · Natural product · Anticancer

6.1 Introduction

Paclitaxel is a potent broad-spectrum anticancer drug. Paclitaxel is a diterpenoid molecule with a central 8-member taxane ring. Paclitaxel was initially isolated from the bark of *Taxus brevifolia,* family: *Taxaciae* (the Western Yew tree). The pure form of paclitaxel was isolated in 1969. Paclitaxel is used in the treatment of breast, ovarian, lung cancer, head and neck cancer. Paclitaxel belongs to class IV of the biopharmaceutical classification system, the most difficult category for drug delivery.

The year-wise publication trend for paclitaxel delivery as per the SciFinder search engine is given (Fig. 6.1). The publication trend reveals that the number of publications have increased 10 times in the last 20 years.

Paclitaxel was extracted from the bark of the western yew tree *Taxus brevifolia.* It was identified as one of the active constituents of this plant in 1967. The structure of the compound was elucidated by Wall and Wani 1996. Its cytotoxic activity was reported against human KB carcinoma cell line and mouse leukemia cells. The availability of the compound was limited because of its low yield from the western yew tree. So, Robert A. Holton and his research group came up with a semi-synthetic pathway by which docetaxel was produced from 10-deacetyl baccatin, which was present in abundant quantities in *Taxus baccata* (Alqahtani et al. 2019). A synthetic method for paclitaxel using C7 protected baccatin III with tricyclic ketone have been reported (Holton et al. 1994).

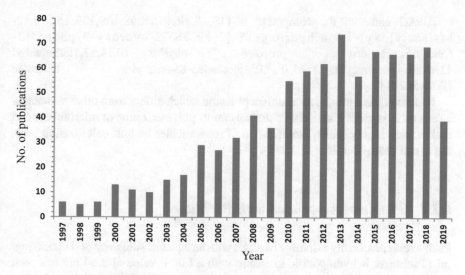

Fig. 6.1 Year-wise publication trend for paclitaxel delivery; this trend has been obtained from SciFinder search on June 24, 2019 using publication year trend; note the increase of publications for paclitaxel delivery over period of 1997–2019; the publication trend signifies demand for newer delivery options for paclitaxel; the existing paclitaxel commercial products lack safety, targeting to specific organs and patient inconvenience due to injections

Fig. 6.2 Chemical structure of paclitaxel and C7 protected position for synthesis of paclitaxel using baccatin III; paclitaxel was extracted from Yew tree initially hence alternative semi-synthetic method using highly abundant baccatin III in the bark of *Taxus wallichiana var. mairei* has been discovered by protecting baccatin III at C7 position and coupling tricyclic ketone to yield paclitaxel

6.1.1 Chemical Structure of Paclitaxel

Paclitaxel is a diterpenoid compound containing a taxane ring along with other hydrophobic substituents. The empirical formula of paclitaxel is $C_{47}H_{51}NO_{14}$. The chemical structure of paclitaxel is given (Fig. 6.2).

IUPAC name of the compound is (1S,2S,3R,4S,7R,9S,10S,12R,15S)-4,12-bis(acetyloxy)-1,9-dihydroxy-15-{[(2R,3S)-2-hydroxy-3-phenyl-3-(phenylformamido)   propa   -noyl]oxy}-10,14,17,1tetramethyl 11-oxo-6-oxatetracyclo[11.3.1.0^3,^{10}.0^4,^{7}]heptadec-13-en-2-yl   benzoate (Weaver 2014).

Paclitaxel has a unique mechanism of action which differs from other anticancer agents such as vinca alkaloids. Paclitaxel causes polymerization of microtubule and makes these highly stable. Stabilization of microtubules inhibits cell division leading to cell death (Schiff et al. 1979).

6.1.2   Problems Associated with Paclitaxel

Paclitaxel is practically insoluble in water with an aqueous solubility of 0.00556 mg/mL. Paclitaxel is hydrophobic in nature with a Log P value of 3.54 but has poor permeability. The molecular weight of paclitaxel is 853.9. The high molecular weight of the drug might be responsible for the poor permeability. The consequence of poor solubility and permeability of the drug is poor oral bioavailability. Paclitaxel is a substrate for P-glycoprotein which is responsible for drug efflux. High molecular weight of the drug has been attributed to drug efflux. Paclitaxel belongs to class IV of the biopharmaceutical classification system (Varma et al. 2005).

6.1.3   Marketed Paclitaxel Products

The commercial paclitaxel formulations include Taxol®, Abraxane®, Genexol PM® and Lipusu® (Nehate et al. 2014). The first marketed product of paclitaxel is Taxol®, which was marketed by Bristol-Myers-Squibb in 1992. Taxol® is available in three strengths i.e. 30 mg, 100 mg and 300 mg per vial. Taxol® consists of purified Cremophor® EL, anhydrous citric acid and dehydrated alcohol. Taxol® is indicated as first-line and subsequent therapy of ovarian cancer, for the adjuvant treatment of node-positive breast cancer administered subsequently to standard doxorubicin-containing combination therapy, for the first-line treatment of non-small cell lung cancer along with cisplatin in patients who are not potential candidates for curative surgery and/or radiation therapy and for the second-line treatment of AIDS-related Kaposi's sarcoma. The major side effects of Taxol® observed in patients are hypersensitivity reaction, low blood counts, hair loss, myalgias, peripheral neuropathy, mouth sores, nausea, vomiting, diarrhoea, swelling of feet, low blood pressure and darkening of the skin. Taxol® injection requires premedication with a steroid to avoid hypersensitivity reaction (Rxlist website 2019).

The main problem associated with Taxol® injection is Cremophor® EL related toxicity. Cremophor® EL is not an inert vehicle and it produces undesirable biological effects. Cremophor® EL can cause a severe anaphylactic reaction (Szebeni et al. 1998). Taxol® needs to be administered using proprietary IVEX-2 filters. IVEX-2 filters avoid leaching of di-(2-ethylhexyl) phthalate in the product. Leaching of di-(2-ethylhexyl) phthalate in the product from polyvinyl chloride container and administration set has been reported. The leaching was attributed to presence of Cremophor® EL and dehydrated alcohol in Taxol® formulation (Kim et al. 2005). The literature has cited a number of research publications for the development of delivery systems without using Cremophor® EL in order to have safer formulation.

Abraxane® is the second commercial product of paclitaxel, which was marketed by Abraxis BioScience in 2005. Abraxane® is composed of drug and human serum albumin and has been prepared by high-pressure homogenization method (Ibrahim et al. 2002). Abraxane® is Cremophor EL®-free formulation and hence does not exhibit hypersensitivity and leaching of plasticizer from polyvinyl chloride. The maximum tolerated dose could be increased from 175 mg/m^2 to 300 mg/m^2 with Abraxane® as compared to that with Taxol® (Gradishar 2006). Lipusu® was the first liposome-based injection of paclitaxel, formulated by Luye Pharmaceutical Co. Ltd. and was marketed in China. Lipusu® contains paclitaxel, lecithin and cholesterol. Lipusu® was effective against breast cancer, ovarian cancer, lung cancer with fewer side effects than Taxol®.

Genexol-PM® is another injectable polymeric micelle based product of paclitaxel, marketed by Samyang Corporation. Genexol-PM® is a poly (ethylene glycol)-b-poly (lactic acid) block co-polymer based micellar formulation. Genexol-PM® does not require steroid premedication and free from the problem of plasticizer leaching from polyvinyl chloride. Genexol-PM® can deliver 300 mg paclitaxel dose without additional toxicity (Kim et al. 2004). Genexol-PM® was stable at 23 °C for 24 h but exhibited precipitation into large needle-like crystals at 40 °C within 2–4 h (Ron et al. 2008). Genexol-PM® was ineffective in cancer patients with multidrug resistance (Fan et al. 2015)

Nanoxel™ is a micellar paclitaxel formulation composed of a copolymer of n-isopropyl acrylamide and n-vinyl pyrrolidone having a mean particle size of 80 nm. The intracellular accumulation of paclitaxel in A549 cell was found to be 2.5-fold and 3.3-fold higher in Nanoxel™ and Abraxane®-treated cells respectively as compared to that of Taxol®. Nanoxel™ at 0.7 mg/mL showed significant aggregation, particle-size growth, and crystallization within 4 h at 40 °C. High-pressure liquid chromatography data revealed slightly lower purity of paclitaxel in Nanoxel™ as compared to both Abraxane® and Genexol PM®. No drug-related mortality was observed following repeated intravenous administration of Abraxane® at dose levels of 5, 15, and 30 mg/kg, while Nanoxel™ exhibited 100% mortality at 15 and 30 mg/kg dose levels in athymic nude mice (Ron et al. 2008; Madaan et al. 2013; Trieu et al. 2008).

6.1.4   Recent Paclitaxel Products Under Clinical Trials

A number of delivery approaches have been reported in the literature to increase the solubility and enhance the antitumor activity of paclitaxel such as micelles, liposomes, nanoparticles, microparticles, solid dispersions, cyclodextrin-drug complex, implants and prodrug. This chapter discusses approaches to paclitaxel delivery. The emphasis is given on choice of material, type of delivery system and performance. Different delivery systems explored for paclitaxel in the literature are discussed in subsequent sections (Fig. 6.3).

DHP107 is a novel lipid-based mucoadhesive oral paclitaxel formulation composed of an edible lipid and FDA approved emulsifiers. Phase II clinical trial is

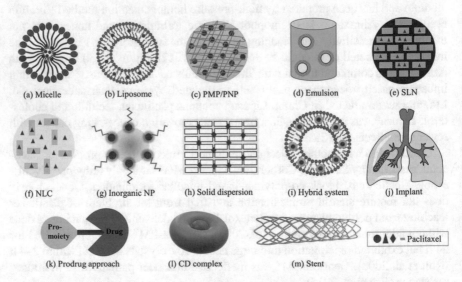

Fig. 6.3 Different delivery systems of paclitaxel (**a**) Micelles: Self-assembled aggregates of polymers or surfactants which have a distinct polar head and non-polar tail region; (**b**) Liposomes: Spherical vesicular structures, made up of one or more phospholipid bilayers and an aqueous core; (**c**) PMP/ PNP: Polymeric microparticles/ Polymeric nanoparticles. These are micron sized or nanosized particles made up of a polymeric matrix in which the drug is dispersed or encapsulated or entrapped; (**d**) Emulsion: Dispersion of two immiscible liquids where the dispersed phase is thoroughly distributed in a dispersion medium with the help of emulsifiers; (**e**) SLN: Solid lipid nanoparticles, submicron carriers made up of high melting point lipid (solid lipid) core and coated by aqueous surfactant layer; (**f**) NLC: Nanostructured lipid carriers, also known as second generation SLN, where the solid lipid core is replaced by a mixture of solid lipid and liquid lipid; (**g**) Inorganic nanoparticle: Inorganic nanoparticles, the core is made up of nanosized inorganic material while the shell is organic, contains drug and/or polymers; (**h**) Solid dispersion: Dispersion of a drug into a solid, inert matrix of carriers; (**i**) Hybrid systems: Combination of more than one systems; (**j**) Implant: Device containing drug, intended to be implanted inside the body; (**k**) Prodrug: Inactive form of a drug, which is converted to a pharmacologically active form upon metabolism inside the body; (**l**) CD complex: Cyclodextrin complex, inclusion complex formed by cyclodextrins with drugs; (**m**) Stent: Metal mesh tube which is implanted in ducts or vessels of the body e.g. coronary artery. Drug-eluting stents are used to deliver drugs to a specific site for a long time

being conducted for this formulation. The study began on July 25, 2018, and it is expected to be completed by September 2019. The use of the formulation is indicated for metastatic breast cancer (Kang et al. 2018). Another oral formulation of paclitaxel is under phase II clinical trial, named as Oraxol®. Oraxol® consists of paclitaxel and HM30181A, a P-glycoprotein inhibitor. Oraxol® has been found to be well tolerated in patients without any notable hypersensitivity reaction (Lee et al. 2015). The toxicity issues in the existing commercial formulations have led to the development of alternate formulations for paclitaxel.

6.2 Micelle-Based Delivery System

Micelles are nanometre sized structures formed by self-assembly of amphiphilic block copolymers. The hydrophilic part is oriented towards water whereas the hydrophobic portion remains inside the micellar structure (Kwon and Okano 1996). The concentration at which the micelles are formed in water is known as critical micellar concentration. Lower the critical micellar concentration better is the stability of the micelles. The micelles need to be intact for solubilising the hydrophobic drug. Upon ingestion, micelles' structure may be collapsed due to dilution with aqueous fluids resulting in precipitation of drug. The breakdown of micelles can be avoided by the use of amphiphilic copolymers having a low critical micellar concentration (Owen et al. 2012).

Different amphiphilic polymers have been used in the past for preparation of drug-loaded micelles. These polymers include diblock copolymers such as poly (ethylene glycol)-b-poly (lactide), poly (ethylene glycol)-b- poly (lactide-co-glycolide), etc. Triblock copolymers have also been used for micelle preparation such as poly (ethylene glycol)-poly (ε-caprolactone-co-L lactide) (Wang et al. 2013), monomethoxy poly (ethylene glycol)- poly (caprolactone)-D-α-tocopheryl polyethylene glycol 1000 succinate (Zhang and Zhang 2015). Chitosan has been modified with hydrophobic moieties to form an amphiphilic polymer. These hydrophobic groups include α-tocopherol succinate (Liang et al. 2016), Pluronic F127 (Xu et al. 2015a, b), N-succinyl palmitoyl group (Yuan et al. 2015), N-octyl N-trimethyl group (Zhang et al. 2016), polycaprolactone (Almeida et al. 2018). Serum albumin has been modified with octyl group to make it amphiphilic for micelle preparation (Liu et al. 2011). The schematic of types of paclitaxel micelles are given (Fig. 6.4).

6.2.1 Simple Micelle-Based Delivery System

Simple micelles are prepared by using an amphiphilic copolymer. The copolymer may be di or triblock copolymer composed of the hydrophilic and hydrophobic component. Hydrophobically modified serum albumin copolymer was synthesized for preparation of paclitaxel micelles. Serum albumin was chosen as it is

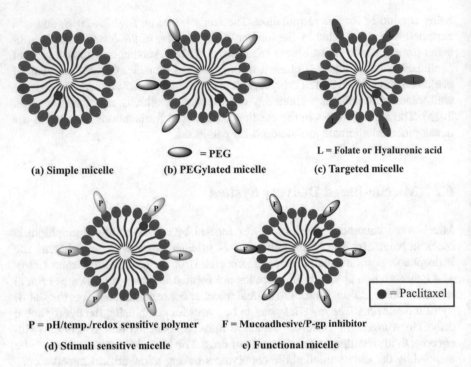

= PEG

L = Folate or Hyaluronic acid

(a) Simple micelle　　　**(b) PEGylated micelle**　　　**(c) Targeted micelle**

= Paclitaxel

P = pH/temp./redox sensitive polymer　　　**F = Mucoadhesive/P-gp inhibitor**

(d) Stimuli sensitive micelle　　　　**e) Functional micelle**

Fig. 6.4 Types of paclitaxel micelles in literature; (**a**) simple micelle, (**b**) poly (ethylene glycol) ylated micelle where poly(ethylene glycol) is coated or conjugated to the micelle, (**c**) targeted micelle; targeting ligands such as folate or hyaluronic acid is conjugated to the micelle, (**d**) stimuli sensitive micelle; stimuli-responsive i.e. pH/temp/redox-sensitive polymer is conjugated to micelle and (**e**) functional micelles; functional molecule such as mucoadhesive agent, P-glycoprotein (P-gp) inhibitor or biodegradable polymer is used

endogenously present in the plasma. Serum albumin is biocompatible and nontoxic protein. The reactive primary amino group of albumin was conjugated to octyl chain. The paclitaxel micelles prepared using synthesized octyl conjugated serum albumin exhibited 1.3 times higher drug loading than that of unmodified serum albumin. Octyl group was chosen to enhance interaction between the hydrophobic drug and micelle. *In vitro* cytotoxicity in Hepg2 cells exhibited a higher percentage of cell viability than Taxol® indicating better safety profile of octyl modified albumin-based paclitaxel micelles (Gong et al. 2009). Lauryl carbamate modified Inulin was used to prepare paclitaxel micelles. Inulin, a hydrophilic carbohydrate offers the advantage of being an alternate for poly(ethylene glycol)ylation, poly(ethylene glycol)ylation and has been found to exhibit production of anti-poly(ethylene glycol) Immunoglobulin M. Paclitaxel micelles of lauryl carbamate modified Inulin exhibited enhanced anticancer activity in mouse melanoma B16F10 cells at half the dose (Muley et al. 2016).

Polyion complex based micelle composed of cationic Pluronic F127 modified chitosan and anionic sodium cholate was reported for paclitaxel delivery. Sodium cholate, a bile salt was chosen to enhance drug loading in the Pluronic based micelles. Pluronic block copolymers are approved by USFDA but have the drawback of low drug loading. The synthesized polymer showed higher drug loading and lower critical micelle concentration. These micelles exhibited inhibition of growth of cancer cells in a drug-resistant MCF7 cell line (Ge et al. 2016). In another report, polyion complex based paclitaxel micelle using Pluronic F127 conjugated chitosan with cysteine was investigated. The polyion complex micelles exhibited four times higher drug loading as compared to Pluronic F127 based micelles. The polyion complex micelles exhibited pH-dependent and sustained release upon oral administration (Xu et al. 2015a, b).

Paclitaxel exhibited poor drug loading in poly (ethylene glycol)-b-poly (ε-caprolactone). Hence, to increase the drug loading, acyl ester bulky prodrugs of paclitaxel were synthesized. The prodrugs were incorporated into poly (ethylene glycol)-b-poly (ε-caprolactone) micelles. The hydrophobic modification of the drug resulted in enhanced solubilization of the prodrug in micelles. These prodrugs exhibited enhancement in drug blood levels as revealed in pharmacokinetics studies in rats. The prodrug micelles not only increased mean residence time and volume of distribution but also sustained the drug release. Chemical modification of the drug resulted in a shift of renal to non-renal clearance of paclitaxel (Forrest et al. 2008).

Paclitaxel micelles prepared using poly (ethylene glycol)-b-poly (lactide) have shown low drug loading. A copolymer of poly (2,4 vinylbenzyloxy)- N, N-diethylnicotinamide) with poly (ethylene glycol) was synthesized and used to prepare paclitaxel micelles. N-N diethylnicotinamide was chosen as a hydrotropic agent. The micelles exhibited 37% by weight drug loading as compared to 27% by weight in poly (ethylene glycol)-b-poly (lactide) micelles. The paclitaxel micelles were stable for 4 weeks at 25 °C whereas there was precipitation of drug in poly (ethylene glycol)-b-poly (lactide) micelles. There was higher antiproliferation of human cancer cells by micelles with synthesized copolymer than poly (ethylene glycol)-b-poly (lactide) micelles (Lee et al. 2007).

6.2.2 Mixed Micelle-Based Delivery System

Mixed micelles utilize two or more polymers to encapsulate the hydrophobic drug in the micelle. The advantages of these over simple micelles include lower critical micelle concentration, higher drug loading and the opportunity to incorporate multiple functionalities in the micellar structure (Attia et al. 2011). Paclitaxel mixed micelles were developed by combining Pluronic P123 with Pluronic F127. The rationale behind such combination relied on co-micellization of these Pluronic grades together during micelle formation. The critical micelle concentration was reduced in mixed micelles as compared to those prepared from Pluronic P123. The mixed micelles reduced half-maximal inhibitory concentration in A549 human lung adenocarcinoma cell lines by 4-fold as compared to Taxol® injection. Mixed micelles exhibited sustained release of paclitaxel. Forty-six percent drug was released after 2 h followed by slow release up to 12 h whereas Taxol® injection released the drug immediately in an aqueous medium containing sodium salicylate by dialysis method at 37 °C and 100 rpm. The mixed micelles exhibited better stability upon 10 times dilution with phosphate buffer solution and storage for 48 h. Ninety percent drug was retained in mixed micelles whereas 35% of the drug was precipitated in Taxol® injection. The precipitation of the drug was attributed to dissociation of the micelles upon dilution in case of Taxol® injection (Wei et al. 2009).

Sterically stabilized mixed micelles of paclitaxel were developed to increase the drug solubilization with lower critical micelle concentration. It was achieved by incorporation of hydrophobic phosphatidylcholine derivative along with poly (ethylene glycol)-g-distearoylphosphatidylethanolamine. The mixed micelles of this graft copolymer exhibited 1000-fold enhancement in paclitaxel solubility. The cytotoxicity study using human breast cancer cell and MCF 7 cell line exhibited comparable cytotoxicity of the mixed micelle with that of 10% paclitaxel solution in dimethyl sulfoxide (Krishnadas et al. 2003). Paclitaxel mixed micelles composed of poly(ethylene glycol-co-lactic acid) and D-α-tocopheryl polyethylene glycol 1000 succinate exhibited higher cellular uptake and effectiveness against multidrug-resistant tumor cells as compared to Genexol PM® (Fan et al. 2015).

Paclitaxel mixed micelles based on Soluplus® and Solutol® HS 15 were prepared for enhancement of drug solubilization. Solutol® HS15 (Murgia et al. 2013) and Soluplus® (Jin et al. 2015) were used in solid dispersions to increase drug solubilization and to prevent drug agglomeration. These mixed micelles exhibited higher plasma concentration in a pharmacokinetic study in Sprague Dawley rats than free paclitaxel. The mixed micelles exhibited higher cytotoxicity than free paclitaxel in MB-231 cell line (Hou et al. 2016). In another study, paclitaxel mixed micelles were prepared by a combination of α- tocopherol polyethylene glycol 1000 succinate and Plasdone® S630. α- tocopherol polyethylene glycol 1000 succinate was chosen for its lower critical micelle concentration of 0.02% w/w and Plasdone® S630® for higher drug solubilization capacity in solid dispersions. The mixed micelles released only 28% drug whereas there was 70% drug release from plain paclitaxel. About 50% drug was released in a sustained manner from mixed micelles upto 50 h. A pharmacokinetic study in Sprague Dawley rats exhibited five times increase in bioavailability of paclitaxel as compared to free paclitaxel. *In vitro* toxicity study in A459 cells revealed enhanced cytotoxicity of mixed micelles as compared to free paclitaxel (Hou et al. 2017).

6.2.3   Stimuli-Responsive Micelle-Based Delivery System

Micelles which can associate or dissociate with respect to change in stimulus are termed as stimuli-responsive micelles. The stimuli such as variation in pH, oxidation-reduction potential, magnetic field, and ultrasound have been used to fabricate stimuli-responsive micelles. pH-sensitive systems are the most popular among these stimuli since there is a variation in pH of fluids present in the milieu of different sites of the gastrointestinal tract, blood, tumor or body organs. The pH is acidic in the stomach which increases to neutral in the intestine. A pH of 7.4 exists in normal cells whereas it is slightly acidic around tumor cells (Shen et al. 2008). Intracellular and extracellular glutathione concentration varies significantly and hence could be used to trigger an oxidation-reduction reaction (Schafer and Buettner 2001). A pH and oxidation-reduction potential based micelles utilize internal milieu of the human body whereas ultrasound and magnetic field are external stimuli to trigger the release of drug from micelles.

pH-sensitive paclitaxel micelles composed of methoxy poly (ethylene glycol)-b-poly (β-aminoester) were reported. These micelles exhibited 13% and 45% drug release in pH 7.4 and pH 6.0 phosphate buffer solutions respectively after 3 days. The micelles with 16 mg/kg dose exhibited the lowest tumor volume after 10 days of treatment with B16F10 tumor-bearing mice as compared to saline solution (Han et al. 2009).

pH-sensitive paclitaxel micelles using methoxy poly (ethylene glycol)-poly (ε-caprolactone)-D-α-tocopheryl polyethylene glycol 1000 succinate were designed. The purpose of use of biodegradable polycaprolactone core was to dissolve the drug in the core whereas polyethylene glycol could bypass reticuloendothelial system uptake. D-α-tocopheryl polyethylene glycol 1000 succinate was chosen to overcome multidrug resistance. These micelles exhibited complete drug release in pH 5.0 whereas only 60% drug was released in pH 7.4 buffer solution after 140 h. The micelles exhibited higher antitumor activity in A549 cells than the free drug. The pharmacokinetic study in Sprague Dawley rats revealed 4-fold higher bioavailability than the free drug (Zhang and Zhang 2015).

N-octyl-N-(2-carboxyl-cyclohexamethenyl) chitosan-based paclitaxel micelles were investigated. High drug loading of 43.25% w/w was obtained in the micelles with a particle size of 145 nm and low critical micellar concentration of 42 μg/ml. The micelles exhibited pH-sensitive drug release. The pharmacokinetic study in Sprague Dawley mice exhibited longer half-life and a larger volume of distribution than Taxol® (Liu et al. 2011).

Polyethylene glycol-fluorenylmethyoxycarbonyl-disulfide-farnesyl thiosalicylic acid-based paclitaxel micelles were developed for intracellular delivery. Disulfide group was introduced to impart redox-sensitivity to the system. The micelles formed filamentous micelles with 10-fold lower critical micelle concentration values. The drug loading was found to be 34% w/w in the micelles. The micelles exhibited 74% drug release with 10 mM glutathione in release medium but micelles could release only 40% drug in release medium sans glutathione. The intracellular concentration of glutathione is in the millimolar scale whereas it is in the micromolar level outside the cells (Schafer and Buettner 2001). Pharmacokinetics and biodistribution studies revealed more retention of paclitaxel in the bloodstream with accumulation in tumor (Xu et al. 2015a, b).

Thermosensitive multi-arm star-shaped copolymer-based folate decorated paclitaxel micelles were formulated. The micelles were designed using 4-arm block copolymer consisting of poly (N-isopropylacrylamide-co-acrylamide), poly (ε-caprolactone) and folate conjugated methoxy poly (ethylene glycol)/polyethylene glycol. These blocks were chosen for thermosensitive, biodegradable and biocompatible properties respectively. The micelles exhibited temperature-sensitive drug release. The drug release from these micelles was faster at 40 °C than at 37 °C at the end of 100 h. The rationale was based on the fact that temperature in normal cells is ~37.5 °C whereas 40 °C in tumor environment (Rezaei et al. 2012).

Magnetically responsive paclitaxel micelles were designed. The magnetic material used as superparamagnetic iron oxide nanoparticles which were incorporated in poly (N-isopropylacrylamide-r-acrylamide) conjugated to an arginine-glycine-aspartic acid peptide. These micelles were formulated to have dual targeting. The peptide ligand was selected due to its affinity for $\alpha_v\beta_3$ integrin on cancer cells (Zitzmann et al. 2002) and magnetic force assisted drug release by virtue of iron oxide nanoparticles. The cellular uptake was improved for paclitaxel micelles with peptide in HeLa cells. Incorporation of magnetically responsive material in the micelles exhibited a synergistic effect on cellular uptake in a cell line (Lin et al. 2015).

Paclitaxel micelles based on monomethoxy poly (ethylene glycol)-b-poly (D, L lactide) have increased the maximum tolerated dose to 50 mg/kg from 20 mg/kg for Taxol® in murine B16 melanoma induced female SPF C57 BL/6 mice. This formulation exhibited poor drug internalization (Kim et al. 2001). Howard and coworkers reported local ultrasound-assisted paclitaxel drug internalization from paclitaxel polymeric micelles. Ultrasound enhanced the drug internalization by 20-fold followed by 90% tumor reduction in drug-resistant MCF7/ADmt cell lines (Howard et al. 2006).

6.2.4 Functional Micelle-Based Delivery System

Paclitaxel micelles having specific functionality have been developed in the past. These functionalities include avoidance of multidrug resistance; p-glycoprotein inhibition, biodegradability, mucoadhesive property and the use of carriers with intrinsic anticancer activity. The purpose of functional micelles were to enhance bioavailability, reduce toxicity, and treatment of multidrug-resistant cancer.

P-glycoprotein inhibitors such as brometetradrine (Zhang et al. 2017), N-octyl-N′-phthalyl-O-phosphoryl chitosan derivative (Qu et al. 2019), elacridar (Sarisozen et al. 2012) and Pluronic F127 (Dahmani et al. 2012) were used in compositions of paclitaxel micelles to overcome multidrug resistance. P-glycoprotein causes multidrug resistance by efflux of drug from cells (Gottesman et al. 1996). Various biodegradable polymers such as poly (ethylene oxide)-b-poly (ε-caprolactone) (Cai et al. 2007), block copolymers of poly (ethylene glycol) and poly (lactide) (Yang et al. 2009a, b) and poly (ethylene glycol)-poly (L-lactide)-poly {3(S) methyl-morpholine −2,5 dione} (Zhao et al. 2012) were used to prepare biodegradable paclitaxel micelles. Thioglycolic acid-modified octyl glycol chitosan (Huo et al. 2018) and chitosan grafted polycaprolactone (Almeida et al. 2018) were used to prepare mucoadhesive paclitaxel micelles. Excipients having intrinsic anticancer activity such as Rhein (Wang et al. 2019), methoxy poly(ethylene glycol) conjugated octacosanol (Chu et al. 2016), ethylene glycol-b-dendritic polylysine conjugated phenethyl isothiocyanate (Xiang et al. 2018) and poly (2-ethyl-2 oxazoline) vitamin E succinate and α-tocopherol polyethylene glycol 1000 succinate (Qu et al. 2018) were exploited for preparation of paclitaxel micelles to further substantiate its anticancer activity.

6.2.5 Targeted Micelle-Based Delivery System

Drug delivery to tumor cells can be improved by passive and active targeting of the nanocarriers. Systems that target the systemic circulation can be kept under a passive targeting system. It is a sort of passive process which exploits the natural biodistribution of the carrier and as a result, it eventually accumulates in certain organs, mainly liver and spleen (Yadav et al. 2019). In passive targeting to the tumor, the drug delivery carriers are exploited for their localization in the tumors by enhanced permeation and retention effects, phagocytosis and/or reticuloendothelial system. Nanocarriers having a size between 100 and 800 nm can accumulate in cancer cells due to leaky vasculature of blood vessels and inadequate lymphatic drainage of cancer cells (Iyer et al. 2006).

Active targeting refers to the attachment of certain components to the drug carrier which can direct the system to the target cells. This is based on molecular recognition phenomenon between ligand and receptor. The ligand possesses high affinity for specific receptor or surface determinant that has been overexpressed in the tumor cells. Overexpression of certain receptors like folate, transferrin, lipoprotein can be exploited as a target for cancer targeting. Certain tumor-associated antigens like melanoma-associated antigen, carcinoembryonic antigen, and other antigens e.g. CD44 are also overexpressed in tumor cells (Thanki et al. 2015). Various peptide-based ligands e.g. albumin, Arg-Gly-Asp peptide; carbohydrate-based ligands e.g. β-galactose; vitamin-based ligands e.g. folate; chimeric proteins and antibodies were used for active targeting. Attachments of these ligands or engineered homing devices or antibodies to drug carrier could facilitate the binding of drug carrier to the target cells. The concentration of these surface modifiers should be optimized in order to avoid reticuloendothelial system uptake or interaction with blood components (Kumar Khanna 2012).

Paclitaxel micelles were reported using mixed micelles composed of N-octyl-N′-trimethyl chitosan, poly (ethylene glycol stearate) and heparin sodium (Zhang et al. 2016), peptide ligand for epidermal growth factor receptor conjugated to poly (ethylene glycol)-distearophosphatidylcholine ethanolamine (Ren et al. 2015). Other systems have used dequalinium (Yao et al. 2011) and prostate carcinoma binding peptide 1 (Chen et al. 2016) for mitochondrial and prostate cancer targeting respectively.

Paclitaxel targeted micelles having dual response were designed wherein any two functionalities like pH sensitivity, biodegradability, active targeting and redox sensitivity have been combined. These micelles include folate conjugated poly (ethyleneimine-pluronic) and Pluronic L121 (Xu et al. 2012), Ala-Pro-Arg-Pro-Gly peptide conjugated poly (lactide-co-glycolide)-b-poly (ethylene glycol) copolymer

(Guo et al. 2015). Other systems include materials such as hyaluronic acid conjugated deoxycholic acid by a redox-sensitive disulfide bond (Li et al. 2015) and E-selectin binding peptide conjugated hyaluronic acid micelles (Han et al. 2017).

6.2.6   Crosslinked Micelle-Based Delivery System

Micelles structure breaks down causing burst release of drug upon dilution in the body. Crosslinking is used to increase the stability of micelles against such breakdown. Crosslinking enables intactness upon dilution of the micelle (O'Reilly et al. 2006). Core crosslinked paclitaxel micelles for improved stability have been reported using micelles composed of poly (ethylene glycol methyl ether acrylate)-b-poly (carboxyethyl acrylate) by disulfide crosslinking (Du et al. 2016) and crosslinked di/triblock copolymers of methoxy poly (ethylene glycol) and poly (ε-caprolactone) by potassium persulfate (Shuai et al. 2004). The details of various paclitaxel micelles are summarized (Table 6.1).

6.3   Liposomal Delivery System

6.3.1   Simple Liposome-Based Delivery System

Liposomes are spherical vesicular delivery systems composed of phospholipids or other amphiphilic lipids. The advantages of liposomes are bio-compatibility, ease of preparation and applicability for targeted delivery. Liposomes can be used as a drug carrier for both hydrophilic and hydrophobic drugs (Akbarzadeh et al. 2013).

Liposomal paclitaxel formulation composed of soy lecithin and cholesterol were reported. The particle size of liposomes and encapsulation efficiency were found to be 131 ± 30.5 nm and $94.5 \pm 3.2\%$ respectively. *In-vitro* drug release in pH 7.4 phosphate buffer saline was 56% after 96 h. Cytotoxicity study using HeLa cell line exhibited enhanced antitumor efficiency (Nguyen et al. 2017).

6.3.2   Polyethylene Glycolated Liposome-Based Delivery System

Polyethylene glycolation involves coating of poly (ethylene glycol) on liposomes to enhance hydrophilicity and blood circulation time. Liposomes with particle size more than 200 nm are directly taken up by the reticuloendothelial system. The larger particle size of liposomes causes faster elimination of liposomes. The circulation time can be prolonged by poly (ethylene glycol)ylation of liposomes (Sharma and Kumar 2015).

Table 6.1 Paclitaxel delivery system based on micelles

Serial number	System detail	Remarks	Reference
1	**P-glycoprotein inhibitor-based micelles**		
A	Brometetradine	Brometetradine was co-encapsulated in Solutol HS15 and TPGS mixed micelles. These micelles exhibited superior cytotoxicity and anticancer effect against MCF7/Adr cell line.	Zhang et al. (2016)
B	OPPC (N-octyl-N′-pthalyl-O-phosphoryl chitosan)	OPPC based self-assembled paclitaxel micelles exhibited higher loading than those with unmodified chitosan. These micelles showed higher cellular uptake in Caco-2 cells than Taxol®.	Qu et al. (2019)
C	Elacridar	Elacridar coloaded 1,2-Distearoyl-*sn*-glycero-3-phosphoethanolamine-N-[methoxy (polyethene glycol) -2000] micelles for intracellular delivery by enhanced permeation and retention effect avoided nonspecific distribution of micelles. This led to P-glycoprotein inhibition of only cancer cells. These micelles exhibited higher cytotoxicity in human cancer cell lines.	Sarisozen et al. (2012)
D	Pluronic F127	Paclitaxel mixed micelles composed of Pluronic F127 and low molecular weight all trans-retinoid acid exhibited higher drug loading and lower critical micellar concentration. *In situ* permeability study through rat intestine exhibited 6-fold enhancement as compared to that with Taxol®.	Dahmani et al. (2012)
2	**Targeted micelle-based delivery system**		
A	Heparin based system	N-octyl-N-trimethyl chitosan-polyethene glycol 100 stearate-based paclitaxel micelles showed high drug loading. Heparin sodium coated micelles exhibited higher cellular uptake with longer circulation time in rats.	Zhang et al. (2016)
B	Dual responsive system	pH-responsive and cancer cell recognition programmed mixed micelles composed of (D- α-tocopheryl poly-ethylene glycol 1000 succinate -b-poly(β-aminoester) and aptamer exhibited stability at pH 7.4 but released the drug at tumor pH.	Zhang et al. (2015)
C	Mitochondria-targeted system	Paclitaxel micelles composed of poly (ethylene glycol-2000)-g-distearoylphosphotidylethanol amine, tocopheryl polyethylene glycol succinate with dequalinium had a higher solubility than 1 mg/ml, exhibited delayed-release and enhanced cellular uptake in MCF7/Adr cells than that of Taxol®.	Yao et al. (2011)

Polyethylene glycolated liposomal formulation of paclitaxel composed of soybean phosphatidylcholine and 1, 2-distearoyl-sn-glycero-3-phosphoethanolamine [methoxy (polyethylene glycol)-2000], Cholesterol, Tween 80 and 3-(4, 5-Dimethyltiazol-2-ly)-2, 5-diphenyl- tetrazolium bromide was reported. This liposomal paclitaxel formulation resulted in enhancement of solubility by 2000 times as compared to conventional liposomes. The pharmacokinetic study on Sprague Dawley rats revealed a 3-fold increase in biological half-life in poly(ethylene glycol)ylated liposomal formulation of paclitaxel (Yang et al. 2007).

6.3.3 Targeted Liposomal Delivery System

The targeted delivery system is a strategy of delivering drugs or medications only to its site of action and not to the other cells, tissues or organs. The concentration of medication is higher in particular cells, tissues or organs as compared to others. Targeting systems have the advantages of improving the therapeutic efficacy and reduction of side effects of drugs (Gupta and Sharma 2011). Passive targeting involves targeting of drugs to the systemic circulation of the body. In this method, drug targeting occurs because of the body's natural response to physicochemical characteristics of the drug or drug carrier system. Enhanced permeability and retention effect passively target the drug to the tumor cells (Bazak et al. 2014).

Active targeting involves the use of a ligand that has an affinity for specific receptors which are over-expressed in cancerous organs, tissues, and cells. Transferrin, peptides, antibodies, nucleic acids, and their fragments are used as a ligand for active targeting (Atar et al. 2018).

Targeting of paclitaxel liposomal formulation to the lungs composed of phospholipon 90H and Tween-80 were reported. The particle size of liposomes and entrapment efficiency were found to be 8.166 ± 0.459 µm and $92.2 \pm 2.56\%$ respectively. These liposomes exhibited 60.26% drug release in 24 h. *In vivo* pharmacokinetic study in rabbits revealed a 1.6-fold increase in half-life and the significant decrease in plasma clearance than Taxol®. Liposomes of size higher than 8 µm could be the potential carrier for paclitaxel for the treatment of lung cancer (Wei et al. 2014).

Targeting of paclitaxel liposomal formulation to mitochondria composed of triphenyl phosphene, 6-bromohexanoic acid, dicyclohexylcarbodiimide, 4-dimethylamino-pyridine were reported. D- α- tocopheryl poly-ethylene glycol 1000 succinate-triphenylphosphine conjugate was used as the mitochondrial targeting material. The particle size of liposomes and entrapment efficiency were found to be 84.67 ± 0.61 nm and $86.27 \pm 3.15\%$ respectively. The liposomes exhibited higher anticancer *in-vitro* activity in human lung cancer A549 cells. The paclitaxel liposomes could enhance the cellular uptake and were selectively accumulated in mitochondria. The increased uptake of the liposomes enhanced the anticancer efficacy (Zhou et al. 2013).

A targeted paclitaxel liposomal formulation composed of dipalmitoyl phosphatidylcholine, dimyristoyl phosphatidylglycerol/monomethoxy, polyethylene glycol 2000, distearoyl phosphatidylethanolamine and folate-poly(ethylene glycol) 3350-distearyl phosphatidylethanolamine was reported. The particle size of liposomes was found to be 97.1 nm. Cytotoxicity study in the human carcinoma cell line, KB exhibited 4-fold and 2.5-fold lower half maximal inhibitory concentration than non-targeted liposomes and Taxol® respectively. The clearance of Taxol® was 16.5 times higher than folate receptor-targeted liposomes (Wu et al. 2006).

A targeted liposomal paclitaxel formulation composed of dipalmitoyl-sn-glycero-3-phosphocholine and 1, 2-dioleoyl-3-trimethylammonium-propane, and hyaluronic acid was developed. The particle size of liposomes and encapsulation efficiency were found to be 106.4 ± 3.2 nm and 92.1 ± 1.7% respectively. *In vitro* drug release in pH 7.4 phosphate buffer saline was 95% in 40 h. *In vivo* antitumor efficacy and biodistribution studies were performed on 4T1 tumour-bearing animal models exhibited higher accumulation of the drug in tumor (Ravar et al. 2016).

6.3.4   Stimuli Sensitive Liposomal Delivery System

Magnetically assisted delivery is a novel approach for delivery of drugs using engineered smart microcarriers which can overcome a number of limitations facing current methods of delivering medicines. The drug is formulated into a pharmaceutically stable formulation which is usually injected through the artery that supplies the target organ or tumor in the presence of an external magnetic field (Koppisetti and Sahiti 2011). Focused ultrasound is a method to increase the permeability of drugs into the blood-brain barrier to promote drug delivery to specific brain regions. It is a potential method of delivery of the drugs into the brain for brain tumors (Burgess et al. 2015).

Focused ultrasound was used to enhance permeation through the blood-brain barrier and blood tumor barrier. 1, 2-dipalmitoyl-sn-glycero-3-phosphocholine, 1, 2-distearoyl-sn-glycero-3-phosphoethanolamine and cholesterol were used to prepare liposomal paclitaxel formulation. Focused ultrasound exposure with a 10 ms pulse length and 1 Hz pulse repetition frequency at 0.64 MPa peak rarefactional pressure was used. *In-vivo* study in rats showed a 3-fold increase in blood drug level from the ultrasound-assisted liposomal formulation as compared to liposomal paclitaxel formulation (Shen et al. 2017). Paclitaxel loaded magneto liposomes composed of 1, 2-dipalmitoyl-sn-glycero-3-phosphocholine and 1-palmitoyl-2-oleoyl-sn-glycero-3-phospho-rac-glycerol were reported. The encapsulation efficiency of 83 ± 3% was obtained. *In-vitro* study under magnetic field on HeLa cell line revealed that 89% of cells were killed (Kulshrestha et al. 2012).

6.3.5   Mucoadhesive Liposomal Delivery System

Mucoadhesion is commonly defined as the adhesion between two materials, at least one of which is a mucosal surface. The mucoadhesive property of a delivery system is dependent upon a variety of factors, including the nature of the mucosal tissue and the physicochemical properties of the polymeric formulation (Shaikh et al. 2011).

Mucoadhesive liposomes composed of chitosan, thioglycolic acid and Pluronic F127 were used for paclitaxel oral delivery. The particle size of liposomes was found to be 121.4 nm. *In-vivo* study on rats revealed a 3-fold higher bioavailability than non-mucoadhesive liposomes. Enhanced mucoadhesion was observed which increased the retention time of the chitosan, thioglycolic acid and Pluronic F127 containing liposomal paclitaxel formulation (Liu et al. 2018).

6.4   Nanoparticle-Based Delivery System

Nanoparticles can be defined as submicron (10–1000 nm) particles or systems derived from polymers, lipids, and/or inorganic materials. Nanoparticle-based systems can overcome some of the limitations associated with conventional cancer therapy (Xin et al. 2017). The limitations include poor solubility of drugs, low specificity to the tumor cells, inability to accumulate in tumor cells, rapid removal of drugs from the systemic circulation and non-specific targeting. Nanoparticles have the potential to increase bioavailability and enhance the delivery of poorly soluble anticancer drugs like paclitaxel (Tran et al. 2017). Due to small size, high surface to volume ratio, ability to modulate drug release and the possibility of surface modification, nanoparticles lead to enhanced accumulation of drugs in tumor cells. Surface modification, use of internal and external stimuli allows selective targeting of drugs to tumor cells. Nanoparticulate systems are also useful to overcome multidrug resistance facilitated by P-glycoprotein efflux pumps in tumor cells (Shi et al. 2017).

6.4.1   Single Nanocarrier Based System

Paclitaxel nanoparticles fabricated using a single polymer have been discussed in this subsection. The carriers can be natural e.g. albumin (Li et al. 2012); semisynthetic e.g. chitosan (Gupta et al. 2017); and synthetic e.g. poly (styrene-co-maleic acid) (Dalela et al. 2015), poly (n-butyl cyanoacrylate) (Huang et al. 2007). Use of biodegradable polymers like poly (lactic-co-glycolic acid) (Le Broc et al. 2013), poly(ε-caprolactone) (Zhu et al. 2010), poly(lactide) (Zhang and Feng 2006) were also reported because of their biocompatibility and non-toxicity. Paclitaxel loaded poly (lactic-co-glycolic acid) nanoparticles were formulated using Poloxamer 188 as a stabilizer. Paclitaxel nanoparticles exhibited sustained release of the drug over a period of 4 days. The formulation was safe for intravenous administration. Pharmacokinetic studies in rats have verified longer retention of drug in the systemic circulation. Mean residence time and plasma half-life were increased by 77 and 45-fold respectively as compared to the pure drug (Mittal et al. 2019).

The chitosan-based biocompatible nanoparticulate system was prepared to achieve better delivery of paclitaxel which has shown burst release followed by sustained release of the drug. The half-maximal inhibitory concentration was reduced by 1.6-fold as compared to the pure drug which was attributed to the higher uptake of nanoparticles. Therapeutic efficacy was induced by apoptosis induction. The drug loading was found to be 11.57% (Gupta et al. 2017). Polygeline has been used as a carrier for paclitaxel delivery. Polygeline based paclitaxel nanoparticles were formulated as a reconstitutable lyophilized powder for intravenous administration. The drug loading was found to be 12.04%. It has shown a relative bioavailability of 89.83% as compared to Abraxane® (Xiong et al. 2019).

6.4.2   Multiple Nanocarriers-Based System

It covers the delivery systems of paclitaxel which contain more than one polymer or carriers. Polymers can remain as separate entities as in the case of layer-by-layer based albumin-bound paclitaxel nanoparticles. The nanoparticle was fabricated by alternate deposition of poly(arginine) and poly (ethylene glycol)-block-poly (L-aspartic acid) onto the drug-albumin conjugate. The drug loading capacity was found to be 48% w/w. The protective layers have improved the colloidal stability as well as biodistribution of albumin-bound paclitaxel nanoparticles. Mean residence time and plasma half-life were increased 5 and 4-fold respectively as compared to albumin-drug conjugate. These multilayer conjugates exhibited enhanced cytotoxicity and apoptosis properties (Ruttala et al. 2017).

Two or more polymers can be copolymerized and loaded with paclitaxel to design paclitaxel loaded nanoparticles. Methoxy poly (ethylene glycol)-poly(ε-caprolactone) nanoparticles loaded with paclitaxel have shown improved anti-glioblastoma activity than Taxol®. The drug loading was found to be 8.2%. High accumulation of the drug in brain cells was observed. The mean survival time was increased to 28 days which was 20 days in Taxol® treated animals. (Xin et al. 2010). Poly (ethylene glycol)ylated poly(ε-caprolactone) based paclitaxel nanoparticles were designed which exhibited similar biodistribution compared to Taxol®. Polyethylene glycol and poly(ε-caprolactone) side chains of the nanoparticles could avoid opsonization (Colombo et al. 2015).

Paclitaxel loaded nanoparticles composed of poly(ε-caprolactone)–co-α-tocopheryl polyethylene glycol 1000 succinate were formulated. The nanoparticles had 6% drug loading. The nanoparticles exhibited sustained drug release over a period of 144 h in pH 7.4 phosphate buffer solution. The half-maximal inhibitory concentration for the nanoparticles was found to be 7.8-fold lower than Abraxane®. *In vivo* pharmacokinetic study in rats revealed that the half-life was increased by 11 and 1.2-fold as compared to Taxol® and Abraxane® respectively. The formulation could increase the systemic circulation and lower the elimination of drug than its pre-existing marketed formulations (Bernabeu et al. 2014).

6.4.3 Stimuli Sensitive Nanosystem

pH sensitivity can be incorporated into either the corona or core of the nanoparticles. The core-shell type nanoparticles are formulated with pH-dependent solubility and lower critical solution temperature which exhibits drug release at tumor pH (Sutradhar and Amin 2014).

Various polymers have been reported to exhibit pH-responsive release of paclitaxel, such as poly [2-(dimethylamino) ethyl methacrylate-co-methacrylic acid] (Lee et al. 2011), O-carboxymethyl chitosan (Sahu et al. 2011), poly (acrylic acid)-poly(ethylene oxide) (Nguyen et al. 2019), poly(4-vinyl pyridine) (Contreras-Cáceres et al. 2017), poly (lactic acid)-block-poly(ethylenimine) and poly (ethylene glycol)-block-poly (l-aspartic acid sodium salt) (Jin et al. 2018a, b). Chitosan modified poly (lactic-co-glycolic acid) nanoparticles of paclitaxel were designed to overcome the initial burst release of poly (lactic-co-glycolic acid) nanoparticles. The drug loading was found to be 6.42%. Chitosan modified nanoparticles were pH-responsive and have shown faster release in pH 5.5 than at pH 7.4. Enhanced cytotoxicity and cellular uptake in MDA-MB-231 cells were due to chitosan modification (Lu et al. 2019).

Thermoresponsive delivery systems have been explored for paclitaxel where a change in temperature had triggered the release of the drug from the system. A novel thermoresponsive triblock copolymer comprised of methoxy poly (ethylene glycol), poly(octadecanedioicanhydride) and D, L-lactic acid oligomer was developed for paclitaxel delivery. The nanoparticles were formulated as a freeze-dried powder which is suitable for peritumoral or intratumoral injection upon redispersion with water at ambient temperature. The drug loading was found to be 0.90%. Paclitaxel accumulation in tumor cells was enhanced whereas systemic exposure was reduced. Plasma half-life and area under the curve were increased by 4.2 and 1.8-fold respectively as compared to Taxol® in the pharmacokinetic study (Liang et al. 2017).

Paclitaxel conjugated recombinant chimeric polypeptide-based nanoparticles have been reported. The nanoparticles self-assemble into spherical nanoparticles of size below 100 nm. The *in vitro* dissolution study of the nanoparticles did not release the drug in pH 7.4 phosphate buffer solution but completely released the drug within 6 h in pH 6.5 carbonate buffer solution. The systemic exposure of paclitaxel was increased by 2-fold as compared to that of Abraxane®. The tumor uptake of the nanoparticles was doubled as compared to *that with* Abraxane®. Abraxane® treated prostate cancer animals could survive less than 60 days, whereas nanoparticle treated animals survived more than 70 days with almost complete tumor reduction (Bhattacharyya et al. 2015).

The reducing environment of tumors serves as a unique internal signal that allows redox-responsive nanoparticles to degrade in tumor cells and release the loaded drug. Rapid drug release can be attributed to the high concentration of glutathione in cancer cells (Guo et al. 2018). Cross-linked polymeric nanocarriers made up of poly (lactic acid) core and glutathione-responsive disulfide cross-linked poly (oligo-ethylene glycol) corona was loaded with paclitaxel. The drug loading capacity was found to be 20%. *In vitro* release of paclitaxel from the nanoparticles was enhanced in the presence of glutathione in acidic pH. The half-maximal inhibitory concentration showed an 11-fold increase activity in OVCAR-3 cells compared to Taxol® which indicated better antitumor efficacy of the nanoparticles (Samarajeewa et al. 2013).

6.4.4 Functional Polymeric Nanoparticles

Multidrug resistance is the major reason for the failure of conventional anticancer therapy. P-glycoprotein mediated drug efflux from cancer cells is one of the important mechanisms of multidrug resistance (Mansoori et al. 2017). Various nanoparticulate systems of paclitaxel have been formulated to overcome this issue. Paclitaxel was loaded in poly [2-(dimethylamino)-ethyl methacrylate-co-methacrylic acid] to obtain water-soluble nanoparticles. These nanoparticles were able to reverse the multidrug resistance in two p-glycoprotein expressing breast cancer cell lines i.e. MCF/ADR, MT3/ADR cell lines compared to pure paclitaxel. This was due to the uptake of nanoparticles by endocytosis, bypassing the P-glycoprotein efflux pump (Lee et al. 2011).

Poly(lactide)-D-α-tocopheryl polyethylene glycol 1000 succinate copolymer-based nanoparticles of paclitaxel were developed, where D-α-tocopheryl polyethylene glycol 1000 succinate acts as a P-glycoprotein inhibitor. The drug loading was found to be 5.2%. These nanoparticles exhibited much higher *in vitro* cytotoxicity in HT-29 cells as compared to Taxol® (Zhang and Feng 2006).

P-glycoprotein inhibition potency of tannic acid was used by formulating tannic acid nanoparticles of paclitaxel. Tannic acid has lowered the P-glycoprotein expression; inhibited metastasis, clonogenic formation, proliferation; decreased expression of multidrug-resistant protein; and increased expression of tumor suppressor proteins of MDA-MB-231 breast cancer cells (Chowdhury et al. 2019).

Mucoadhesive polymeric nanoparticles relish localization and prolonged residence at the site of absorption (Boddupalli et al. 2010). Paclitaxel conjugated trimethyl chitosan nanoparticles were developed for oral and intravenous administration of the drug. Intestinal transport of paclitaxel was promoted by the mucoadhesive property of trimethyl chitosan. A pharmacokinetic study in tumor-bearing mice revealed prolong blood retention and improved tumor accumulation of paclitaxel. Further modification of nanoparticles by folic acid has improved the overall antitumor efficacy (He and Yin 2017).

6.4.5 Targeted Polymeric Nanoparticles

Chitosan modified poly (lactic-co-glycolic acid) nanoparticles were loaded with paclitaxel. Chitosan modification produced a positively charged surface which resulted in increased uptake of the nanoparticles into lung cancer cell line A549.

Moreover, a lung-specific biodistribution was achieved as compared to Taxol®. The transient formation of aggregates in plasma increased the size of the nanoparticles upon intravenous administration. Electrical interaction between positively charged nanoparticles and negatively charged tumor vasculature enhanced the accumulation of nanoparticles in lung tumor (Yang et al. 2009a, b).

Various active targeting based polymeric nanoparticles have been reported for paclitaxel. Paclitaxel loaded cationized polyacrylamide nanoparticles engineered with recognition peptide VRPMPLQ were designed. The presence of positive charge and the recognition peptide directed the nanoparticles to dysplasia regions in the colon where sialic acid was overexpressed. The uptake of the peptide containing nanoparticles was increased 10-fold as compared to non-peptide nanoparticles (Tiwari et al. 2017). CD44 is a non-kinase transmembrane glycoprotein that is over-expressed in certain cancers like gallbladder cancer, breast cancer and ovarian cancer (Chen et al. 2018). Hyaluronic acid serves as a targeting ligand for overexpressed CD44. Hyaluronic acid decorated serum albumin-based paclitaxel nanoparticles were designed for targeting ovarian cancer cells. The encapsulation efficiency was found to be 90%. The uptake of the nanoparticles was attributed to receptor-mediated endocytosis. A fluorescent dye fluorescein isothiocyanate was labelled to the nanoparticles for *in vitro* imaging of the tumor. The half-maximal inhibitory concentration was decreased by 4.35-fold in hyaluronic acid conjugated paclitaxel as compared to free paclitaxel (Edelman et al. 2019).

Ganipineni and co-workers have compared passive, active, magnetic and hybrid targeting strategies of paclitaxel for glioblastoma treatment. For passive targeting, paclitaxel and superparamagnetic iron oxide loaded poly (lactic-co-glycolic acid) nanoparticles were formulated. Accumulation of nanoparticles in the U87MG tumor model was due to enhanced permeability and retention effect. Active targeting was based on the fact that $\alpha_v\beta_3$ integrin is overexpressed in glioblastoma and Arg-Gly-Asp tripeptide was served as a targeting ligand. So, the surface of the nanoparticles was modified with Arg-Gly-Asp to achieve active targeting. The inherent magnetic property of superparamagnetic iron oxide was explored for magnetic targeting. Hybrid targeting was achieved by adding active and magnetic targeting. The best therapeutic effect was obtained in magnetic targeting followed by the hybrid targeting (Ganipineni et al. 2019). The other polymeric nanocarrier based targeted delivery systems of paclitaxel is given (Table 6.2).

6.4.6   Inorganic Nanoparticle-Based System

Inorganic nanoparticles are developed lately in addition to polymeric or organic nanoparticles. Typically, the inorganic nanoparticles are made up of inorganic core and organic shell (Bayda et al. 2018). The core contains metals like iron oxide, gold, aluminium or non-metals like carbon or silica. The shell protects the core from chemical interactions during circulation and/or acts as a substrate for conjugation with biomolecules, such as antibodies, proteins, and oligonucleotides. Inorganic

Table 6.2 Polymeric nanocarrier based targeted delivery of paclitaxel

Serial number	Systems	Remarks	References
1	**Passive targeting based systems**		
A	Paclitaxel loaded poly (ethylene glycol)-block poly (trimethylene carbonate – nanoparticle	The nanoparticles exhibited sustained release and slightly higher *in vivo* antitumor efficacy than Taxol®. The uptake of nanoparticles to glioblastoma was enhanced due to passive targeting.	Jiang et al. (2011)
B	Association of poly (ethylene glycol) -paclitaxel with free paclitaxel [poly (ethylene glycol)-paclitaxel/paclitaxel]	*In vitro* cytotoxicity was more in poly (ethylene glycol)-paclitaxel/paclitaxel than poly (ethylene glycol)-poly(lactide)/paclitaxel. *In vivo* cellular uptake and antitumor efficacy was more in poly (ethylene glycol)-paclitaxel/ paclitaxel than poly (ethylene glycol)- poly(lactide)/paclitaxel and Taxol® due to efficient passive targeting.	Lu et al. (2014)
2	**Active targeting-based systems**		
A	Wheat germ agglutinin conjugated isopropyl myristate incorporated poly(lactide-co-glycolide) nanoparticles	The nanoparticles exhibited stronger *in vitro* cytotoxicity due to efficient cellular uptake via Wheat germ agglutinin receptor-mediated endocytosis and isopropyl myristate-facilitated release of paclitaxel from the nanoparticles.	Mo and Lim (2005)
B	Poly (vinyl benzyl lactonamide) incorporated paclitaxel loaded poly(lactide-co-glycolide) nanoparticles	Enhanced cellular uptake and cytotoxicity were observed due to receptor-mediated endocytosis of nanoparticles to hepatic cancer cells. Poly (vinyl benzyl lactonamide) carried galactose residues which were served as a ligand.	Wang et al. (2012)
C	Follicle-stimulating hormone β 81–95 peptide conjugated poly (ethylene glycol)-poly(lactide) nanoparticles	Specific paclitaxel uptake by follicle-stimulating hormone receptor expressing ovarian cancer cells. Side effects were reduced as compared to commercial paclitaxel.	Zhang et al. (2013)
D	15F and regulon-peptide modified polyester based nanoparticles	Internalization of nanoparticles was mediated by low-density lipoprotein receptor-1 in the brain. Cellular uptake by glioma cells was improved due to peptide functionalization.	Di Mauro et al. (2018)
E	Alendronate modified paclitaxel-polydopamine nanoparticles	The nanoparticles exhibited specificity towards osteosarcoma cells due to alendronate modification. *In vivo* antitumor activity was better as compared to Taxol® with lesser side effects.	Zhao et al. (2019)

nanoparticles offer good stability and biocompatibility. These can alter the drug release profile and provide the site of attachment for targeting molecules. Additionally, few metallic nanoparticles can also be used for contrast imaging and photodynamic therapy (Núñez et al. 2018).

Paclitaxel was loaded into mesoporous silica nanoparticles of three different pore sizes and was evaluated for *in vitro* and *in vivo* antitumor activity. The *in vitro* release of paclitaxel was dependent on the pore size. Early and late apoptosis was directly proportional to pore size and was boosted as compared to free paclitaxel. Pharmacokinetic parameters were similar to Taxol® (Jia et al. 2013). Paclitaxel was loaded into mesoporous carbon spheres and surface modified by the folate-polyethyleneimine function. The enhanced internalization of nanoparticles by Caco-2 cell lines was due to folate conjugation. The nanoparticles also demonstrated improved oral bioavailability and diminished gastrointestinal toxicity (Wan et al. 2015). Other inorganic nanoparticle-based delivery systems of paclitaxel are given (Table 6.3).

6.4.7   Nanocrystal-Based System

Drug nanocrystals are nanosized, carrier-free, crystalline particles of the drug, usually produced in the form of nanosuspensions and stabilized by surfactants or polymers. Nanocrystals have several advantages over other nanoparticulate systems such as high drug loading capacity, improved solubility, dissolution, stability, and long circulation time. Nanocrystals also enjoy excellent commercialization potentials. So, development of nanocrystals of paclitaxel has gained the interest of researchers in recent years (Lu et al. 2015).

Paclitaxel nanocrystals were prepared using Pluronic F127 as a stabilizer for hyperthermic intraperitoneal chemotherapy of ovarian cancer. The *in vitro* cytotoxicity against human ovarian carcinoma cell line, SKOV-3, was equivalent to Taxol®. The maximum tolerated dose of the nanocrystals was similar to Taxol® however; the rats treated with the nanocrystals recovered faster after hyperthermic intraperitoneal chemotherapy treatment (De Smet et al. 2012). In another study, Pluronic F127 grafted chitosan copolymer was used as a stabilizer for the development of paclitaxel nanocrystals. Enhanced accumulation of paclitaxel in the Caco-2 cell line as a result of P-glycoprotein inhibitory potential of the stabilizer was observed. The nanocrystals have shown 12.6-fold enhanced absorption as compared to Taxol® (Sharma et al. 2015).

Polyethylene glycol stabilized and Arg-Gly-Asp peptide-functionalized paclitaxel nanocrystals were reported. Nanocrystals were coated with polydopamine which gave a site of attachment to polyethylene glycol and Arg-Gly-Asp peptide. Due to surface modification, the uptake and accumulation of nanocrystals were increased in adenocarcinomic pulmonary cells. *In vivo* study in pulmonary tumor-bearing mice resulted in enhanced antitumor efficacy in RGD peptide-functionalized paclitaxel nanocrystals as compared to that with Taxol® (Huang et al. 2019).

Table 6.3 Inorganic nanoparticle-based delivery systems of paclitaxel

Serial number	Systems	Remarks	References
1	Paclitaxel loaded chitosan oligosaccharide stabilized gold nanoparticles	The gold nanoparticles have shown sustained drug release at acidic pH and a strong cytotoxic effect against MDA-MB-231 cells. Gold nanoparticles also have the potential to be used as an optical contrast agent for photoacoustic imaging.	Manivasagan et al. (2016)
2	Poly (ethylene glycol)-carboxyl–poly(ε-caprolactone) magnetic nanoparticles	In the presence of a magnetic field, the nanoparticles were targeted to tumor cells. The nanoparticles were less toxic to normal cells and found to be biocompatible.	Li et al. (2017a, b, c)
3	Gadolinium arsenite nanoparticles	Conjugations of arsenic trioxide and paclitaxel have shown a synergistic effect against paclitaxel-resistant HCT 166. Intracellular uptake of paclitaxel by resistant cells was enhanced.	Chen et al. (2017a, b)
4	Pectin conjugated magnetic graphene oxide nanoparticles	The nanoparticles have shown high drug loading capacity i.e. 36% and compatibility. *In vitro* release was more in endosomal cancer cell medium than normal physiological pH.	Hussien et al. (2018)
5	Hyaluronic acid (HA) functionalized MHANs (mesoporous hollow alumina nanoparticles)	Due to HA modification, MHAN was targeted to liver cancer cells. *In vivo* cytotoxicity in nude mice was better than pure paclitaxel and non-functionalized MHANs	Gao et al. (2019)

6.5   Lipid-Based Delivery System

Lipid-based systems possess the advantage of incorporation of the hydrophobic drug in amorphous form. It helps to enhance the solubility of the drugs. It also helps improving permeability by providing lipophilicity to the delivery system. Lipid systems can be prepared using solid lipids or a mixture of solid lipids and liquid lipids. The former is referred to as solid lipid nanoparticles and the later one as nanostructured lipid carriers. Solid lipid nanoparticles, as well as nanostructured lipid carriers of paclitaxel, have been formulated. These include simple lipid-based systems, poly (ethylene glycol)ylated lipid-based systems, oral systems, local systems, pulmonary systems, targeted systems and stimuli-responsive systems.

6.5.1   Simple Lipid-Based Delivery System

Simple lipid-based systems of paclitaxel have been prepared using lipid and surfactants. Paclitaxel nanostructured lipid particles were prepared using fluorescein isothiocynate-octadecylamine using 0.1% Poloxamer 188 aqueous solution by hot melt high-pressure homogenization. The nanoparticles were freeze-dried and were found to have 89 nm size and more than 30 mV zeta potential. The nanoparticles released ~ 80% drug within 48 h but it was about 50% from nanoparticles prepared without octadecyamine. *In vitro* cellular uptake in A549 cells was found to be higher with octadecylamine nanoparticles (Miao et al. 2015).

In another study, paclitaxel solid lipid nanoparticles composed of trilaurin, phosphotidylcholine were prepared using hot homogenization method. The nanoparticles had a size of ~160 nm. These nanoparticles exhibited sustained release in human plasma over 24 h whereas burst release was observed in Taxol® (Xu et al. 2013).

6.5.2   Lipid-Based Oral Drug Delivery System

Glyceryl monostearate based solid lipid nanoparticles of paclitaxel were prepared by emulsification and evaporation method. Wheat germ agglutinin was conjugated to these solid lipid nanoparticles which exhibited higher cytotoxicity than non-conjugated nanoparticles in A549 cell line. The conjugated nanoparticles exhibited a 2-fold increase of area under the plasma curve and mean residence time during *in vivo* pharmacokinetic study in rats as compared to the plain drug (Pooja et al. 2016).

6.5.3   Lipid-Based Topical Drug Delivery System

Paclitaxel loaded solid lipid nanoparticles composed of Carbapol 940, stearic acid, egg lecithin and Pluronic PF 68 were reported for topical delivery. Particle size and pH of the formulation were found to be 78 nm and 5.4 respectively. The gel released 70% drug in 24 h in pH 7.4 phosphate buffer solution. The nanoparticle-based gel showed higher permeability through the dialysis membrane than that of plain drug-loaded gel. *In vivo* study in mice exhibited higher antitumor efficacy (Bharadwaj et al. 2016). Another report for topical delivery of paclitaxel was based on nanostructured lipid carriers composed of glyceryl behenate, capric-caprylic triglycerides mixture with a surfactant cetylpyridinium chloride. The particle size and drug loading were found to be 270–315 nm and 3% respectively. The higher zeta potential of 20 mV was observed. The positive zeta potential helped to increase the permeability of drug through the skin (Tosta et al. 2014).

6.5.4   Lipid-Based Pulmonary Drug Delivery System

Pulmonary lipid nanoparticles are meant for lung cancer treatment. Nanostructure lipid carriers of paclitaxel were designed with a surfactant that can inhibit P-glycoprotein mediated drug efflux. These nanoparticles exhibited sustained release in pH 7.4 phosphate buffer solution. Nanoparticles of Tween 20 exhibited the highest cellular uptake in Caco-2 cell line. *In vivo* studies of nanoparticles exhibited better therapeutic activity and lung deposition (Kaur et al. 2016).

6.5.5   Lipid-Based Targeted Delivery System

Passive and active targeted lipid-based systems were reported for paclitaxel delivery. Positively charged solid lipid nanoparticles of paclitaxel composed of stearylamine, sodium behenate and polyvinyl alcohol were prepared. These nanoparticles exhibited only 1% drug release after 12 h whereas paclitaxel solution released the drug completely in a phosphate buffer solution with 0.1% v/v Tween 80 solution. Cytotoxicity study in hCMEC/D3 cells exhibited better anticancer activity with improved blood-brain barrier permeation (Chirio et al. 2014). Solid lipid nanoparticles of paclitaxel were prepared using tristearin, hydrogenated soy phosphatidylcholine. The surface of the solid lipid nanoparticles was modified with lactoferrin. The size of nanoparticles and zeta potential were found to be 250 nm and 3.7 mV respectively. Cytotoxicity study in BEAS-2B cell line exhibited a 4-fold reduction in half-maximal inhibitory concentration in lactoferrin conjugate nanoparticles as compared to non-conjugated nanoparticles. *In vivo* distribution study in albino rats exhibited 1.6-fold higher uptake of the drug as compared to non-targeted nanoparticles (Pandey et al. 2015).

Folate-conjugated poly (ethylene glycol)-cholesteryl hemisuccinate was used to prepare a paclitaxel targeted delivery system. Paclitaxel nanostructured lipid carrier was prepared using oleic acid and stearic acid from the conjugated polymer. Transmission electron microscopy revealed 100 nm size for these carriers. The biodistribution study in albino rats exhibited higher deposition of targeted particles in the kidney (Ucar et al. 2017). Paclitaxel solid lipid nanoparticles prepared using conjugated polymer of Pluronic P85, 1, 2 distearoyl–sn-glycero-3-phosphatidylethanolamine and hyaluronic acid by hot homogenization method were developed. *In vitro* drug release study revealed complete drug release from the plain drug but the nanoparticles exhibited sustained release up to 48 h in a phosphate buffer solution with 0.5% w/v Tween 80. The targeted nanoparticles exhibited a 4-fold increase in drug concentration in tumor than from free drug in balb/c mice (Wang et al. 2017).

Paclitaxel brain targeted solid lipid nanoparticles were developed. These were composed of transferrin conjugated polyethylene glycol, oleic acid, compritol ATO 888 and cholesterol. These were prepared by the solvent evaporation method. The size of the nanoparticles and drug loading were found to be 160 nm and 8.6%

respectively. *In vitro* release study showed that the drug release was in a sustained manner up to 75 h in pH 7.4 phosphate buffer saline. 6-fold reduction of half maximal inhibitory concentration in HL 60 cells was shown by targeted nanoparticles as compared to non-targeted nanoparticles (Dai et al. 2018).

6.5.6 Lipid-Based Stimuli Sensitive Delivery System

Lipid-based stimuli sensitive delivery systems have been utilized for paclitaxel delivery such as pH-sensitive system, magnetically assisted system, enzyme-dependent system and photosensitive system. Arginine lauryl ester-based cationic nanostructured lipid carriers for paclitaxel were developed. The incorporation of arginine was attributed to the pH-sensitive behaviour of the system. These cationic nanocarriers were further coated with bovine serum albumin to increase their blood circulation time by charge masking. The size of these coated carriers was below 100 nm. These carriers exhibited 60–65% drug release at 5.4 pH buffer but released only ~35% drug at physiological pH of 7.4 buffer saline with 0.1% Tween 80. Biodistribution studies in tumor-bearing mice revealed higher retention time and tumor targeting from these carriers as compared to Taxol® injection (Li et al. 2012).

Magnetically assisted paclitaxel solid lipid nanoparticles containing magnetite and glyceryl monostearate were investigated. The melting point reduction from 56 to 43 °C was observed after the formation of solid lipid nanoparticles. Magnetic hyperthermia assisted drug release occurs as a response to altered temperature upon magnetic field change (Moros et al. 2019). The size of these nanoparticles was found to be 277 nm. The nanoparticles released drug faster at 43 °C but only 20% at room temperature. Magnetic hyperthermia-induced more than 75% drug release from these nanoparticles whereas only 10% drug was released from normal nanoparticles (Oliveira et al. 2018).

Enzyme cleavable paclitaxel solid lipid nanoparticles were prepared using soy phosphatidylcholine and glyceryl monostearate. These were fabricated by conjugation of polyethylene glycol-conjugated peptide that is cleaved by a metalloprotease. These nanoparticles exhibited sustained drug release over a period of 200 h in pH 7.4 phosphate buffer with 0.1% Tween 80 at 37 °C. These nanoparticles exhibited higher cytotoxicity in the HT1080 cell line as compared to Taxol®. *In vivo* pharmacokinetic study of these nanoparticles in C57BL/6 N mice after incorporation of a fluorophore, dioctadecyl-3, 3, 3′, 3′-tetramethylindodicarbocyanine and 4-chlorobenzesulfonate salt exhibited 50% fluorescence after 10 h post-injection. The unmodified solid lipid nanoparticles were cleared within 4 h of injection in mice (Zheng et al. 2014).

Photosensitive paclitaxel solid lipid nanoparticles composed of a photosensitive material aluminum 1,8,15,22 tetrakis (phenylthio) 29H, 31H-phthalocyanine soy phosphatidylcholine, Pluronic F68 and ascorbic acid were prepared. These nanoparticles exhibited drug release as a consequence of the breakdown of lipids by near-infrared light of 730 nm wavelength. The half-maximal inhibitory concentration of these nanoparticles was 20-fold reduced than Taxol® in A549 cells (Meerovich et al. 2019).

6.6 Microparticulate Delivery System

Microparticles are defined as particulate dispersions or solid particles with a size in the range of 1–1000 μm. The drug is entrapped or encapsulated in a micro-particulate matrix. Microparticles are prepared by spray drying, emulsion polymerization, solvent evaporation, and fluidized bed coating. This microencapsulation strategy gives protection of the drug from the environment, stabilization of sensitive drug substances and masking of unpleasant taste. Hence, microparticles can play an essential role for sustained release, controlled release, enhancing bioavailability and reducing side effects of drugs (Madhav and Kala 2011).

6.6.1 Biodegradable Microparticle-Based Delivery System

Biodegradable polymers comprised of monomers linked to one another through functional groups and have unstable links in the backbone. They are degraded into biologically acceptable molecules in the body. Some common examples of biodegradable polymers are poly (glycolic acid), poly (ε-caprolactone), poly (lactic-co-glycolic acid), chitin, cellulose and alginic acid (Vroman and Tighzert 2009).

Biodegradable paclitaxel microparticles composed of poly (D, L-lactide) were reported. The particle size of microparticles and encapsulation efficiency were found to be 4.6 μm and 90% respectively. Cytotoxicity study using U251 human glioma cells exhibited a 70% reduction in cell viability after 120 h (Song et al. 2010).

Biodegradable paclitaxel composite microparticles composed of poly (lactic-co-glycolic acid) have been reported. The yield and entrapment efficiency were found to be 73.06 ± 1.94% and 29.27 ± 1.09% respectively. Cytotoxicity study revealed that paclitaxel loaded Poly (lactic-co-glycolic acid)-silica microparticles exhibited higher cytotoxicity than those without silica in HeLa cancer cells (Nanaki et al. 2017).

6.6.2 Targeted Microparticles-Based Delivery System

Paclitaxel microparticles composed of Pluronic F-127, Poly (ethylene glycol)-2000 and stearic acid were reported for passive targeting. The particle size of microparticles and encapsulation efficiency were found to be 1.76 ± 0.37 μm and 94.73% respectively. The microparticles exhibited higher cytotoxicity in an *in vitro* study using SKOC-3 ovarian cancer cells than Taxol® (Han et al. 2019). Paclitaxel loaded microparticles composed of folic acid, chitosan and 3-(4, 5-dimethylthiazol-2-yl)-2, 5-diphenyl tetrazolium bromide were used for active targeting. Folic acid-containing microparticles of paclitaxel showed higher cytotoxicity than unmodified microparticles in L929 cells (Wang et al. 2016).

6.6.3   Stimuli Sensitive Microparticles-Based Delivery System

Stimuli are a state of responsiveness to sensory stimulation or excitability. Stimuli sensitive systems deal with the changes in the physiology of the body with respect to the environment changes. These systems are beneficial for the controlled and sustained delivery of the drug in the body (Bhardwaj et al. 2015). Stimuli responsible drug release can be changed in pH, temperature and magnetic field. Magnetically assisted delivery is a novel approach for delivery of drugs using engineered smart microcarriers which overcome a number of limitations facing current methods of delivering medicines. The drug is formulated into a pharmaceutically stable formulation which is usually injected through the artery that supplies the target organ or tumor in the presence of an external magnetic field (Koppisetti and Sahiti 2011).

Thermosensitive paclitaxel microparticles composed of chitosan, poly (lactide-co-glycolide), polyvinyl alcohol were developed. The particle size of microparticles was 6.38 ± 0.13 μm. These microparticles showed 63.0% reduction in tumor volume in M-234p BALB/c tumor model in comparison with non-thermosensitive microparticles (Pesoa et al. 2018). Magnetic assisted paclitaxel delivery from microparticles composed of magnetite, poly (lactic-co-glycolic acid) and polyvinyl alcohol was investigated. The particle size of microparticles and drug loading were found to be 0.7 to 5 μm and 38% respectively. *In-vitro* study in MESSA human uterine sarcoma cells showed 5% enhancement in cell killing from magnetic assisted microparticles as compared to Taxol® (Hamoudeh et al. 2008).

6.7   Emulsion-Based Delivery System

Emulsions are liquid biphasic systems in which one phase is dispersed in another immiscible phase. Emulsions are stabilized by the aid of emulsifying agents. This delivery system can be utilized for administration of poorly water-soluble drugs (Goodarzi and Zendehboudi 2019). Microemulsions can be defined as a thermodynamically stable, isotropically clear dispersion of two immiscible liquids. Globule size of microemulsion ranges from 10 to 200 nm. Microemulsions are stabilized by a mixture of surfactants namely, surfactant and co-surfactant (Lawrence and Rees 2000). On the other hand, nanoemulsions are kinetically stable, isotropically clear systems containing two immiscible liquids. The globule size ranges from 50 to 500 nm (Jaiswal et al. 2015). Microemulsions are formed by self-assembly whereas nanoemulsions are formed by mechanical shear. Emulsions can be administrated by oral or parenteral routes.

6.7.1   Simple Emulsion-Based Delivery System

Simple emulsions are made up of a single lipid or a blend of lipid or oil, an emulsifying agent and water. Paclitaxel oleate, a prodrug of paclitaxel, was incorporated in a phospholipid-based oil-in-water emulsion. The oil phase was comprised of egg phosphatidylcholine, triolein, dipalmitoyl phosphatidylethanolamine and stabilized by polysorbate 80 and oleyl chloride. Upon intravenous administration of the emulsion in rabbits, improved biodistribution and cytotoxicity was obtained as compared to Taxol® (Lundberg et al. 2003).

Two microemulsions of paclitaxel were prepared composed of lecithin, butanol, myvacet oil, water and capmul, myvacet oil, water respectively. Both of the systems were less cytotoxic and less hemolytic as compared to commercial Taxol®. A comparatively higher amount of drug i.e. 12 mg paclitaxel per gram of emulsion could be loaded in the microemulsions due to the high solubility of paclitaxel in the oil phase (Nornoo and Chow 2008). The microemulsions have shown the slow and sustained release of the drug as compared to Taxol®. *In vivo* pharmacokinetic study in rats after intravenous administration revealed that plasma half-life and biodistribution were improved as compared to Taxol® (Nornoo et al. 2008).

Paclitaxel incorporated nanoemulsion of paclitaxel was designed using medium-chain triglyceride as oil phase and Tween 80 as a surfactant. The half-maximal inhibitory concentration of paclitaxel was reduced by 18.82 times in drug-resistant MCF7 cells which indicated that paclitaxel resistance was reversed due to P-glycoprotein inhibition. Tumor volume became 10.06% in microemulsion treated animal models, as compared to paclitaxel solution (Bu et al. 2014).

Tocosol™ is a tocopherol-based formulation of paclitaxel, manufactured by Bayer Schering Pharma. The product failed in phase III clinical trial. Abu-Fayyad and his co-workers have designed an emulsion-based formulation, similar to Tocosol™ but substituting the tocopherol with tocoretinol. The shell of the emulsion was made up of polyethylene glycol and tocoretinol. Cytotoxicity of the nanoemulsion was checked on PANC-1 cells and Bx-PC-3 cells. The new system displayed better results than Tocosol™ along with a reduction in the half-maximal inhibitory concentration value (Abu-Fayyad et al. 2018).

6.7.2   Emulsion-Based System for Targeted Delivery

Emulsion-based systems can be passively or actively targeted to tumor cells. A series of lipid nanoemulsion of paclitaxel was prepared by taking medium and long-chain glycerides, oleic acid as oil phase, Tween 80 and polyethylene glycol as a surfactant. The nanoemulsions have shown stronger cytotoxicity against MCF7 cells. *In vivo* study in mice revealed that the nanoemulsion had improved biodistribution and intratumor accumulation than Taxol®. The improved activity was correlated to enhanced permeability and retention effect (Chen et al. 2017a, b).

Hyaluronic acid serves as a targeting ligand for CD44 overexpressing cancer cells. Hyaluronic acid-coated nanoemulsion, loaded with paclitaxel was prepared for the treatment of ovarian cancer. D, L-α-tocopheryl acetate, soybean oil was taken as oil phase whereas polysorbate 80 was used as a surfactant. The drug loading was found to be 3.75%. Cell affinity studies in SK-OV-3 and OVCAR-3 cells revealed that hyaluronic acid coated nanoemulsion had 10 times higher targeting ability than uncoated nanoemulsion (Kim and Park 2017).

6.7.3   Self Emulsifying Drug Delivery System

Self emulsifying drug delivery systems are made up of a mixture of oil, surfactant, solvents and can be administrated orally by encapsulating in gelatin capsules. After reaching the gastrointestinal tract, these systems form macro or microemulsions. These systems can protect drugs which are prone to enzymatic hydrolysis in the gastrointestinal tract. It results in the enhanced oral bioavailability of poorly soluble drugs (Gursoy et al. 2003).

A series of self-microemulsifying oily formulations were evaluated for their potential for oral delivery of paclitaxel. The self-emulsifying formulations were made up of different concentrations of paclitaxel, D-alpha tocopheryl polyethylene glycol 1000 succinate, tyloxapol, ethanol, docusate sodium and cyclosporin A. *In vivo* study in P-glycoprotein knockout mice and wild type mice revealed that bioavailability of paclitaxel was similar to Taxol® after oral administration of the formulations. Efficacy of the formulations was improved due to the presence of cyclosporin A and a P-glycoprotein inhibitor (Oostendorp et al. 2011).

Paclitaxel loaded self-nanoemulsifying drug delivery system was developed using tocopheryl polyethylene glycol succinate as oil; labrasol as a surfactant; lauroglycol 90 as co-surfactant and ethanol as co-solvent. It showed higher cytotoxicity in MDA-MB-231 cells than Taxol®. Drug release from the self-nano emulsifying system has shown sustained release. Survivin, a member of the inhibitor of apoptosis family, was downregulated by the system due to the presence of tocopheryl polyethylene glycol succinate. Its oral bioavailability was 4-fold higher than Taxol® (Meher et al. 2018).

6.8   Solid Dispersion-Based Delivery System

Solid dispersions are a dispersion of drug in the amorphous polymeric matrix. The purpose of solid dispersions is to enhance drug solubility and to stabilize the drug (Huang and Dai 2014). Paclitaxel solid dispersion composed of hydroxypropyl β-cyclodextrin, polyvinyl pyrrolidone, α-tocopherol and polyoxyl 40 hydrogenated castor oil prepared by a supercritical antisolvent process which exhibited a 3-fold increase in solubility as compared to Taxol®. This solid dispersion exhibited higher tumor growth inhibition as compared to Taxol® up to 90 days in female athymic nude mice bearing MDA-MB-231 cancer (Shanmugam et al. 2011). In another study, solid dispersion of paclitaxel referred to as Modrapac001 formulation which was prepared from polyvinyl pyrrolidone K30 and sodium lauryl sulfate. This ternary solid dispersion exhibited a multifold increase in drug dissolution (Moes et al. 2013).

6.9   Cyclodextrin-Based Delivery System

Cyclodextrins are cyclic oligosaccharides of glucose having α-1, 4 glycosidic bonds. It has 6, 7 or 8 units present in α, β and γ derivatives respectively. Cyclodextrins have truncated cone shape wherein the inner cavity is hydrophobic and the outer surface is hydrophilic. Water-insoluble molecules can be incorporated into the hydrophobic cavity of cyclodextrins and they enhance the solubility and/or stability of the molecule. pH-sensitive paclitaxel nanoparticles were prepared using acetylated α-cyclodextrin by oil/water solvent evaporation method. These nanoparticles exhibited faster release i.e. 83% in 8 h in pH 5.0 but slower drug release in pH 7.4 phosphate buffer. The cellular uptake in B16F10 cells exhibited complete cellular uptake within 4 h from pH-sensitive nanoparticles. These nanoparticles exhibited better effectiveness in tumor inhibition *in vivo* in the melanoma-bearing nude mouse model than the plain drug itself (He et al. 2013).

Poly (anhydride) oral nanoparticles containing a complex of paclitaxel in β-cyclodextrin were developed. The drug loading was negligible with the drug per se but it was increased 2 to 4-fold after complexation of the drug in the nanoparticles. These nanoparticles exhibited 83% relative bioavailability with respect to Taxol® in rats. Hence, these nanoparticles could be useful for parenteral to oral switch of paclitaxel (Agüeros et al. 2010). In another study, when paclitaxel was incorporated into 2, 6 dimethyl β-cyclodextrin complex, the solubility was increased to 2.3 mM (Hamada et al. 2006).

Poly (acrylic acid)-co-β-cyclodextrin was used to prepare paclitaxel nanoparticles. The nanoparticles exhibited 170 nm size and 100-fold solubility enhancement. The nanoparticles showed sustained release at pH 7.4 phosphate buffer over a period of 120 days. *In vivo* imaging in mice for these nanoparticles revealed their accumulation in the tumor for 144 h due to enhanced permeability and retention effect. *In vivo* antitumor study in H22 tumor-bearing mice exhibited higher antitumor activity and 50 days of survival time than that of Taxol® (Yuan et al. 2016).

Folate-conjugated cyclodextrin nanoparticles of paclitaxel were used for targeted delivery to tumor cells. Cytotoxicity study in the 4T cell line exhibited higher cytotoxicity than that of Taxol®. The survival rate of these nanoparticles was longer than that of Taxol® in 4T cell-bearing mice (Erdoğar et al. 2018).

6.10   Implant-Based Drug Delivery System

Implants are sterile solid masses consisting of a drug made by compression or moulding. These provide a controlled delivery of drug over a period of time at the site of implantation. Implantation can be done under the skin into the subcutaneous tissue called subcutaneous implant. Two types of implants are available such as biodegradable and non-biodegradable implants. Implants are prepared by compression, moulding or extrusion. The advantages are controlled drug delivery with the minimized side effects (Zaki et al. 2012).

6.10.1   Biodegradable Implant

The advantages of biodegradable systems include polymers used in implants are inert and are converted into nontoxic components such as carbon dioxide and water. There is no need for surgical removal of implants after the therapy. The polymers used in biodegradable implants are polylactic acid, poly (lactide-co-glycolide) and chitosan (Rajgor et al. 2011).

Biodegradable implants of paclitaxel composed of poly (ε-caprolactone), polyethylene glycol 6000 were reported. The drug loading was found to be 93%. *In vitro* study showed 79% drug release after 30 days in pH 7.4 phosphate buffer saline (Hiremath et al. 2013). Biodegradable chitosan-based film implants of paclitaxel composed of chitosan, lysozyme, Tween 80 and Poloxamer 407 were reported. The drug loading and *in vitro* release in pH 7.4 phosphate buffer saline after 6 h were 31% and 10% respectively. *In vivo* study using Swiss mice revealed that the implant was biodegradable as it lost the integrity and identity after 50 days (Dhanikula and Panchagnula 2004). Biodegradable microfiber implants of paclitaxel composed of poly (D, L lactide-co-glycolide) were investigated. The drug loading and encapsulation efficiency were found to be 8.92% and 98% respectively. *In vitro* release study showed 16% drug release after 9 days in pH 7.4 phosphate buffer saline (Ranganath and Wang 2008).

Biodegradable stent coating to paclitaxel implants composed of poly (lactic-co-glycolic acid) and poly (vinyl alcohol)-graft-poly (lactic-co-glycolic acid) were reported. The drug loading was found to be 10%. *In vitro* release study showed 64% drug release after 48 days in pH 7.4 phosphate buffer saline (Westedt et al. 2006).

6.10.2   Non-biodegradable Implant

Non-biodegradable implants are used to administer medications. It is useful for patients who need treatment for chronic diseases. These implants need to be removed surgically once the medication has been released. These implants are not designed to be metabolized by the body. These implants start to work immediately upon insertion and can continue releasing drug doses in a controlled manner over the years. The polymers used in non-biodegradable implants are poly (methyl methacrylate), thermoplastic polyurethane and poly (ethylene vinyl acetate (Stewart et al. 2018).

Non-biodegradable stent implants of paclitaxel composed of poly (methyl methacrylate), 316 L stainless steel stents; poly (n-butyl methacrylate) and poly (ethylene-co-vinyl acetate) were reported. *In vitro* study showed 60% drug release after 14 days in pH 7.4 phosphate buffer solution (Shaulov et al. 2009). In another study, non-biodegradable stent implants of paclitaxel composed of 316 L stainless steel stents, N-(2-carboxyethyl) pyrrole and butyl ester of N-(2-carboxyethyl) pyrrole were reported. *In vitro* study showed 50% drug release after 30 days in pH 7.4 phosphate buffer solution (Okner et al. 2009).

6.10.3   Site-Specific Implant-Based Delivery System

Targeting to a particular organ or site is a smart approach for delivery of biologically active agents. It enhances the therapeutic efficacy of the drugs and reduces the drug concentration in non-target organs or sites. It can be very useful for the delivery of anticancer drugs to a particular site (Himri and Guaadaoui 2018).

Urinary tract tumor-specific implant of paclitaxel composed of poly(ε-caprolactone) resin PCL 787, alginic acid sodium salt, and gelatin were reported. *In vitro* release study exhibited complete drug release after 72 h in pH 5.5 artificial urine solution. *In-vitro* study in T24 cell line revealed the half-maximal inhibitory concentration of 7.3 ng/ml after 72 h. The tumor growth inhibition from paclitaxel implant was 75% in T24 cells (Barros et al. 2016).

Ovarian tumor-specific implant of paclitaxel composed of chitosan, phosphatidylcholine and 3-(4, 5- dimethyl-2-thiazolyl)-2, 5-diphenyl-[2H]-tetrazolium bromide) was reported. *In-vitro* study in SKOV-3 cells showed half-maximal inhibitory concentration of 0.211 μm/mL of these paclitaxel implant (Ho et al. 2005).

The tracheal tumor-specific implant of paclitaxel composed of Carbopol 974P, ethylene-vinyl acetate was reported. Carbapol 974P was chosen for mucoadhesive effect in the trachea. *In vitro* release study showed 77% drug release after 85 days in pH 7.8 phosphate buffer solution with 0.5% sodium dodecyl sulfate (Jin et al. 2018a, b).

6.10.4   Stimuli Sensitive Implant

Stimuli sensitive paclitaxel implants composed of poly (D, L lactide-co-glycolide), polyethylene glycol and Tween 80 were reported. The irradiation was performed using Cobalt 60 isotope to enhance the drug release. *In vitro* study showed 10% drug release after 6 h in pH 7.2 phosphate buffer solution (Wang et al. 2003).

6.10.5   In-Situ Forming Implants

In-situ implants are a potential alternative strategy to preformed implants and they avoid surgery. They are injected as a low viscous solution and transform in the body to a gel or solid depot due to phase separation or lowered solubility (Kempe and Mäder 2012). The *in-situ* forming implant of paclitaxel composed of polyethylene glycol, trimethylene carbonate and glycolide was reported. *In vivo* study of this implant in the mouse flank model exhibited higher antitumor efficacy than that of Taxol® (Olbrich et al. 2012). *In situ* forming injectable paclitaxel implant composed of n-butyl-2-cyanoacrylate and ethyl oleate was reported. It was based on the sol-gel transition. *In vitro* drug release in pH 7.4 sodium salicylate solution exhibited 25% from the implant versus 91% from Taxol® after 6 days. Fifty percent of the implant was degraded in female ICR mice after 8 weeks and hence it was biodegradable. The tumor inhibition in mice bearing human MDA-MB-231 was 80% whereas it was 44% in Taxol® treated mice (Wu et al. 2017).

6.11   Prodrug-Based Delivery System

The prodrug is defined as an inactive form of a drug, which is converted to a pharmacologically active form upon metabolism inside the body. Various prodrugs of paclitaxel have been designed in order to alter its physicochemical properties to make it suitable for delivery systems.

6.11.1   Small Molecule-Based Prodrug Delivery System

Small molecules such as malic acid, squalene and silicate-based prodrugs of paclitaxel have been developed in the past. The hydrophilic prodrug of paclitaxel was synthesized by incorporation of dihydroxyl derivative using solketal chloroformate. The prodrug exhibited pH-sensitive hydrolysis at the acidic condition. The prodrug exhibited 50 times higher water solubility than the unmodified drug. The prodrug exhibited a 2.5-fold higher maximum tolerable dose as compared to the unmodified drug in Balb/c mice (Niethammer et al. 2001). Silicate ester prodrugs of paclitaxel were synthesized and loaded into poly (ethylene glycol)-b-poly(lactide-co-glycolide) nanoparticles by using flash nanoprecipitation method. The drug loading was found to be 74%. The silicate ester was sensitive to acidic condition leading to hydrolysis followed by drug release. These prodrugs exhibited a 10-fold reduction in half-maximal inhibitory concentration in the cell lines (Han et al. 2015).

The lipidic prodrug of paclitaxel was prepared with lipophilic squalenoyl derivative to increase its lipid affinity. The prodrug was further converted into multilamellar vesicles using 1,2-dimyristoyl-sn-glycero-3-phosphocholine. The prodrug could enhance the loading of the drug into vesicles (Sarpietro et al. 2012).

6.11.2 Peptide-Based Prodrug Delivery System

Peptides such as octreotide and Ac-Cys-Arg-Gly-Asp-Arg-NH$_2$ (Papas et al. 2007) have been used to prepare paclitaxel prodrugs. Octreotide based paclitaxel prodrug was prepared to increase the hydrophilicity of the drug that resulted in 20,000 times solubility enhancement in water than an unmodified drug. The prodrug exhibited sustained drug release over a period of 168 h and faster drug release in acidic pH. *In vitro* cytotoxicity study in NCI-H446 cells revealed 2.6-fold lower half maximal inhibitory concentration for the octreotide prodrug as compared to the unmodified drug. *In vivo* antitumor growth inhibition study in NCI-H446 xenograft model exhibited 1.22 times higher antitumor effect of peptide prodrug than that of Taxol® (Huo et al. 2015).

6.11.3 Polymer-Based Prodrug Delivery System

The drug is attached to the polymer chemically either directly or via a spacer. The nature of the polymer and spacer determines the site of release of the drug from the prodrug and its release kinetics (Ringsdorf 1975). The molecular weight of the polymer should be up to 40,000 Da to aid the filtration by kidneys (Hoste et al. 2004). Poly-(γ-L-glutamyl glutamine) conjugate of paclitaxel was developed which showed a 3-fold reduction of tumor growth in NCI-H460 human lung cancer cells than Abraxane® (Feng et al. 2010). Oligo (lactide)$_8$-paclitaxel prodrug composed of poly (ethylene glycol)-b-poly (D, L lactic acid) was reported. Paclitaxel loaded into poly (ethylene glycol)-b-poly (lactic acid) exhibited the issue of stability and faster elimination from the body. *In vitro* cytotoxicity study in 4T1-Luc breast cancer cells for the drug-loaded synthesized polymer exhibited lower half maximal inhibitory concentration than the plain drug (Tam et al. 2019).

6.11.4 Targeted Prodrug Delivery System

The targeted prodrug of paclitaxel composed of folic acid, distearoyl phosphatidyl-choline, triolein cholesteryl oleate and polyethylene glycol was reported. *In vitro* study showed 10% drug release after 24 h in pH 7.4 phosphate buffer solution. *In-vitro* study in KB, M109 and CHO cell line revealed higher cytotoxicity compared to Taxol®. *In vivo* antitumor efficacy in Balb/c mice revealed that folic acid conjugated with paclitaxel containing cholesteryl oleate exhibited higher antitumor efficacy compared to Taxol® (Stevens et al. 2004).

The targeted prodrug of paclitaxel composed of hyaluronic acid, diphenylphos-phinic chloride, adipic acid dihydrazide and succinic anhydride was synthesized. *In vitro* study in H22 cell line revealed higher cytotoxicity compared to paclitaxel alone. *In vivo* antitumor study in H22 tumor-bearing mice revealed a 4-fold decrease in tumor growth compared to paclitaxel alone (Xu et al. 2015a, b).

Arg-Gly-Asp peptide-based targeting of paclitaxel prodrug composed of poly (lactic-co-glycolic acid), 1-ethyl-3- (3-dimethylaminopropyl) carbodiimide, Soybean lecithin and 3, 3′-dithiodipropionic acid was prepared. *In vivo* antitumor study in lung tumor-bearing mice revealed peptide modified paclitaxel had higher antitumor efficacy compared to free paclitaxel (Wang et al. 2018).

6.11.5   Stimuli-Sensitive Prodrug Delivery System

Prodrug based pH-sensitive paclitaxel micelles prepared using poly (ethylene glycol), 2-hydroxyethyl vinyl ether and acryloyl chloride was reported. The drug loading capacity was found to be 60.3%. *In vitro* study in HeLa and MDA-MB-231 cells exhibited higher antitumor activity in poly (ethylene glycol) conjugated paclitaxel compared to paclitaxel alone (Huang et al. 2018).

The photosensitive prodrug of paclitaxel composed of folic acid and polyethylene glycol was reported. The prodrug of paclitaxel was activated by the far-red light. *In vitro* study in SKOV-3 cells exhibited folic acid conjugated and photosensitive prodrug of paclitaxel with polyethylene glycol exhibited more cytotoxicity than non-conjugated paclitaxel (Thapa et al. 2017).

The redox-sensitive prodrug of paclitaxel composed of methoxy polyethylene glycol amine, N-hydroxysuccinimide, 4-dimethylaminopyridine and 3, 3′-dithiodi-propionic acid was reported. These polymeric paclitaxel conjugates were structurally confirmed by ^{1}H NMR and exhibited an approximately 23,000-fold increase in water solubility over parent paclitaxel. *In vivo* studies on NCI-H466 tumor-bearing nude mice exhibited that redox-sensitive prodrug of paclitaxel conjugates possessed superior tumor targeting ability and antitumor activity compared to Taxol® (Yin et al. 2015).

6.12   Hybrid Delivery System

Liposomes, micelles, polymeric nanoparticles, and lipid-based nanoparticles are some of the most explored delivery systems for paclitaxel. Each of these systems exhibit some inherent problems. Difficulties allied to these systems can be minimized by combining two or more of these systems which can be defined as a hybrid drug delivery system.

Bao and co-workers have developed such a hybrid system by combining gold nanoparticles and liposomes. A thiol-terminated polyethylene glycol-paclitaxel derivative was covalently attached on the surface of gold nanoparticles followed by the incorporation of the conjugate in the phospholipid bilayer region of the liposomes. Due to this, the drug loading was increased to 50.4%. *In vitro* study exhibited 50% drug release in 7.4 pH phosphate buffer solutions. The hybrid system exhibited prolonged circulation of the drug and increased accumulation in the liver as compared to Taxol®. Improved pharmacokinetic properties of the system were attributed to the protective shell of the liposomes (Bao et al. 2014).

In another study, hybrid liposomes were prepared by incorporating paclitaxel loaded polyionic micelles. The micelles were made up of positively charged Pluronic F127- poly (ethyleneimine) copolymer and negatively charged sodium cholate whereas the liposomes were composed of phospholipid. The drug loading of the final system was found to be 5.45%. It was able to sustain the release of paclitaxel. Cellular uptake of the system by multidrug-resistant breast cancer cells was significantly higher which can be attributed to the P-glycoprotein inhibitory potential of Pluronic F127. Formation of liposome not only allowed intestinal absorption of paclitaxel but also stabilized the drug from gastric degradation. The absolute oral bioavailability of paclitaxel in hybrid liposomes was found to be 37.91% (Li et al. 2017a, b, c).

Silica coated liposomes of paclitaxel were developed by Ingle and co-workers, named as liposils. Stability study of 6 months revealed that the liposils were more stable than corresponding liposomes. *In vitro* release study showed silica coating was able to retard the drug release from the liposils. The liposils also exhibited *in vivo* release of paclitaxel over a longer duration than corresponding liposomes and Taxol®. Biodistribution study in B16F10 cells revealed that accumulation of liposils and liposomes was significantly higher in the tumor cells as compared to that of Taxol® (Ingle et al. 2018).

6.13 Miscellaneous Delivery System

Miscellaneous drug delivery systems for paclitaxel based on dendrimer, exosomes, dry powder inhalations, and hydrogels are discussed in this subsection.

6.13.1 Dendrimer-Based Delivery System

Dendrimer-based paclitaxel delivery composed of methoxy polyethylene glycol-succinimidyl carboxymethyl ester, N-ethyl diisopropylamine, poly (amidoamine), α-tocopheryl succinate and polyethylene glycol was reported. The diameter and entrapment efficiency were found to be 31.19 ± 0.07 nm and $78.33 \pm 2.81\%$ respectively. *In vitro* study showed 63.97% drug release after 36 h in pH 7.4 phosphate buffer saline. *In vitro* cytotoxicity study in B16F10 and MDA MB231 cells demonstrated that the paclitaxel loaded methoxy polyethylene glycol-succinimidyl carboxymethyl ester, N-ethyl diisopropylamine, poly (amidoamine), α-tocopheryl succinate and polyethylene glycol exhibited higher cytotoxicity compared to free paclitaxel (Bhatt et al. 2019).

Dendrimer-based paclitaxel delivery composed of azido-cyanine 5.5, N, N-diisopropylethylamine, and N, N, N′, N′-tetramethyl – (1H-benzotriazol-1-yl) uranium hexafluorophosphate was reported. The diameter and drug content were found to be 74 nm and 20.9% respectively. *In vitro* cytotoxicity study in 4T1, 3T3 and C2C12 exhibited higher cytotoxicity than paclitaxel injection (Li et al. 2017a, b, c).

6.13.2 Dry Powder Inhalation-Based Drug Delivery

Dry powder inhalation formulation based on paclitaxel micelles composed of folic acid, polyethylene glycol and dextran was reported. The particle size and entrapment efficiency of the micelles were found to be 57 ± 2 nm and 99 ± 2% respectively. *In vitro* release study from the micelles showed 70% drug release after 24 h in pH 5.0 phosphate buffer. *In-vitro* study of the micelles in M109-HiFR cell line revealed 2-fold reduction of the half-maximal inhibitory concentration as compared to that of Taxol®. The formulation exhibited a fine particle fraction upto 50% with good redispersibility in physiological buffer (Rosière et al. 2016).

6.13.3 Exosomes

Exosomes of paclitaxel isolated from raw milk of Jersey cows were reported. The particle size and loading efficiency were found to be 108 nm and 8% respectively. *In vitro* study showed 90% drug release after 48 h in pH 5.0 fed-state simulated-gastric fluid. *In vivo* antitumor efficacy of exosomes of paclitaxel using athymic nude mice bearing subcutaneous lung cancer A549 xenografts exhibited higher cytotoxicity compared to paclitaxel alone (Agrawal et al. 2017).

6.13.4 Hydrogel-Based Delivery System

Thermosensitive hydrogel formulation of paclitaxel composed of chitosan, β-glycero phosphate and sodium dodecyl sulfate. *In vitro* release study showed 92% drug release after 30 days in pH 7.4 phosphate buffer saline. *In vivo* antitumor efficacy of hydrogel formulation of paclitaxel in EMT-6 tumors implanted subcutaneously on Balb/c mice exhibited 4-fold higher efficacy than Taxol® (Ruel-Gariépy et al. 2004).

Self-assembled paclitaxel hydrogel of composed of folic acid, hydroxyl benzo-triazole, alkaline phosphatise and 3-(4, 5-dimethylthiazol-2-yl)-2, 5-diphenyltetra-zolium bromide was reported. The hydrodynamic diameter and zeta potential were found to be 137.3 ± 15.2 nm and 43.8 ± 3.2 mV respectively. *In vitro* study in HepG2 cells exhibited higher cytotoxicity compared to free paclitaxel. *In vivo* anti-tumor study in nude mouse model exhibited enhanced anticancer efficacy compared to free paclitaxel. *In vivo* real-time imaging further demonstrated that the hydrogel preferentially accumulated in tumor site (Shu et al. 2017).

6.14 Conclusion

Poor solubility and permeability of Paclitaxel have compelled formulation scientists to commercialize it as injectable products such as Taxol®, Genexol PM®, and Abraxane®. Newer delivery systems like micelles, nanoparticles, liposomes, lipid systems that can overcome the issues associated with the commercial products of paclitaxel have been discussed. Some of these systems have utilized specific materials to impart stimuli sensitivity, targeting or functionality. The highest solubility enhancement has been shown by prodrugs, mixed micelles, cyclodextrin complexes and solid dispersions for paclitaxel. Drug loading could be increased in implants, nanoparticles, emulsions and lipid systems of paclitaxel. Implant and cyclodextrin complexes of paclitaxel exhibited sustained release characteristics for a prolonged period of time. The half-maximal inhibitory concentration of paclitaxel was reduced substantially in the case of micelles, emulsions, nanoparticles. Enhancement in oral bioavailability was found to be best in nanoparticles. More efficient tumor growth inhibition was found in cases of emulsion, liposome, prodrug and nanoparticulate delivery system as compared to that with Taxol®. Non-invasive oral systems of paclitaxel such as DHP107 and Oraxol® are under clinical evaluation.

Acknowledgements The authors PA, SK, KV acknowledge scholarship provided by Department of Pharmaceuticals, Government of India.

References

Abu-Fayyad A, Kamal MM, Carroll JL, Dragoi AM, Cody R, Cardelli J, Nazzal S (2018) Development and in-vitro characterization of nanoemulsions loaded with paclitaxel/γ-tocotrienol lipid conjugates. Int J Pharm 536(1):146–157. https://doi.org/10.1016/j.ijpharm.2017.11.062

Agrawal AK, Aqil F, Jeyabalan J, Spencer WA, Bcck J, Gachuki BW, Alhakeem SS, Oben K, Munagala R, Bondada S, Gupta RC (2017) Milk-derived exosomes for oral delivery of paclitaxel. Nanomedicine 13(5):1627–1636. https://doi.org/10.1016/j.nano.2017.03.001

Agüeros M, Zabaleta V, Espuelas S, Campanero MA, Irache JM (2010) Increased oral bioavailability of paclitaxel by its encapsulation through complex formation with cyclodextrins in poly (anhydride) nanoparticles. J Control Release 145(1):2–8. https://doi.org/10.1016/j.jconrel.2010.03.012

Akbarzadeh A, Rezaei-Sadabady R, Davaran S, Joo SW, Zarghami N, Hanifehpour Y, Samiei M, Kouhi M, Nejati-Koshki K (2013) Liposome: classification, preparation, and applications. Nanoscale Res Lett 8(1):102. https://doi.org/10.1186/1556-276X-8-102

Almeida A, Silva D, Gonçalves V, Sarmento B (2018) Synthesis and characterization of chitosan-grafted-polycaprolactone micelles for modulate intestinal paclitaxel delivery. Drug Deliv Transl Res 8(2):387–397. https://doi.org/10.1007/s13346-017-0357-8

Alqahtani FY, Aleanizy FS, El ET, Alkahtani HM, AlQuadeib BT (2019) Paclitaxel. Profiles Drug Subst Excip Relat Methodol 44:205–238. https://doi.org/10.1016/bs.podrm.2018.11.001

Atar K, Eroğlu H, Çalış S (2018) Novel advances in targeted drug delivery. J Drug Target 26(8):633–642. https://doi.org/10.1080/1061186X.2017.1401076

Attia ABE, Ong ZY, Hedrick JL, Lee PP, Ee PLR, Hammond PT, Yang YY (2011) Mixed micelles self-assembled from block copolymers for drug delivery. Curr Opin Colloid Interface Sci 16(3):182–194. https://doi.org/10.1016/j.cocis.2010.10.003

Bao Y, Guo Y, Zhuang X, Li D, Cheng B, Tan S, Zhang Z (2014) D-α-tocopherol polyethylene glycol succinate-based redox-sensitive paclitaxel prodrug for overcoming multidrug resistance in cancer cells. Mol Pharm 11(9):3196–3209. https://doi.org/10.1021/mp500384d

Barros AA, Browne S, Oliveira C, Lima E, Duarte ARC, Healy KE, Reis RL (2016) Drug-eluting biodegradable ureteral stent: new approach for urothelial tumors of upper urinary tract cancer. Int J Pharm 513(1–2):227–237. https://doi.org/10.1016/j.ijpharm.2016.08.061

Bayda S, Hadla M, Palazzolo S, Riello P, Corona G, Toffoli G, Rizzolio F (2018) Inorganic nanoparticles for cancer therapy: a transition from lab to clinic. Curr Med Chem 25(34):4269–4303. https://doi.org/10.2174/0929867325666171229141156

Bazak R, Houri M, El Achy S, Hussein W, Refaat T (2014) Passive targeting of nanoparticles to cancer: a comprehensive review of the literature. Mol Clin Oncol 2(6):904–908. https://doi.org/10.3892/mco.2014.356

Bernabeu E, Helguera G, Legaspi MJ, Gonzalez L, Hocht C, Taira C, Chiappetta DA (2014) Paclitaxel-loaded PCL–TPGS nanoparticles: in vitro and in vivo performance compared with Abraxane®. Colloids Surf B: Biointerfaces 113:43–50. https://doi.org/10.1016/j.colsurfb.2013.07.036

Bharadwaj R, Das PJ, Pal P, Mazumder B (2016) Topical delivery of paclitaxel for treatment of skin cancer. Drug Dev Ind Pharm 42(9):1482–1494. https://doi.org/10.3109/03639045.2016.1151028

Bhardwaj A, Kumar L, Mehta S, Mehta A (2015) Stimuli-sensitive Systems-an emerging delivery system for drugs. Artif Cells Nanomed Biotechnol 43(5):299–310. https://doi.org/10.3109/21691401.2013.856016

Bhatt H, Rompicharla SVK, Ghosh B, Biswas S (2019) α-tocopherol succinate anchored poly(ethylene glycol)ylated poly (Amidoamine) Dendrimer for the delivery of paclitaxel: assessment of in vitro and in-vivo therapeutic efficacy. Mol Pharm. https://doi.org/10.1021/acs.molpharmaceut.8b01232

Bhattacharyya J, Bellucci JJ, Weitzhandler I, McDaniel JR, Spasojevic I, Li X, Lin CC, Chi JTA, Chilkoti A (2015) A paclitaxel-loaded recombinant polypeptide nanoparticle outperforms Abraxane in multiple murine cancer models. Nat Commun 6:7939. https://doi.org/10.1038/ncomms8939

Boddupalli BM, Mohammed ZN, Nath RA, Banji D (2010) Mucoadhesive drug delivery system: an overview. J Adv Pharm Technol Res 1(4):381–387. https://doi.org/10.4103/0110-5558.76436

Bu H, He X, Zhang Z, Yin Q, Yu H, Li Y (2014) A TPGS-incorporating nanoemulsion of paclitaxel circumvents drug resistance in breast cancer. Int J Pharm 471(1–2):206–213. https://doi.org/10.1016/j.ijpharm.2014.05.039

Burgess A, Shah K, Hough O, Hynynen K (2015) Focused ultrasound-mediated drug delivery through the blood–brain barrier. Expert Rev Neurother 15(5):477–491. https://doi.org/10.1586/14737175.2015.1028369

Cai S, Vijayan K, Cheng D, Lima EM, Discher DE (2007) Micelles of different morphologies—advantages of worm-like filomicelles of PEO-PCL in paclitaxel delivery. Pharm Res 24(11):2099–2109. https://doi.org/10.1007/s11095-007-9335-z

Chen H, Wu F, Li J, Jiang X, Cai L, Li X (2016) DUP1 peptide modified micelle efficiently targeted delivery paclitaxel and enhance mitochondrial apoptosis on PSMA-negative prostate cancer cells. Springerplus 5(1):362. https://doi.org/10.1186/s40064-016-1992-0

Chen FY, Zhang Y, Chen XY, Li JQ, Xiao XP, Yu LL, Tang Q (2017a) Development of a hybrid paclitaxel-loaded arsenite nanoparticle (HPAN) delivery system for synergistic combined therapy of paclitaxel-resistant cancer. J Nanopart Res 19(4):155. https://doi.org/10.1007/s11051-017-3848-0

Chen L, Chen B, Deng L, Gao B, Zhang Y, Wu C, Yu N, Zhou Q, Yao J, Chen J (2017b) An optimized two-vial formulation lipid nanoemulsion of paclitaxel for targeted delivery to tumor. Int J Pharm 534(1–2):308–315. https://doi.org/10.1016/j.ijpharm.2017.10.005

Chen C, Zhao S, Karnad A, Freeman JW (2018) The biology and role of CD44 in cancer progression: therapeutic implications. J Hematol Oncol 11(1):64. https://doi.org/10.1186/s13045-018-0605-5

Chirio D, Gallarate M, Peira E, Battaglia L, Muntoni E, Riganti C, Biasibetti E, Capucchio MT, Valazza A, Panciani P, Lanotte M (2014) Positive-charged solid lipid nanoparticles as paclitaxel drug delivery system in glioblastoma treatment. Eur J Pharm Biopharm 88(3):746–758. https://doi.org/10.1016/j.ejpb.2014.10017.

Chowdhury P, Nagesh PK, Hatami E, Wagh S, Dan N, Tripathi MK, Khan S, Hafeez BB, Meibohm B, Chauhan SC, Jaggi M (2019) Tannic acid-inspired paclitaxel nanoparticles for enhanced anticancer effects in breast cancer cells. J Colloid Interface Sci 535:133–148. https://doi.org/10.1016/j.jcis.2018.09.072

Chu B, Qu Y, Huang Y, Zhang L, Chen X, Long C, He Y, Ou C, Qian Z (2016) Poly(ethylene glycol)-derivatized octacosanol as micellar carrier for paclitaxel delivery. Int J Pharm 500(1–2):345–359. https://doi.org/10.1016/j.ijpharm.2016.01.030.

Colombo C, Morosi L, Bello E, Ferrari R, Licandro SA, Lupi M, Ubezio P, Morbidelli M, Zucchetti M, D'Incalci M, Moscatelli D (2015) PEGylated nanoparticles obtained through emulsion polymerization as paclitaxel carriers. Mol Pharm 13(1):40–46. https://doi.org/10.1021/acs.molpharmaceut.5b00383

Contreras-Cáceres R, Leiva MC, Ortiz R, Díaz A, Perazzoli G, Casado-Rodríguez MA, Melguizo C, Baeyens JM, López-Romero JM, Prados J (2017) Paclitaxel-loaded hollow-poly (4-vinyl-pyridine) nanoparticles enhance drug chemotherapeutic efficacy in lung and breast cancer cell lines. Nano Res 10(3):856–875. https://doi.org/10.1007/s12274-016-1340-2

Dahmani FZ, Yang H, Zhou J, Yao J, Zhang T, Zhang Q (2012) Enhanced oral bioavailability of paclitaxel in pluronic/LHR mixed polymeric micelles: preparation, in vitro and in vivo evaluation. Eur J Pharm Sci 47(1):179–189. https://doi.org/10.1016/j.ejps.2012.05.015

Dai Y, Huang J, Xiang B, Zhu H, He C (2018) Antiproliferative and apoptosis triggering potential of paclitaxel-based targeted-lipid nanoparticles with enhanced cellular internalization by transferrin receptors—a study in leukemia cells. Nanoscale Res Lett 13(1):271. https://doi.org/10.1186/s11671-018-2688-x

Dalela M, Shrivastav TG, Kharbanda S, Singh H (2015) pH-sensitive biocompatible nanoparticles of paclitaxel-conjugated poly (styrene-co-maleic acid) for anticancer drug delivery in solid tumors of syngeneic mice. ACS Appl Mater Interfaces 7(48):26530–26548. https://doi.org/10.1021/acsami.5b07764

De Smet L, Colin P, Ceelen W, Bracke M, Van Bocxlaer J, Remon JP, Vervaet C (2012) Development of a nanocrystalline paclitaxel formulation for HIPEC treatment. Pharm Res 29(9):2398–2406. https://doi.org/10.1007/s11095-012-0765-x

Dhanikula AB, Panchagnula R (2004) Development and characterization of biodegradable chitosan films for local delivery of paclitaxel. AAPS J 6(3):88–99. https://doi.org/10.1208/aapsj060327

Di Mauro PP, Cascante A, Vilà PB, Gómez-Vallejo V, Llop J, Borrós S (2018) Peptide-functionalized and high drug loaded novel nanoparticles as dual-targeting drug delivery system for modulated and controlled release of paclitaxel to brain glioma. Int J Pharm 553(1–2):169–185. https://doi.org/10.1016/j.ijpharm.2018.10.022

Du AW, Lu H, Stenzel M (2016) Stabilization of paclitaxel-conjugated micelles by cross-linking with cystamine compromises the antitumor effects against two-and three-dimensional tumor cellular models. Mol Pharm 13(11):3648–3656. https://doi.org/10.1021/acs.molpharmaceut.6b00410

Edelman R, Assaraf YG, Slavkin A, Dolev T, Shahar T, Livney YD (2019) Developing body-components-based theranostic nanoparticles for targeting ovarian cancer. Pharmaceutics 11(5):216. https://doi.org/10.3390/pharmaceutics11050216

Erdoğar N, Esendağlı G, Nielsen TT, Esendağlı-Yılmaz G, Yöyen-Ermiş D, Erdoğdu B, Sargon MF, Eroğlu H, Bilensoy E (2018) Therapeutic efficacy of folate receptor-targeted amphiphilic cyclodextrin nanoparticles as a novel vehicle for paclitaxel delivery in breast cancer. J Drug Target 26(1):66–74. https://doi.org/10.1080/1061186X.2017.1339194

Fan Z, Chen C, Pang X, Yu Z, Qi Y, Chen X, Liang H, Fang X, Sha X (2015) Adding vitamin E-TPGS to the formulation of Genexol-PM: specially mixed micelles improve drug-loading ability and cytotoxicity against multidrug-resistant tumors significantly. PLoS One 10(4):e0120129

Feng Z, Zhao G, Yu L, Gough D, Howell SB (2010) Preclinical efficacy studies of a novel nanoparticle-based formulation of paclitaxel that out-performs Abraxane. Cancer Chemother Pharmacol 65(5):923–930. https://doi.org/10.1007/s00280-009-1099-1

Forrest ML, Yáñez JA, Remsberg CM, Ohgami Y, Kwon GS, Davies NM (2008) Paclitaxel prodrugs with sustained release and high solubility in poly (ethylene glycol)-b-poly (ε-caprolactone) micelle nanocarriers: pharmacokinetic disposition, tolerability, and cytotoxicity. Pharm Res 25(1):194–206. https://doi.org/10.1007/s11095-007-9451-9

Ganipineni LP, Ucakar B, Joudiou N, Riva R, Jérôme C, Gallez B, Danhier F, Préat V (2019) Paclitaxel-loaded multifunctional nanoparticles for the targeted treatment of glioblastoma. J Drug Target:1–10. https://doi.org/10.1080/1061186X.2019.1567738

Gao Y, Hu L, Liu Y, Xu X, Wu C (2019) Targeted delivery of paclitaxel in liver cancer using hyaluronic acid functionalized mesoporous hollow alumina nanoparticles. Biomed Res Int 2019:10. https://doi.org/10.1155/2019/2928507

Ge Y, Zhao Y, Li L (2016) Preparation of sodium cholate-based micelles through non-covalent bonding interaction and application as oral delivery systems for paclitaxel. Drug Deliv 23(7):2555–2565. https://doi.org/10.3109/10717544.2015.1028604.

Gong J, Huo M, Zhou J, Zhang Y, Peng X, Yu D, Zhang H, Li J (2009) Synthesis, characterization, drug-loading capacity and safety of novel octyl modified serum albumin micelles. Int J Pharm 376(1–2):161–168. https://doi.org/10.1016/j.ijpharm.2009.04.033

Goodarzi F, Zendehboudi S (2019) A comprehensive review on emulsions and emulsion stability in chemical and energy industries. Can J Chem Eng 97(1):281–309. https://doi.org/10.1002/cjce.23336

Gottesman MM, Pastan I, Ambudkar SV (1996) P-glycoprotein and multidrug resistance. Curr Opin Genet Dev 6(5):610–617

Gradishar WJ (2006) Albumin-bound paclitaxel: a next-generation taxane. Expert Opin Pharmacother 7(8):1041–1053. https://doi.org/10.1517/14656566.7.8.1041

Guo P, Song S, Li Z, Tian Y, Zheng J, Yang X, Pan W (2015) In vitro and in vivo evaluation of APRPG-modified angiogenic vessel targeting micelles for anticancer therapy. Int J Pharm 486(1–2):356–366. https://doi.org/10.1016/j.ijpharm.2015.03.067

Guo X, Cheng Y, Zhao X, Luo Y, Chen J, Yuan WE (2018) Advances in redox-responsive drug delivery systems of tumor microenvironment. J Nanobiotechnol 16(1):74. https://doi.org/10.1186/s12951-018-0398-2

Gupta M, Sharma V (2011) Targeted drug delivery system: a review. Res J Chem Sci 1(2):135–138. https://doi.org/10.22159/ijap.2017v9i6.22556.

Gupta U, Sharma S, Khan I, Gothwal A, Sharma AK, Singh Y, Chourasia MK, Kumar V (2017) Enhanced apoptotic and anticancer potential of paclitaxel loaded biodegradable nanoparticles based on chitosan. Int J Biol Macromol 98:810–819. https://doi.org/10.1016/j.ijbiomac.2017.02.030

Gursoy N, Garrigue JS, Razafindratsita A, Lambert G, Benita S, Bloom DR (2003) Excipient effects on in vitro cytotoxicity of a novel paclitaxel self-emulsifying drug delivery system. J Pharm Sci 92(12):2411–2418. https://doi.org/10.1002/jps.10501

Hamada H, Ishihara K, Masuoka N, Mikuni K, Nakajima N (2006) Enhancement of water-solubility and bioactivity of paclitaxel using modified cyclodextrins. J Biosci Bioeng 102(4):369–371. https://doi.org/10.1263/jbb.102.369

Hamoudeh M, Diab R, Fessi H, Dumontet C, Cuchet D (2008) Paclitaxel-loaded microparticles for intratumoral administration via the TMT technique: preparation, characterization, and preliminary antitumoral evaluation. Drug Dev Ind Pharm 34(7):698–707. https://doi.org/10.1080/03639040701842444

Han JK, Kim MS, Lee DS, Kim YS, Park RW, Kim K, Kwon IC (2009) Evaluation of the antitumor effects of paclitaxel-encapsulated pH-sensitive micelles. Macromol Res 17(2):99–103. https://doi.org/10.1007/BF03218661

Han J, Michel AR, Lee HS, Kalscheuer S, Wohl A, Hoye TR, McCormick AV, Panyam J, Macosko CW (2015) Nanoparticles containing high loads of paclitaxel-silicate prodrugs: formulation, drug release, and anticancer efficacy. Mol Pharm 12(12):4329–4335. https://doi.org/10.1021/acs.molpharmaceut.5b00530

Han X, Dong X, Li J, Wang M, Luo L, Li Z, Lu X, He R, Xu R, Gong M (2017) Free paclitaxel-loaded E-selectin binding peptide modified micelle self-assembled from hyaluronic acid-paclitaxel conjugate inhibit breast cancer metastasis in a murine model. Int J Pharm 528(1–2):33–46. https://doi.org/10.1016/j.ijpharm.2017.05.063

Han S, Dwivedi P, Mangrio FA, Dwivedi M, Khatik R, Cohn DE, Si T, Xu RX (2019) Sustained release paclitaxel-loaded core-shell-structured solid lipid microparticles for intraperitoneal chemotherapy of ovarian cancer. Artif Cells Nanomed Biotechnol 47(1):957–967. https://doi.org/10.1080/21691401.2019.1576705

He R, Yin C (2017) Trimethyl chitosan based conjugates for oral and intravenous delivery of paclitaxel. Acta Biomater 53:355–366. https://doi.org/10.1016/j.actbio.2017.02.012

He H, Chen S, Zhou J, Dou Y, Song L, Che L, Zhou X, Chen X, Jia Y, Zhang J, Li S (2013) Cyclodextrin-derived pH-responsive nanoparticles for delivery of paclitaxel. Biomaterials 34(21):5344–5358. https://doi.org/10.1016/j.biomaterials.2013.03.068

Himri I, Guaadaoui A (2018) Cell and organ drug targeting: types of drug delivery systems and advanced targeting strategies. In: Nanostructures for the engineering of cells, tissues and organs, pp 1–66. https://doi.org/10.1016/B978-0-12-813665-2.00001-6

Hiremath JG, Khamar NS, Palavalli SG, Rudani CG, Aitha R, Mura P (2013) Paclitaxel loaded carrier based biodegradable polymeric implants: preparation and in vitro characterization. Saudi Pharm J 21(1):85–91. https://doi.org/10.1016/j.jsps.2011.12.002

Ho EA, Vassileva V, Allen C, Piquette-Miller M (2005) In vitro and in vivo characterization of a novel biocompatible polymer–lipid implants system for the sustained delivery of paclitaxel. J Control Release 104(1):181–191. https://doi.org/10.1016/j.jconrel.2005.02.008

Holton RA, Somoza C, Kim HB, Liang F, Biediger RJ, Boatman PD, Shindo M, Smith CC, Kim S (1994) First total synthesis of taxol. 1. Functionalization of the B ring. J Am Chem Soc 116(4):1597–1598. https://doi.org/10.1021/ja00083a066

Hoste K, De Winne K, Schacht E (2004) Polymeric prodrugs. Int J Pharm 277(1–2):119–131. https://doi.org/10.1016/j.ijpharm.2003.07.016

Hou J, Sun E, Sun C, Wang J, Yang L, Jia XB, Zhang ZH (2016) Improved oral bioavailability and anticancer efficacy on breast cancer of paclitaxel via Novel Soluplus®—Solutol® HS15 binary mixed micelles system. Int J Pharm 512(1):186–193. https://doi.org/10.1016/j.ijpharm.2016.08.045

Hou J, Sun E, Zhang ZH, Wang J, Yang L, Cui L, Ke ZC, Tan XB, Jia XB, Lv H (2017) Improved oral absorption and anti-lung cancer activity of paclitaxel-loaded mixed micelles. Drug Deliv 24(1):261–269. https://doi.org/10.1080/10717544.2016.1245370

Howard B, Gao Z, Lee SW, Seo MH, Rapoport N (2006) Ultrasound-enhanced chemotherapy of drug-resistant breast cancer tumors by micellar-encapsulated paclitaxel. Am J Drug Deliv 4(2):97–104

Huang Y, Dai WG (2014) Fundamental aspects of solid dispersion technology for poorly soluble drugs. Acta Pharm Sin B 4(1):18–25. https://doi.org/10.1016/j.apsb.2013.11.001

Huang CY, Chen CM, Lee YD (2007) Synthesis of high loading and encapsulation efficient paclitaxel-loaded poly (n-butyl cyanoacrylate) nanoparticles via miniemulsion. Int J Pharm 338(1–2):267–275. https://doi.org/10.1016/j.ijpharm.2007.01.052

Huang D, Zhuang Y, Shen H, Yang F, Wang X, Wu D (2018) Acetal-linked poly(ethylene glycol) ylated paclitaxel prodrugs forming free-paclitaxel-loaded pH-responsive micelles with high drug loading capacity and improved drug delivery. Mater Sci Eng C 82:60–68. https://doi.org/10.1016/j.msec.2017.08.063

Huang ZG, Lv FM, Wang J, Cao SJ, Liu ZP, Liu Y, Lu WY (2019) RGD-modified poly(ethylene glycol)ylated paclitaxel nanocrystals with enhanced stability and tumor-targeting capability. Int J Pharm 556:217–225. https://doi.org/10.1016/j.ijpharm.2018.12.023

Huo M, Zhu Q, Wu Q, Yin T, Wang L, Yin L, Zhou J (2015) Somatostatin receptor–mediated specific delivery of paclitaxel prodrugs for efficient cancer therapy. J Pharm Sci 104(6):2018–2028. https://doi.org/10.1002/jps.24438

Huo M, Fu Y, Liu Y, Chen Q, Mu Y, Zhou J, Li L, Xu W, Yin T (2018) N-mercapto acetyl-N′-octyl-O, N ″-glycol chitosan as an efficiency oral delivery system of paclitaxel. Carbohydr Polym 181:477–488. https://doi.org/10.1016/j.carbpol.2017.10.066

Hussien NA, Işıklan N, Türk M (2018) Pectin-conjugated magnetic graphene oxide nanohybrid as a novel drug carrier for paclitaxel delivery. Artif Cells Nanomed Biotechnol 46(sup1):264–273. https://doi.org/10.1080/21691401.2017.1421211

Ibrahim NK, Desai N, Legha S, Soon-Shiong P, Theriault RL, Rivera E, Esmaeli B, Ring SE, Bedikian A, Hortobagyi GN, Ellerhorst JA (2002) Phase I and pharmacokinetic study of ABI-007, a Cremophor-free, protein-stabilized, nanoparticle formulation of paclitaxel. Clin Cancer Res 8(5):1038–1044

Ingle SG, Pai RV, Monpara JD, Vavia PR (2018) Liposils: an effective strategy for stabilizing paclitaxel loaded liposomes by surface coating with silica. Eur J Pharm Sci 122:51–63. https://doi.org/10.1016/j.ejps.2018.06.02.

Iyer AK, Khaled G, Fang J, Maeda H (2006) Exploiting the enhanced permeability and retention effect for tumor targeting. Drug Discov Today 11(17–18):812–818. https://doi.org/10.1016/j.drudis.2006.07.005

Jaiswal M, Dudhe R, Sharma PK (2015) Nanoemulsion: an advanced mode of drug delivery system. 3 Biotech 5(2):123–127. https://doi.org/10.1007/s13205-014-0214-0

Jia L, Shen J, Li Z, Zhang D, Zhang Q, Liu G, Zheng D, Tian X (2013) In vitro and in vivo evaluation of paclitaxel-loaded mesoporous silica nanoparticles with three pore sizes. Int J Pharm 445(1–2):12–19. https://doi.org/10.1016/j.ijpharm.2013.01.058

Jiang X, Xin H, Sha X, Gu J, Jiang Y, Law K, Chen Y, Chen L, Wang X, Fang X (2011) Poly(ethylene glycol)ylated poly (trimethylene carbonate) nanoparticles loaded with paclitaxel for the treatment of advanced glioma: in vitro and in vivo evaluation. Int J Pharm 420(2):385–394. https://doi.org/10.1016/j.ijpharm.2011.08.052.

Jin X, Zhou B, Xue L, San W (2015) Soluplus® micelles as a potential drug delivery system for reversal of resistant tumor. Biomed Pharmacother 69:388–395. https://doi.org/10.1016/j.biopha.2014.12.028

Jin M, Jin G, Kang L, Chen L, Gao Z, Huang W (2018a) Smart polymeric nanoparticles with pH-responsive and poly(ethylene glycol)-detachable properties for co-delivering paclitaxel and survivin siRNA to enhance antitumor outcomes. Int J Nanomedicine 13:2405–2426. https://doi.org/10.2147/IJN.S161426

Jin Z, Chen Z, Wu K, Shen Y, Guo S (2018b) Investigation of migration-preventing tracheal stent with high dose of 5-fluorouracil or paclitaxel for local drug delivery. ACS Appl Bio Mater 1(5):1328–1336. https://doi.org/10.1021/acsabm.8b00290

Kang YK, Ryu MH, Park SH, Kim JG, Kim JW, Cho SH, Park YI, Park SR, Rha SY, Kang MJ, Cho JY (2018) Efficacy and safety findings from DREAM: a phase III study of DHP107 (oral paclitaxel) versus iv paclitaxel in patients with advanced gastric cancer after failure of first-line chemotherapy. Ann Oncol 29(5):1220–1226. https://doi.org/10.1093/annonc/mdy055

Kaur P, Garg T, Rath G, Murthy RR, Goyal AK (2016) Development, optimization and evaluation of surfactant-based pulmonary nanolipid carrier system of paclitaxel for the management of drug resistance lung cancer using Box-Behnken design. Drug Deliv 23(6):1912–1925. https://doi.org/10.3109/10717544.2014.993486

Kempe S, Mäder K (2012) In situ forming implants—an attractive formulation principle for parenteral depot formulations. J Control Release 161(2):668–679. https://doi.org/10.1016/j.jconrel.2012.04.016

Kim JE, Park YJ (2017) High paclitaxel-loaded and tumor cell-targeting hyaluronan-coated nanoemulsions. Colloids Surf B: Biointerfaces 150:362–372. https://doi.org/10.1016/j.colsurfb.2016.10.050

Kim SC, Kim DW, Shim YH, Bang JS, Oh HS, Kim SW, Seo MH (2001) In vivo evaluation of polymeric micellar paclitaxel formulation: toxicity and efficacy. J Control Release 72(1–3):191–202. https://doi.org/10.1016/S0168-3659(01)00275-9

Kim TY, Kim DW, Chung JY, Shin SG, Kim SC, Heo DS, Kim NK, Bang YJ (2004) Phase I and pharmacokinetic study of Genexol-PM, a cremophor-free, polymeric micelle-formulated paclitaxel, in patients with advanced malignancies. Clin Cancer Res 10(11):3708–3716

Kim SC, Yoon HJ, Lee JW, Yu J, Park ES, Chi SC (2005) Investigation of the release behavior of DEHP from infusion sets by paclitaxel-loaded polymeric micelles. Int J Pharm 293(1–2):303–310. https://doi.org/10.1016/j.ijpharm.2005.01.011

Koppisetti V, Sahiti B (2011) Magnetically modulated drug delivery systems. Int J Drug Dev Res 3(1):260–266

Krishnadas A, Rubinstein I, Önyüksel H (2003) Sterically stabilized phospholipid mixed micelles: in vitro evaluation as a novel carrier for water-insoluble drugs. Pharm Res 20(2):297–302. https://doi.org/10.1023/A:1022243709003.

Kulshrestha P, Gogoi M, Bahadur D, Banerjee R (2012) In vitro application of paclitaxel loaded magnetoliposomes for combined chemotherapy and hyperthermia. Colloids Surf B: Biointerfaces 96:1–7. https://doi.org/10.1016/j.colsurfb.2012.02.029

Kumar Khanna V (2012) Targeted delivery of nanomedicines. ISRN Pharmacol 2012. https://doi.org/10.5402/2012/571394

Kwon GS, Okano T (1996) Polymeric micelles as new drug carriers. Adv Drug Deliv Rev 21(2):107–116. https://doi.org/10.1016/S0169-409X(96)00401-2

Lawrence MJ, Rees GD (2000) Microemulsion-based media as novel drug delivery systems. Adv Drug Deliv Rev 45(1):89–121. https://doi.org/10.1016/S0169-409X(00)00103-4

Le Broc-Ryckewaert D, Carpentier R, Lipka E, Daher S, Vaccher C, Betbeder D, Furman C (2013) Development of innovative paclitaxel-loaded small poly(lactide-co-glycolide) nanoparticles: study of their antiproliferative activity and their molecular interactions on prostatic cancer cells. Int J Pharm 454(2):712–719. https://doi.org/10.1016/j.ijpharm.2013.05.018

Lee MK, Lim SJ, Kim CK (2007) Preparation, characterization and in vitro cytotoxicity of paclitaxel-loaded sterically stabilized solid lipid nanoparticles. Biomaterials 28(12):2137–2146. https://doi.org/10.1016/j.biomaterials.2007.01.014

Lee Y, Graeser R, Kratz F, Geckeler KE (2011) Paclitaxel-loaded polymer nanoparticles for the reversal of multidrug resistance in breast cancer cells. Adv Funct Mater 21(22):4211–4218. https://doi.org/10.1002/adfm.201100853

Lee KW, Lee KH, Zang DY, Park YI, Shin DB, Kim JW, Im SA, Koh SA, Yu KS, Cho JY, Jung JA (2015) Phase I/II study of weekly oraxol for the second-line treatment of patients with metastatic or recurrent gastric cancer. Oncologist 20(8):896–897. https://doi.org/10.1634/theoncologist.

Li S, Su Z, Sun M, Xiao Y, Cao F, Huang A, Li H, Ping Q, Zhang C (2012) An arginine derivative contained nanostructure lipid carriers with pH-sensitive membranolytic capability for lysoso-

molytic anti-cancer drug delivery. Int J Pharm 436(1–2):248–257. https://doi.org/10.1016/j.
ijpharm.2012.06.040

Li J, Yin T, Wang L, Yin L, Zhou J, Huo M (2015) Biological evaluation of redox-sensitive micelles
based on hyaluronic acid-deoxycholic acid conjugates for tumor-specific delivery of paclitaxel.
Int J Pharm 483(1–2):38–48. https://doi.org/10.1016/j.ijpharm.2015.02.002

Li X, Yang Y, Jia Y, Pu X, Yang T, Wang Y, Ma X, Chen Q, Sun M, Wei D, Kuang Y (2017a)
Enhanced tumor targeting effects of a novel paclitaxel-loaded polymer: poly(ethylene glycol)–
PCCL-modified magnetic iron oxide nanoparticles. Drug Deliv 24(1):1284–1294. https://doi.
org/10.1080/10717544.2017.1373167

Li Y, Chen Z, Cui Y, Zhai G, Li L (2017b) The construction and characterization of hybrid
paclitaxel-in-micelle-in-liposome systems for enhanced oral drug delivery. Colloids Surf B:
Biointerfaces 160:572–580. https://doi.org/10.1016/j.colsurfb.2017.10.016

Li N, Cai H, Jiang L, Hu J, Bains A, Hu J, Gong Q, Luo K, Gu Z (2017c) Enzyme-sensitive
and amphiphilic PEGylated dendrimer-paclitaxel prodrug-based nanoparticles for enhanced
stability and anticancer efficacy. ACS Appl Mater Interfaces 9(8):6865–6877. https://doi.
org/10.1021/acsami.6b15505

Liang N, Sun S, Hong J, Tian J, Fang L, Cui F (2016) In vivo pharmacokinetics, biodistribution
and antitumor effect of paclitaxel-loaded micelles based on α-tocopherol succinate-modified
chitosan. Drug Deliv 23(8):2651–2660. https://doi.org/10.3109/10717544.2015.1045103.

Liang Y, Dong C, Zhang J, Deng L, Dong A (2017) A reconstituted thermosensitive hydrogel system based on paclitaxel-loaded amphiphilic copolymer nanoparticles and antitumor efficacy.
Drug Dev Ind Pharm 43(6):972–979. https://doi.org/10.1080/03639045.2017.1287718

Lin MM, Kang YJ, Sohn Y, Kim DK (2015) Dual targeting strategy of magnetic nanoparticle-
loaded and RGD peptide-activated stimuli-sensitive polymeric micelles for delivery of paclitaxel. J Nanopart Res 17(6):248. https://doi.org/10.1007/s11051-015-3033-2

Liu J, Li H, Chen D, Jin X, Zhao X, Zhang C, Ping Q (2011) In vivo evaluation of novel chitosan
graft polymeric micelles for delivery of paclitaxel. Drug Deliv 18(3):181–189. https://doi.org/1
0.3109/10717544.2010.520355

Liu Y, Yang T, Wei S, Zhou C, Lan Y, Cao A, Yang J, Wang W (2018) Mucus adhesion-and pene-
tration-enhanced liposomes for paclitaxel oral delivery. Int J Pharm 537(1–2):245–256. https://
doi.org/10.1016/j.ijpharm.2017.12.044

Lu J, Chuan X, Zhang H, Dai W, Wang X, Wang X, Zhang Q (2014) Free paclitaxel loaded
poly(ethylene glycol)ylated-paclitaxel nanoparticles: preparation and comparison with other
paclitaxel systems in vitro and in vivo. Int J Pharm 471(1–2):525–535. https://doi.org/10.1016/j.
ijpharm.2014.05.032

Lu Y, Chen Y, Gemeinhart RA, Wu W, Li T (2015) Developing nanocrystals for cancer treatment.
Nanomedicine 10(16):2537–2552. https://doi.org/10.2217/NNM.15.73

Lu B, Lv X, Le Y (2019) Chitosan-modified poly(lactide-co-glycolide) nanoparticles for control-
released drug delivery. Polymers 11(2):304. https://doi.org/10.3390/polym11020304

Lundberg BB, Risovic V, Ramaswamy M, Wasan KM (2003) A lipophilic paclitaxel derivative
incorporated in a lipid emulsion for parenteral administration. J Control Release 86(1):93–100.
https://doi.org/10.1016/S0168-3659(02)00323-1

Madaan A, Singh P, Awasthi A, Verma R, Singh AT, Jaggi M, Mishra SK, Kulkarni S, Kulkarni H
(2013) Efficiency and mechanism of intracellular paclitaxel delivery by novel nanopolymer-
based tumor-targeted delivery system, Nanoxel TM. Clin Transl Oncol 15(1):26–32

Madhav NS, Kala S (2011) Review on microparticulate drug delivery system. Int J PharmTech
Res 3(3):1242–1244

Manivasagan P, Bharathiraja S, Bui NQ, Lim IG, Oh J (2016) Paclitaxel-loaded chitosan oligosac-
charide-stabilized gold nanoparticles as novel agents for drug delivery and photoacoustic imaging of cancer cells. Int J Pharm 511(1):367–379. https://doi.org/10.1016/j.ijpharm.2016.07.025

Mansoori B, Mohammadi A, Davudian S, Shirjang S, Baradaran B (2017) The different mechanisms of cancer drug resistance: a brief review. Adv Pharm Bull 7(3):339–348. https://doi.
org/10.15171/apb.2017.041

Meerovich I, Nichols MG, Dash AK (2019) Low-intensity light-induced paclitaxel release from lipid-based nano-delivery systems. J Drug Target:1–13. https://doi.org/10.108 0/1061186X.2019.1571066

Meher JG, Dixit S, Pathan DK, Singh Y, Chandasana H, Pawar VK, Sharma M, Bhatta RS, Konwar R, Kesharwani P, Chourasia MK (2018) Paclitaxel-loaded TPGS enriched self-emulsifying carrier causes apoptosis by modulating survivin expression and inhibits tumour growth in syngeneic mammary tumours. Artif Cells Nanomed Biotechnol:1–15. https://doi.org/10.108 0/21691401.2018.1492933.

Miao J, Du Y, Yuan H, Zhang X, Li Q, Rao Y, Zhao M, Hu F (2015) Improved cytotoxicity of paclitaxel loaded in nanosized lipid carriers by intracellular delivery. J Nanopart Res 17(1):10. https://doi.org/10.1007/s11051-014-2852-x

Mittal P, Vardhan H, Ajmal G, Bonde GV, Kapoor R, Mittal A, Mishra B (2019) Formulation, optimization, hemocompatibility and pharmacokinetic evaluation of poly(lactide-co-glycolide) nanoparticles containing paclitaxel. Drug Dev Ind Pharm 45(3):365–378. https://doi.org/1 0.1080/03639045.2018.1542706

Mo Y, Lim LY (2005) Preparation and in vitro anticancer activity of wheat germ agglutinin (WGA)-conjugated poly(lactide-co-glycolide) nanoparticles loaded with paclitaxel and isopropyl myristate. J Control Release 107(1):30–42. https://doi.org/10.1016/j.jconrel.2004.06.024

Moes J, Koolen S, Huitema A, Schellens J, Beijnen J, Nuijen B (2013) Development of an oral solid dispersion formulation for use in low-dose metronomic chemotherapy of paclitaxel. Eur J Pharm Biopharm 83(1):87–94. https://doi.org/10.1016/j.ejpb.2012.09.016

Moros M, Idiago-López J, Asín L, Moreno-Antolín E, Beola L, Grazú V, Fratila RM, Gutiérrez L, de la Fuente JM (2019) Triggering antitumoural drug release and gene expression by magnetic hyperthermia. Adv Drug Deliv Rev 138:326–343. https://doi.org/10.1016/j.addr.2018.10.004

Muley P, Kumar S, El Kourati F, Kesharwani SS, Tummala H (2016) Hydrophobically modified inulin as an amphiphilic carbohydrate polymer for micellar delivery of paclitaxel for intravenous route. Int J Pharm 500(1–2):32–41. https://doi.org/10.1016/j.ijpharm.2016.01.005

Murgia S, Fadda P, Colafemmina G, Angelico R, Corrado L, Lazzari P, Monduzzi M, Palazzo G (2013) Characterization of the Solutol® HS15/water phase diagram and the impact of the Δ9-tetrahydrocannabinol solubilization. J Colloid Interface Sci 390(1):129–136. https://doi.org/10.1016/j.jcis.2012.08.068

Nanaki S, Siafaka PI, Zachariadou D, Nerantzaki M, Giliopoulos DJ, Triantafyllidis KS, Kostoglou M, Nikolakaki E, Bikiaris DN (2017) Poly(lactide-co-glycolide)/SBA-15 mesoporous silica composite microparticles loaded with paclitaxel for local chemotherapy. Eur J Pharm Sci 99:32–44. https://doi.org/10.1016/j.ejps.2016.12.010.

Nehate C, Jain S, Saneja A, Khare V, Alam N, Dhar Dubey R, Gupta PN (2014) Paclitaxel formulations: challenges and novel delivery options. Current drug delivery 11(6):666–686. https://doi.org/10.2174/156720181166614

Nguyen TL, Nguyen TH, Nguyen DH (2017) Development and in vitro evaluation of liposomes using soy lecithin to encapsulate paclitaxel. Int J Biomater 2017. https://doi.org/10.1155/2017/8234712

Nguyen HT, Soe ZC, Yang KY, Dai Phung C, Nguyen LTT, Jeong JH, Jin SG, Choi HG, Ku SK, Yong CS, Kim JO (2019) Transferrin-conjugated pH-sensitive platform for effective delivery of porous palladium nanoparticles and paclitaxel in cancer treatment. Colloids Surf B: Biointerfaces 176:265–275. https://doi.org/10.1016/j.colsurfb.2019.01.010

Niethammer A, Gaedicke G, Lode HN, Wrasidlo W (2001) Synthesis and preclinical characterization of a paclitaxel prodrug with improved antitumor activity and water solubility. Bioconjug Chem 12(3):414–420. https://doi.org/10.1021/bc000122g

Nornoo AO, Chow DSL (2008) Cremophor®free intravenous microemulsions for paclitaxel: II. Stability, in vitro release and pharmacokinetics. Int J Pharm 349(1–2):117–123. https://doi.org/10.1016/j.ijpharm.2007.07.043

Nornoo AO, Osborne DW, Chow DSL (2008) Cremophor®-free intravenous microemulsions for paclitaxel: I: formulation, cytotoxicity and hemolysis. Int J Pharm 349(1–2):108–116. https://doi.org/10.1016/j.ijpharm.2007.07.042

Núñez C, Estévez SV, del Pilar Chantada M (2018) Inorganic nanoparticles in diagnosis and treatment of breast cancer. JBIC J Biol Inorg Chem 23(3):331–345. https://doi.org/10.1007/s00775-018-1542-z

Okner R, Shaulov Y, Tal N, Favaro G, Domb AJ, Mandler D (2009) Electropolymerized tricopolymer based on N-pyrrole derivatives as a primer coating for improving the performance of a drug-eluting stent. ACS Appl Mater Interfaces 1(4):758–767. https://doi.org/10.1021/am800139s

Olbrich JM, Tate PL, Corbett JT, Lindsey JM III, Nagatomi SD, Shalaby WS, Shalaby SW (2012) Injectable in situ forming controlled release implant composed of a poly-ether-ester-carbonate and applications in the field of chemotherapy. J Biomed Mater Res A 100(9):2365–2372. https://doi.org/10.1002/jbm.a.34179.

Oliveira RR, Carrião MS, Pacheco MT, Branquinho LC, de Souza ALR, Bakuzis AF, Lima EM (2018) Triggered release of paclitaxel from magnetic solid lipid nanoparticles by magnetic hyperthermia. Mater Sci Eng C 92:547–553. https://doi.org/10.1016/j.msec.2018.07.011

Oostendorp RL, Buckle T, Lambert G, Garrigue JS, Beijnen JH, Schellens JH, van Tellingen O (2011) Paclitaxel in self-micro emulsifying formulations: oral bioavailability study in mice. Investig New Drugs 29(5):768–776. https://doi.org/10.1007/s10637-010-9421-7

O'Reilly RK, Hawker CJ, Wooley KL (2006) Cross-linked block copolymer micelles: functional nanostructures of great potential and versatility. Chem Soc Rev 35(11):1068–1083. https://doi.org/10.1039/b514858h

Owen SC, Chan DP, Shoichet MS (2012) Polymeric micelle stability. Nano Today 7(1):53–65. https://doi.org/10.1016/j.nantod.2012.01.002

Pandey V, Gajbhiye KR, Soni V (2015) Lactoferrin-appended solid lipid nanoparticles of paclitaxel for effective management of bronchogenic carcinoma. Drug Deliv 22(2):199–205. https://doi.org/10.3109/10717544.2013.877100

Papas S, Akoumianaki T, Kalogiros C, Hadjiarapoglou L, Theodoropoulos PA, Tsikaris V (2007) Synthesis and antitumor activity of peptide-paclitaxel conjugates. J Pept Sci 13(10):662–671. https://doi.org/10.1002/psc.899.

Pesoa JI, Rico MJ, Rozados VR, Scharovsky OG, Luna JA, Mengatto LN (2018) Paclitaxel delivery system based on poly (lactide-co-glycolide) microparticles and chitosan thermo-sensitive gel for mammary adenocarcinoma treatment. J Pharm Pharmacol 70(11):1494–1502. https://doi.org/10.1111/jphp.13006.

Pooja D, Kulhari H, Kuncha M, Rachamalla SS, Adams DJ, Bansal V, Sistla R (2016) Improving efficacy, oral bioavailability, and delivery of paclitaxel using protein-grafted solid lipid nanoparticles. Mol Pharm 13(11):3903–3912. https://doi.org/10.1021/acs.molpharmaceut.6b00691

Qu X, Zou Y, He C, Zhou Y, Jin Y, Deng Y, Wang Z, Li X, Zhou Y, Liu Y (2018) Improved intestinal absorption of paclitaxel by mixed micelles self-assembled from vitamin E succinate-based amphiphilic polymers and their transcellular transport mechanism and intracellular trafficking routes. Drug Deliv 25(1):210–225. https://doi.org/10.1080/10717544.2017.1419513

Qu G, Hou S, Qu D, Tian C, Zhu J, Xue L, Ju C, Zhang C (2019) Self-assembled micelles based on N-octyl-N'-phthalyl-O-phosphoryl chitosan derivative as an effective oral carrier of paclitaxel. Carbohydr Polym 207:428–439. https://doi.org/10.1016/j.carbpol.2018.11.099

Shaikh R, Singh TRR, Garland MJ, Woolfson AD, Donnelly RF (2011) Mucoadhesive drug delivery systems. J Pharm Bioallied Sci 3(1):89. https://doi.org/10.4103/0975-7406.76478

Rajgor N, Patel M, Bhaskar VH (2011) Implantable drug delivery systems: an overview. Surg Neurol Int 2(2). https://doi.org/10.4103/0975-8453.86297

Ranganath SH, Wang CH (2008) Biodegradable microfiber implants delivering paclitaxel for postsurgical chemotherapy against malignant glioma. Biomaterials 29(20):2996–3003. https://doi.org/10.1016/j.biomaterials.2008.04.002

Ravar F, Saadat E, Gholami M, Dehghankelishadi P, Mahdavi M, Azami S, Dorkoosh FA (2016) Hyaluronic acid-coated liposomes for targeted delivery of paclitaxel, in-vitro characterization and in-vivo evaluation. J Control Release 229:10–22. https://doi.org/10.1016/j.jconrel.2016.03.012

Ren H, Gao C, Zhou L, Liu M, Xie C, Lu W (2015) EGFR-targeted poly (ethylene glycol)-distearoylphosphatidylethanolamine micelle loaded with paclitaxel for laryngeal cancer: preparation, characterization and in vitro evaluation. Drug Deliv 22(6):785–794. https://doi.org/10.310 9/10717544.2014.896057

Rezaei SJT, Nabid MR, Niknejad H, Entezami AA (2012) Folate-decorated thermoresponsive micelles based on star-shaped amphiphilic block copolymers for efficient intracellular release of anticancer drugs. Int J Pharm 437(1–2):70–79. https://doi.org/10.1016/j.ijpharm.2012.07.069

Ringsdorf H (1975) Structure and properties of pharmacologically active polymers. In: Journal of Polymer Science, Polymer Symposia, vol 51, no 1. Wiley Subscription Services, Inc., A Wiley Company, New York, pp 135–153. https://doi.org/10.1002/polc.5070510111

Ron N, Cordia J, Yang A, Ci S, Nguyen P, Hughs M, Desai N (2008) Comparison of physicochemical characteristics and stability of three novel formulations of paclitaxel: Abraxane, Nanoxel, and Genexol PM

Rosière R, Van Woensel M, Mathieu V, Langer I, Mathivet T, Vermeersch M, Amighi K, Wauthoz N (2016) Development and evaluation of well-tolerated and tumor-penetrating polymeric micelle-based dry powders for inhaled anti-cancer chemotherapy. Int J Pharm 501(1–2):148–159

Ruel-Gariépy E, Shive M, Bichara A, Berrada M, Le Garrec D, Chenite A, Leroux JC (2004) A thermosensitive chitosan-based hydrogel for the local delivery of paclitaxel. Eur J Pharm Biopharm 57(1):53–63. https://doi.org/10.1016/S0939-6411(03)00095-X

Ruttala HB, Ramasamy T, Shin BS, Choi HG, Yong CS, Kim JO (2017) Layer-by-layer assembly of hierarchical nanoarchitectures to enhance the systemic performance of nanoparticle albumin-bound paclitaxel. Int J Pharm 519(1–2):11–21. https://doi.org/10.1016/j.ijpharm.2017.01.011

Rxlist website (2019) https://www.rxlist.com/taxol-drug.html, July 04, 2019

Sahu SK, Maiti S, Maiti TK, Ghosh SK, Pramanik P (2011) Hydrophobically modified carboxymethyl chitosan nanoparticles targeted delivery of paclitaxel. J Drug Target 19(2):104–113. https://doi.org/10.3109/10611861003733987

Samarajeewa S, Shrestha R, Elsabahy M, Karwa A, Li A, Zentay RP, Kostelc JG, Dorshow RB, Wooley KL (2013) In vitro efficacy of paclitaxel-loaded dual-responsive shell cross-linked polymer nanoparticles having orthogonally degradable disulfide cross-linked corona and polyester core domains. Mol Pharm 10(3):1092–1099. https://doi.org/10.1021/mp3005897

Sarisozen C, Vural I, Levchenko T, Hincal AA, Torchilin VP (2012) Long circulating poly(ethylene glycol)-PE micelles co-loaded with paclitaxel and elacridar (GG918) overcome multidrug resistance. Drug Deliv 19(8):363–370. https://doi.org/10.3109/10717544.2012.724473

Sarpietro MG, Ottimo S, Paolino D, Ferrero A, Dosio F, Castelli F (2012) Squalenoyl prodrug of paclitaxel: synthesis and evaluation of its incorporation in phospholipid bilayers. Int J Pharm 436(1–2):135–140. https://doi.org/10.1016/j.ijpharm.2012.06.034

Schafer FQ, Buettner GR (2001) Redox environment of the cell as viewed through the redox state of the glutathione disulfide/glutathione couple. Free Radic Biol Med 30(11):1191–1212

Schiff PB, Fant J, Horwitz SB (1979) Promotion of microtubule assembly in vitro by taxol. Nature 277(5698):665. https://doi.org/10.1038/277665a0

Shanmugam S, Park JH, Chi SC, Yong CS, Choi HG, Woo JS (2011) Antitumor efficacy of solid dispersion of paclitaxel prepared by supercritical antisolvent process in human mammary tumor xenografts. Int J Pharm 403(1–2):130–135. https://doi.org/10.1016/j.ijpharm.2010.10.033

Sharma NK, Kumar V (2015) Liposomal paclitaxel: recent trends and future perspectives. Int J Pharm Sci Rev Res 31(1):205–211

Sharma S, Verma A, Pandey G, Mittapelly N, Mishra PR (2015) Investigating the role of Pluronic-g-Cationic polyelectrolyte as functional stabilizer for nanocrystals: impact on paclitaxel oral bioavailability and tumor growth. Acta Biomater 26:169–183. https://doi.org/10.1016/j.actbio.2015.08.005

Shaulov Y, Okner R, Levi Y, Tal N, Gutkin V, Mandler D, Domb AJ (2009) Poly (methyl methacrylate) grafting onto stainless steel surfaces: application to drug-eluting stents. ACS Appl Mater Interfaces 1(11):2519–2528. https://doi.org/10.1021/am900465t

Shen Y, Tang H, Radosz M, Van Kirk E, Murdoch WJ (2008) pH-responsive nanoparticles for cancer drug delivery. Methods Mol Biol 437:183–216. https://doi.org/10.1007/978-1-59745-210-6_10

Shen Y, Pi Z, Yan F, Yeh CK, Zeng X, Diao X, Hu Y, Chen S, Chen X, Zheng H (2017) Enhanced delivery of paclitaxel liposomes using focused ultrasound with microbubbles for treating nude mice bearing intracranial glioblastoma xenografts. Int J Nanomedicine 12:5613. https://doi.org/10.2147/IJN.S136401

Shi J, Kantoff PW, Wooster R, Farokhzad OC (2017) Cancer nanomedicine: progress, challenges and opportunities. Nat Rev Cancer 17(1):20–37. https://doi.org/10.1038/nrc.2016.108

Shu C, Sabi-mouka EM, Wang X, Ding L (2017) Self-assembly hydrogels as multifunctional drug delivery of paclitaxel for synergistic tumour-targeting and biocompatibility in vitro and in vivo. J Pharm Pharmacol 69(8):967–977. https://doi.org/10.1111/jphp.12732

Shuai X, Merdan T, Schaper AK, Xi F, Kissel T (2004) Core-cross-linked polymeric micelles as paclitaxel carriers. Bioconjug Chem 15(3):441–448. https://doi.org/10.1021/bc034113u

Song TT, Yuan XB, Sun AP, Wang H, Kang CS, Ren Y, He B, Sheng J, Pu PY (2010) Preparation of injectable paclitaxel sustained release microspheres by spray drying for inhibition of glioma in vitro. J Appl Polym Sci 115(3):1534–1539. https://doi.org/10.1002/app.31105

Stevens PJ, Sekido M, Lee RJ (2004) A folate receptor–targeted lipid nanoparticle formulation for a lipophilic paclitaxel prodrug. Pharm Res 21(12):2153–2157. https://doi.org/10.1007/s11095-004-7667-5

Stewart S, Domínguez-Robles J, Donnelly R, Larrañeta E (2018) Implantable polymeric drug delivery devices: classification, manufacture, materials, and clinical applications. Polymers 10(12):1379. https://doi.org/10.3390/polym10121379

Sutradhar KB, Amin ML (2014) Nanotechnology in cancer drug delivery and selective targeting. ISRN Nanotechnol 2014:1–12. https://doi.org/10.1155/2014/939378

Szebeni J, Alving CR, Muggia FM (1998) Complement activation by Cremophor EL as a possible contributor to hypersensitivity to paclitaxel: an in vitro study. JNCI: J Natl Canc Inst 90(4):300–306. https://doi.org/10.1093/jnci/90.4.300

Tam YT, Shin DH, Chen KE, Kwon GS (2019) Poly (ethylene glycol)-block-poly (d, l-lactic acid) micelles containing oligo (lactic acid) 8-paclitaxel prodrug: in vivo conversion and antitumor efficacy. J Control Release 298:186–193. https://doi.org/10.1016/j.jconrel.2019.02.017

Thanki K, Kushwah V, Jain S (2015) Recent advances in tumor targeting approaches. In: Targeted drug delivery: concepts and design. Springer, Cham, pp 41–112. https://doi.org/10.1007/978-3-319-11355-5_2

Thapa P, Li M, Karki R, Bio M, Rajaputra P, Nkepang G, Woo S, You Y (2017) Folate-poly(ethylene glycol) conjugates of a far-red light-activatable paclitaxel prodrug to improve selectivity toward folate receptor-positive cancer cells. ACS Omega 2(10):6349–6360. https://doi.org/10.1021/acsomega.7b01105

Tiwari S, Tirosh B, Rubinstein A (2017) Increasing the affinity of cationized polyacrylamide-paclitaxel nanoparticles towards colon cancer cells by a surface recognition peptide. Int J Pharm 531(1):281–291. https://doi.org/10.1016/j.ijpharm.2017.08.092

Tosta FV, Andrade LM, Mendes LP, Anjos JLV, Alonso A, Marreto RN, Lima EM, Taveira SF (2014) Paclitaxel-loaded lipid nanoparticles for topical application: the influence of oil content on lipid dynamic behavior, stability, and drug skin penetration. J Nanopart Res 16(12):2782. https://doi.org/10.1007/s11051-014-2782-7

Tran S, DeGiovanni PJ, Piel B, Rai P (2017) Cancer nanomedicine: a review of recent success in drug delivery. Clin Transl Med 6(1):44. https://doi.org/10.1186/s40169-017-0175-0

Trieu V, D'Cruz O, Desai, N (2008) Comparison of antitumor activity of three cremophor-free paclitaxel formulations, Abraxane, Nanoxel, and Genexol PM

Ucar E, Teksoz S, Ichedef C, Kilcar AY, Medine EI, Ari K, Parlak Y, Bilgin BES, Unak P (2017) Synthesis, characterization and radiolabeling of folic acid modified nanostructured lipid carriers as a contrast agent and drug delivery system. Appl Radiat Isot 119:72–79. https://doi.org/10.1016/j.apradiso.2016.11.002

Varma MV, Sateesh K, Panchagnula R (2005) Functional role of P-glycoprotein in limiting intestinal absorption of drugs: contribution of passive permeability to P-glycoprotein mediated efflux transport. Mol Pharm 2(1):12–21. https://doi.org/10.1021/mp0499196

Vroman I, Tighzert L (2009) Biodegradable polymers. Materials 2(2):307–344. https://doi.org/10.3390/ma2020307

Wall ME, Wani MC (1996) Camptothecin and taxol: from discovery to clinic. J Ethnopharmacol 51(1–3):239–254. https://doi.org/10.1016/0378-8741(95)01367-9

Wan L, Wang X, Zhu W, Zhang C, Song A, Sun C, Jiang T, Wang S (2015) Folate-polyethyleneimine functionalized mesoporous carbon nanoparticles for enhancing oral bioavailability of paclitaxel. Int J Pharm 484(1–2):207–217. https://doi.org/10.1016/j.ijpharm.2015.02.054

Wang J, Ng CW, Win KY, Shoemakers P, Lee TKY, Feng SS, Wang CH (2003) Release of paclitaxel from polylactide-co-glycolide (poly(lactide-co-glycolide)) microparticles and discs under irradiation. J Microencapsul 20(3):317–327. https://doi.org/10.1080/0265204021000058401.

Wang Y, Jiang G, Qiu T, Ding F (2012) Preparation and evaluation of paclitaxel-loaded nanoparticle incorporated with galactose-carrying polymer for hepatocyte targeted delivery. Drug Dev Ind Pharm 38(9):1039–1046. https://doi.org/10.3109/03639045.2011.637052

Wang F, Shen Y, Xu X, Lv L, Li Y, Liu J, Li M, Guo A, Guo S, Jin F (2013) Selective tissue distribution and long circulation endowed by paclitaxel loaded poly(ethylene glycol)ylated poly (ε-caprolactone-co-l-lactide) micelles leading to improved anti-tumor effects and low systematic toxicity. Int J Pharm 456(1):101–112. https://doi.org/10.1016/j.ijpharm.2013.08.008

Wang F, Yang S, Hua D, Yuan J, Huang C, Gao Q (2016) A novel preparation method of paclitaxccl-loaded folate-modified chitosan microparticles and in vitro evaluation. J Biomater Sci Polym Ed 27(3):276–289. https://doi.org/10.1080/09205063.2015.1121366

Wang F, Li L, Liu B, Chen Z, Li C (2017) Hyaluronic acid decorated pluronic P85 solid lipid nanoparticles as a potential carrier to overcome multidrug resistance in cervical and breast cancer. Biomed Pharmacother 86:595–604. https://doi.org/10.1016/j.biopha.2016.12.041

Wang G, Wang Z, Li C, Duan G, Wang K, Li Q, Tao T (2018) RGD peptide-modified, paclitaxel prodrug-based, dual-drugs loaded, and redox-sensitive lipid-polymer nanoparticles for the enhanced lung cancer therapy. Biomed Pharmacother 106:275–284. https://doi.org/10.1016/j.biopha.2018.06.137

Wang X, Guo Y, Qiu L, Wang X, Li T, Han L, Ouyang H, Xu W, Chu K (2019) Preparation and evaluation of carboxymethyl chitosan-rhein polymeric micelles with synergistic antitumor effect for oral delivery of paclitaxel. Carbohydr Polym 206:121–131. https://doi.org/10.1016/j.carbpol.2018.10.096

Weaver BA (2014) How Taxol®/paclitaxel kills cancer cells. Mol Biol Cell 25(18):2677–2681. https://doi.org/10.1091/mbc.e14-04-0916

Wei Z, Hao J, Yuan S, Li Y, Juan W, Sha X, Fang X (2009) Paclitaxel-loaded Pluronic P123/F127 mixed polymeric micelles: formulation, optimization and in vitro characterization. Int J Pharm 376(1–2):176–185. https://doi.org/10.1016/j.ijpharm.2009.04.030

Wei Y, Xue Z, Ye Y, Huang Y, Zhao L (2014) Paclitaxel targeting to lungs by way of liposomes prepared by the effervescent dispersion technique. Arch Pharm Res 37(6):728–737. https://doi.org/10.1007/s12272-013-0181-8

Westedt U, Wittmar M, Hellwig M, Hanefeld P, Greiner A, Schaper AK, Kissel T (2006) Paclitaxel releasing films consisting of poly (vinyl alcohol)-graft-poly (lactide-co-glycolide) and their potential as biodegradable stent coatings. J Control Release 111(1–2):235–246. https://doi.org/10.1016/j.jconrel.2005.12.012

Wu J, Liu Q, Lee RJ (2006) A folate receptor-targeted liposomal formulation for paclitaxel. Int J Pharm 316(1–2):148–153. https://doi.org/10.1016/j.ijpharm.2006.02.027

Wu Y, Wang L, Zhang K, Zhou L, Zhang X, Jiang X, Zhu C (2017) N-Butyl-2-cyanoacrylate-based injectable and in situ-forming implants for efficient intratumoral chemotherapy. Drug Deliv 24(1):729–736. https://doi.org/10.1080/10717544.2017.1309478

Xiang J, Wu B, Zhou Z, Hu S, Piao Y, Zhou Q, Wang G, Tang J, Liu X, Shen Y (2018) Synthesis and evaluation of a paclitaxel-binding polymeric micelle for efficient breast cancer therapy. Sci China Life Sci:1–12. https://doi.org/10.1007/s11427-017-9274-9.

Xin H, Chen L, Gu J, Ren X, Luo J, Chen Y, Jiang X, Sha X, Fang X (2010) Enhanced anti-glioblastoma efficacy by paclitaxel-loaded poly(ethylene glycol)ylated poly (ε-caprolactone) nanoparticles: in vitro and in vivo evaluation. Int J Pharm 402(1–2):238–247. https://doi.org/10.1016/j.ijpharm.2010.10.005.

Xin Y, Yin M, Zhao L, Meng F, Luo L (2017) Recent progress on nanoparticle-based drug delivery systems for cancer therapy. Cancer Biol Med 14(3):228–241. https://doi.org/10.20892/j.issn.2095-3941.2017.0052

Xiong K, Wu J, Liu Y, Wu N, Ruan J (2019) Drug carrier-oriented polygeline for preparing novel polygeline-bound paclitaxel nanoparticles. J Pharm Sci 108(6):2012–2021. https://doi.org/10.1016/j.xphs.2019.01.005

Xu W, Cui Y, Ling P, Li LB (2012) Preparation and evaluation of folate-modified cationic pluronic micelles for poorly soluble anticancer drug. Drug Deliv 19(4):208–219. https://doi.org/10.310 9/10717544.2012.690005

Xu W, Lim SJ, Lee MK (2013) Cellular uptake and antitumour activity of paclitaxel incorporated into trilaurin-based solid lipid nanoparticles in ovarian cancer. J Microencapsul 30(8):755–761. https://doi.org/10.3109/02652048.2013.788083

Xu W, Fan X, Zhao Y, Li L (2015a) Cysteine modified and bile salt based micelles: preparation and application as an oral delivery system for paclitaxel. Colloids Surf B: Biointerfaces 128:165–171. https://doi.org/10.1016/j.colsurfb.2015.02.031

Xu C, He W, Lv Y, Qin C, Shen L, Yin L (2015b) Self-assembled nanoparticles from hyaluronic acid–paclitaxel prodrugs for direct cytosolic delivery and enhanced antitumor activity. Int J Pharm 493(1–2):172–181. https://doi.org/10.1016/j.ijpharm.2015.07.069

Yadav KS, Mishra DK, Deshpande A, Pethe AM (2019) Levels of drug targeting. In: Basic fundamentals of drug delivery. Academic, pp 269–305. https://doi.org/10.1016/B978-0-12-817909-3.00007-8.

Yang T, Cui FD, Choi MK, Cho JW, Chung SJ, Shim CK, Kim DD (2007) Enhanced solubility and stability of poly(ethylene glycol)ylated liposomal paclitaxel: in vitro and in vivo evaluation. Int J Pharm 338(1–2):317–326. https://doi.org/10.1016/j.ijpharm.2007.02.011

Yang L, Wu X, Liu F, Duan Y, Li S (2009a) Novel biodegradable polylactide/poly (ethylene glycol) micelles prepared by direct dissolution method for controlled delivery of anticancer drugs. Pharm Res 26(10):2332–2342. https://doi.org/10.1007/s11095-009-9949-4

Yang R, Yang SG, Shim WS, Cui F, Cheng G, Kim IW, Kim DD, Chung SJ, Shim CK (2009b) Lung-specific delivery of paclitaxel by chitosan-modified poly(lactide-co-glycolide) nanoparticles via transient formation of microaggregates. J Pharm Sci 98(3):970–984. https://doi.org/10.1002/jps.21487.

Yao HJ, Ju RJ, Wang XX, Zhang Y, Li RJ, Yu Y, Zhang L, Lu WL (2011) The antitumor efficacy of functional paclitaxel nanomicelles in treating resistant breast cancers by oral delivery. Biomaterials 32(12):3285–3302. https://doi.org/10.1016/j.biomaterials.2011.01.038

Yin T, Wu Q, Wang L, Yin L, Zhou J, Huo M (2015) Well-defined redox-sensitive polyethene glycol–paclitaxel prodrug conjugate for tumor-specific delivery of paclitaxel using octreotide for tumor targeting. Mol Pharm 12(8):3020–3031. https://doi.org/10.1021/acs.molpharmaceut.5b00280

Yuan ZQ, Li JZ, Liu Y, Chen WL, Zhang CG, Zhu WJ, Zhou XF, Liu C, Zhang XN (2015) Systemic delivery of micelles loading with paclitaxel using N-succinyl-palmitoyl-chitosan decorated with cRGDyK peptide to inhibit non-small-cell lung cancer. Int J Pharm 492(1–2):141–151. https://doi.org/10.1016/j.ijpharm.2015.07.022

Yuan S, Chen J, Sheng J, Hu Y, Jiang Z (2016) Paclitaxel loaded β-cyclodextrin modified poly(acrylic acid) nanoparticles through multivalent inclusion for anticancer therapy. Macromol Biosci 16:341–349. https://doi.org/10.1002/mabi.201500302

Zaki M, Patil SK, Baviskar DT, Jain DK (2012) Implantable drug delivery system: a review. Int J PharmTech Res 4(1):280–292

Zhang Z, Feng SS (2006) In vitro investigation on poly (lactide)-tween 80 copolymer nanoparticles fabricated by dialysis method for chemotherapy. Biomacromolecules 7(4):1139–1146. https://doi.org/10.1021/bm050953v

Zhang XY, Zhang YD (2015) Enhanced antiproliferative and apoptosis effect of paclitaxel-loaded polymeric micelles against non-small cell lung cancers. Tumor Biol 36(7):4949–4959. https://doi.org/10.1007/s13277-015-3142-7

Zhang X, Chen J, Kang Y, Hong S, Zheng Y, Sun H, Xu C (2013) Targeted paclitaxel nanoparticles modified with follicle-stimulating hormone β peptide show effective antitumor activity against ovarian carcinoma. Int J Pharm 453(2):498–505. https://doi.org/10.1016/j.ijpharm.2013.06.038

Zhang J, Chen R, Fang X, Chen F, Wang Y, Chen M (2015) Nucleolin targeting AS1411 aptamer modified pH-sensitive micelles for enhanced delivery and antitumor efficacy of paclitaxel. Nano Res 8(1):201–218. https://doi.org/10.1007/s12274-014-0619-4

Zhang F, Fei J, Sun M, Ping Q (2016) Heparin modification enhances the delivery and tumor targeting of paclitaxel-loaded N-octyl-N-trimethyl chitosan micelles. Int J Pharm 511(1):390–402. https://doi.org/10.1016/j.ijpharm.2016.07.020

Zhang T, Luo J, Fu Y, Li H, Ding R, Gong T, Zhang Z (2017) Novel oral administrated paclitaxel micelles with enhanced bioavailability and antitumor efficacy for resistant breast cancer. Colloids Surf B: Biointerfaces 150:89–97. https://doi.org/10.1016/j.colsurfb.2016.11.024

Zhao Y, Li J, Yu H, Wang G, Liu W (2012) Synthesis and characterization of a novel polydepsipeptide contained tri-block copolymer (mpoly(ethylene glycol)–PLLA–PMMD) as self-assembly micelle delivery system for paclitaxel. Int J Pharm 430(1–2):282–291. https://doi.org/10.1016/j.ijpharm.2012.03.043

Zhao L, Bi D, Qi X, Guo Y, Yue F, Wang X, Han M (2019) Polydopamine-based surface modification of paclitaxel nanoparticles for osteosarcoma targeted therapy. Nanotechnology 30(25):255101. https://doi.org/10.1088/1361-6528/ab055f

Zheng J, Wan Y, Elhissi A, Zhang Z, Sun X (2014) Targeted paclitaxel delivery to tumors using cleavable poly(ethylene glycol)-conjugated solid lipid nanoparticles. Pharm Res 31(8):2220–2233. https://doi.org/10.1007/s11095-014-1320-8

Zhou J, Zhao WY, Ma X, Ju RJ, Li XY, Li N, Sun MG, Shi JF, Zhang CX, Lu WL (2013) The anticancer efficacy of paclitaxel liposomes modified with mitochondrial targeting conjugate in resistant lung cancer. Biomaterials 34(14):3626–3638. https://doi.org/10.1016/j.biomaterials.2013.01.078

Zhu Z, Li Y, Li X, Li R, Jia Z, Liu B, Guo W, Wu W, Jiang X (2010) Paclitaxel-loaded poly (N-vinylpyrrolidone)-b-poly (ε-caprolactone) nanoparticles: preparation and antitumor activity in vivo. J Control Release 142(3):438–446. https://doi.org/10.1016/j.jconrel.2009.11.002

Zitzmann S, Ehemann V, Schwab M (2002) Arginine-glycine-aspartic acid (RGD)-peptide binds to both tumor and tumor-endothelial cells in vivo. Cancer Res 62(18):5139–5143

Chapter 7
Phytosomes as Emerging Nanotechnology for Herbal Drug Delivery

Dinesh Kumar, Nitin Vats, Kamal Saroha, and Avtar Chand Rana

Abstract Herbal medicines or phyto-pharmaceuticals have a prominent history in medicine due to their numerous therapeutic significance in healthcare. Even though plant extracts and their phytoconstituents show brilliant *in-vitro* bio-activity, they still possesses poor *in-vivo* actions owing to their high molecular sizes and low lipid solubility or both, rendering them less absorbable with lower bioavailability. Phytosome is novel technology which is known to placate all above problems or limitations of conventional systems of herbal drug delivery. Phytosome are formed by a method where a plant extract or its components get anchored to phospholipids (chiefly phosphotidylcholine), forming a lipid responsive complex. Phospholipid structure comprises of a hydrophilic head and two lipophilic tails and due to their duos nature, phospholipids behaves like an active emulsifier. By conjoining the emulsifying property of phospholipids with the herbal extract/phytoactive constituent, the phytosome affords considerably improved bioavailability and deliver faster and enhanced absorption in intestinal tract.

Since all phytoconstituents are not highly bioavailable, combining them with phospholipid yields a proficient carrier for increased absorption of the key components of herbal extract or medicine. Recently, phytosome methodology has been fruitfully applied over several well-known natural drugs such as ginseng, green tea hawthorn, olive oil and grape seed etc. Indeed, phytosomes are superior to conventional drug delivery systems in terms of the pharmacodynamic and pharmacokinetic properties. This chapter discusses the various aspects, components, methods of preparation of phytosomes and their marketed formulations, therapeutic applications along with the recent research work reported on this technology.

D. Kumar (✉)
Institute of Pharmaceutical Sciences, Kurukshetra University, Kurukshetra, Haryana, India

Department of Clinical Biochemistry & Pharmacology, Ben Gurion University of the Negev, Beer sheva, Israel
e-mail: dineshbarbola@kuk.ac.in

N. Vats · K. Saroha · A. C. Rana
Institute of Pharmaceutical Sciences, Kurukshetra University, Kurukshetra, Haryana, India

© The Editor(s) (if applicable) and The Author(s), under exclusive license to
Springer Nature Switzerland AG 2020
A. Saneja et al. (eds.), *Sustainable Agriculture Reviews 43*, Sustainable
Agriculture Reviews 43, https://doi.org/10.1007/978-3-030-41838-0_7

Keywords Phytosome · Nano drug delivery · Phosphatidylcholine · Natural products

Abbreviations

NDDS   Novel drug delivery system
PC   Phosphatidylcholine
PS   Phosphatidylserine
VDDS   Vesicular drug delivery system

7.1 Introduction

Natural products are the treasurable blessings given by nature to the human being. Natural products have been used for their therapeutic purposes since ages. Hundreds of phytoconstituents present in single natural product work together and impart therapeutic benefits. Natural products are not only being the oldest mode of treatment but the safest, effective and easily available mode of treatment also (De Smet 1997). Although having so many insignificant properties, natural products rely the most on basis of knowledge and clinical experience of the practitioners and formulators.

Natural products are usually obtained by different processes such as extraction, expression distillation, fractionation, purification etc. from whole plant or its parts. However, the particular part or the method to be implemented for specific treatment entirely depends on the manufacturer. Allopathic medicines are currently more popular than traditional ones, especially in developed countries as there are no any standardized natural products having definite therapeutic value. However, most developing countries are continuously using the natural medicines, probably because of high cost of synthetic drugs (Handa et al. 2008; Bonifacio et al. 2014). This ever-increasing dependency over the modern drugs and failure in treatment of various diseases motivates the scientists to find a better alternative.

Low solubility is among the main constraints for using the herbal products for therapeutic purposes. On the other hand, despite the potential efficacy of herbal medicines, they are usually not preferred because of lack of standardization and poor apparent quality. While dealing with herbal extracts, several components are predisposed to degradation in the acidic gastric medium whereas majority of them get broken down inside liver ahead of entering into systemic circulation. Furthermore, natural extracts are oftenly hygroscopic, less compressible and show poor powder flowability. Since plant extracts have plentiful therapeutic significance, now the efforts are being focused on developing their newer carrier systems to tackle the limitations of conventional dosage forms with reduced efficacy of herbal medicines (Aqil et al. 2013; Byeon et al. 2019).

Considering the above mentioned facts, there is an urgent need of such herbal formulations which not only overcome the problems associated with modern medicines but also provide the standardized herbal products with calculative therapeutic effects. From the very beginning, the most important parameters are solubility and absorption to increase the bioavailability of phytoconstituents to obtain more efficacies of the herbal drugs. Previous studies reveal that many therapeutically active phytocomponents like flavonoids, terpenoids and aglycone glycosides are highly polar molecules and hence, they are water soluble molecules. But, they have low absorption due to their poor lipid solubility (Cott 1995) owing to their big molecular size which hinders their absorption by passive diffusion. This limits their ability to cross the lipid-rich bio-membrane and thus, results in poor bioavailability (Manach et al. 2004).

The traditional herbal formulations typically consist of crude extracts of different herbs which also contains undesirable and sometimes, toxic compounds along with the active constituents. Owing to the latest upgradation in the field of herbal products, particular ingredient or a group of analogous constituents from plants are now isolated, purified and characterized for a number of significant uses (Khan et al. 2013). The process of extraction/isolation and purification of particular constituent from plant extract generally results in fractional or total loss of therapeutic activity. Instead of displaying brilliant bioactivity *in vitro*, herbal extracts often exhibit poor effectiveness *in vivo* or in animal models. The key reasons for low bioavailability of crude plant extracts are (i) the bioactive constituents have large multi-ring molecular structures which cannot be absorbed into the blood by simple passive diffusion and, (ii) the bioactive constituents are mostly hydrophilic in nature, hence, poor lipid solubility restricts their ability to cross lipid bio-membranes (Dewan et al. 2016).

Pioneering pre-clinical and biochemical studies have evidenced the ability of natural/plant based compounds in the management of topical disorders, anti-aging, different types of carcinoma and several other fields of therapeutics and precautionary medicine. The water-loving nature and complex structure of these compounds present vital problems because they cause low bioavailability across the gut or skin. Also, some phytomolecules owing to their lipoidal nature, are often unable to pass the small intestine (Sharma and Sikarwar 2005). Various tactics are devised to augment the bioavailability, such as inclusion by solubility and bioavailability enhancers, structural alteration and entrapment with the lipophilic transporters (Venkatesan et al. 2000).

7.2 Approaches to Improve Bioavailability

The efficacy of any natural/herbal formulation solely depends on its ability to deliver an optimum level of the active constituent and different tactics have been employed to improve the bioavailability of natural products (Bhattacharya 2009;

Giriraj 2011). Some of the strategies anticipated to advance the bioavailability are listed below:

(a) Chemical derivatization of the natural products leads to formation of several chemical derivatives (analogues) of natural product which require to be properly screened.
   So, this approach has been designed to increase the bioavailability of natural products by transforming them into a product with similar structure.
(b) Combining natural products with other compounds as adjuvants encouraging the absorption of active constituent.
(c) This approach suggests detailed formulation research of structures, capable of both stabilizing the natural molecules and augmenting their intestinal absorption.

Natural products delivery approach must preferably convey the active constituent at a rate as per the requirements of the body, during the treatment duration and it must guide the desired constituent of herbal extract to the site of action (Saraf 2010). Whereas, conventional drug delivery systems are unable to attain these targets, therefore, significant efforts should be made for developing novel herbal drug delivery system. Incorporating herbal extract/phytoactive constituents into such delivery systems not only decreases the frequency of dosing to attain patient compliance, but also favours the improvement in therapeutic effect by lowering toxicity or augmenting the absorption (Musthaba et al. 2009; Obeid et al. 2017). Thus, novel drug delivery approach is designed to either maintain drug action at a predecided rate or by ensuring a proper and continual effective drug concentration in the body with lowering of undesirable side effects.

Phytopharmaceuticals with greater therapeutic potential may be explored by means of several novel formulation techniques. Complexation of standardized herbal extracts, or mainly, polar phytoactive constituents with lipids especially phosphatidylcholine, results into new drug delivery technique known as phytolipid complexes/phytosomes/herbosomes that produces a greater uptake after the oral route due to higher lipid solubility and permits crossing of the cell barrier consequentially resulting in improved bioavailability. This indicates that a large number of active constituents are working at the site of action in equal or a smaller amount than conventional herbal extract. Hence, the therapeutic efficacy turns into superior, extra measurable and long-lasting outcome (Talware et al. 2018). Thus, phytosome is an herbal drug delivery system which is known to placate all above problems or limitations of conventional systems of herbal drug delivery.

7.3 Phytosomes

The term "Phytosomes" comprises of two words i.e. 'Phyto' meaning *Plant* and 'Some' meaning *cell-like*. They are micelles obtained by complexation of phospholipids with water and this type of complexation is exaggerated by the adding polyphenolic plant extract. The phytosome technique was first established by Indena, an Italian Company. Bombardelli and Spelta in 1991 developed a novel drug delivery system called *Phytosomes* (Bombardelli et al. 1991). This technique is categorized under VDDS which encompasses aquasomes, ethosomes, liposomes, niosomes, and phytosomes. VDDS are systems consisting of a hydrophilic core and outer lipid bilayer shell (Azeez et al. 2017). The fatty bilayer of phytosomes assists 'contact-facilitated drug delivery' which comprises of lipid-lipid interaction between the cell membrane and carrier, causing diffusion of phytoactive constituents inside the cell (Bhupinder et al. 2018).

Phytosomes, also known as *herbosomes*, is a novel phytolipid formulation technology attempting to break through the barriers put forth by the conventional drug delivery systems with respect to bioavailability and stability of plant-derived drugs. It is vesicular drug delivery system where phytoactive constituents of natural product are surrounded by lipid, consequently forming a formulation with better absorption than the conventional ones (Pawar and Bhangale 2015). It is a biocompatible and biodegradable delivery system formed via complexation in a stoichiometric ratio of a phytoactive chemical, or a mixture of phytochemicals, with a phospholipid, mainly phosphatidylcholine or phosphatidylserine in an aprotic solvent, as depicted in Fig. 7.1 (Karatas and Turhan 2015). The spectral values discloses the phospholipid-phytoconstituent interactions due to hydrogen bond formation between the hydrophilic part and the polar parts of the phytoactive constituent (Tripathy et al. 2013).

The average size of a phytosome differs from 50 nanometres to some hundred micrometres (Patel et al. 2013). Phytosome preserves the key components of natural product extract from damage by digestive chemicals and bacteria, as a result of which, improved absorption and bioavailability lead to elevated biological and pharmacokinetic parameters as compared to conventional herbal extracts (Lu et al. 2019).

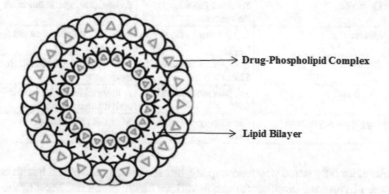

Fig. 7.1 General structure of phytosomes depicting the drug-phospholipid complex in which choline head of phosphatidylcholine (PC) binds to the phytoactive constituent and the tail encapsulates the polar portion of complex to provide a hydrophobic surface

Table 7.1 Phytosome based products developed by various manufacturers alongwith their biological source and therapeutic significance

Sr. No.	Product name	Botanical origin & plant part used	Therapeutic significance
1.	18 Beta-glycyrrhetinic acid phytosome	*Glycyrrhiza glabra* L. root	Lenitive, soothing
2.	Bosexil	*Boswellia serrata* Roxb. root	Anti-photo ageing, soothing
3.	Casperome		Joint health
4.	*Centella asiatica* triterpenes phytosome	*Centella asiatica* L. leaf	Anti-wrinkles, collagen remodelling & reorganization
5	Escin Beta-sitosterol phytosome	*Aesculus hippocastanum* L. seed	Vasokinetic action, shining of skin
6.	*Ginkgo biloba* dimeric flavonoids phytosome	*G. biloba* L. leaf	Antilipase, phosphodiesterase inhibitor
7.	*G. biloba* extract phytosome	*G. biloba* L. leaf	Anti-allergic agent
8.	Ginkgoselect phytosome	*G. biloba* L. leaf	Free radical scavenger, improves cognition and blood circulation
9.	Virtiva	*G. biloba* L. leaf	Cognitive enhancer
10.	Ginseng phytosome	*Panax ginseng* C.A. root	Improves elastic strength of skin
11.	Greenselect phytosome	*Camellia sinensis* L. Kuntze young leaf	Free radical scavenger, weight regulator, whitening agent
12.	Hawthorn phytosome	*Crataegus* spp. flowering tops	Antioxidant, improve cardiovascular fitness
13.	Leucoselect phytosome	*Vitis vinifera* L. seed	Antioxidant, cardio-protective, ultra-violet protective
14.	Meriva	*Curcuma longa* L. rhizome	Joint strengthner, soothing
15.	Proanthocyanidin A_2 phytosome	*Aesculus hippocastanum* L. bark	Ultra-violet protective, promotes skin repair and skin firmness
16.	Quercefit	*Sophora japonica* L. flower buds	Antioxidant, sports nutrition
17.	Quercevita	*S. japonica* L. flower buds	Soothing, lenitive, first aid for skin challenges
18.	Rexatrol	*Polygonum cuspidatum* Sieb. rhizome	Anti-ageing, antioxidant, Sirt1 modulator
19.	Siliphos	*Silybum marianum* L. fruit	Anti-wrinkles, retinoic acid like activity, hepatoprotective
20.	Bergamot phytosome	*Citrus bergamia* fruit juice	Maintains blood levels

Indena, an Italy based pharmacompany, has manufactured several health care & skin care phytosomal products for use in various health problems. Most of the standardized herbal extracts containing polyphenolics and terpenoid fractions are extensively formulated as phytosomes, some of them are summarized in Table 7.1 (Kumar et al. 2019).

7.3.1   Mechanism of Drug Complexation

Phytosomes are the product obtained by the interaction of a phospholipid (stoichiometric quantity) with the plant extract or phytoactive constituent inside an aprotic solvent (Patel et al. 2009). Phosphatidylcholine is a bi-functional compound in which, the choline part of molecule binds to the phytoactive constituent whereas the tail (phosphatidyl part comprising of two long lipophilic chains) encapsulates the polar portion of complex to provide a hydrophobic surface (Amit et al. 2013). Spectral techniques have shown that the phytoactive constituents bind with the choline head through H-bonds thus forming stable and more bioavailable form of herbal drug delivery (Bombardelli et al. 1991).

7.3.2   Components of Phytosomes

7.3.2.1   Phyto-Active Constituents

Herbal drugs are selected on various basis primarily on utility of drug in therapeutics, their nature, availability, estimation method and stability. Nature of phytoactive constituent in the herbal extract is a crucial element in drug selection. Herbal extracts being multicomponent mixture, generally possess multiple ring molecules, owing to large size, are tough to be diffused by simple passive diffusion and depict poor penetrability through the cellular lining of the intestine (Sarika et al. 2016). For producing a phytosomal complex, the drug moiety must have active hydrogens like −NH$_2$, −NH, −OH, −COOH, having the capacity of estabilishing H-bonds between the drug and N-methyl groups of PC (Semalty et al. 2009). Previous work has reported that molecules with conjugated systems of pie electrons are proficient in making various types of complexes with phospholipids.

7.3.2.2   Phospholipid

Phospholipids being amphiphilic and zwitterionic molecules, are considered as a potential component of cell membrane (Olsson and Salem 1997; Singh et al. 2017). They extensively exists in plants and mammals as they are among the necessary constituents of cellular membrane (Seimiya and Ohki 1973). Main sources of phospholipids include rape seed, sunflower seed, soya bean, vegetable oils and cotton whereas mammalian tissues include bovine brain and egg yolk (Chaurio et al. 2009). Phosphatidylcholine (PC), phosphatidylethanolamine (PE), and Phosphatidylserine (PS) are the chief phospholipids largely employed to formulate complexes consisting of an aqueous head group and two lipophilic chains (Suriyakala et al. 2014). However, PC is most widely employed phospholipid, due to its amphipathic nature which imparts it adequate solubility in water and lipid medium. Furthermore, being

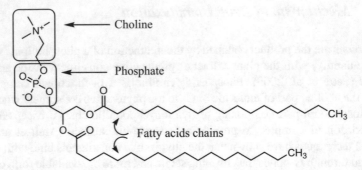

Fig. 7.2 Structure of phosphatidylcholine consisting of a hydrophilic head group and two long lipophilic fatty acid chains

vital component of cell membranes, PC displays strong biocompatibility and little toxicity (Li et al. 2015). Phosphatidylcholine consists of two aliphatic chains surrounding the choline-phytoactive drug complex as depicted in Fig. 7.2.

7.3.2.3 Solvent

During the phospholipid complex formation, the selection of organic solvent plays an important role and relies on the solubilisation parameters of both drug and phospholipids. Previous studies suggested that improved solubility can be achieved by using both protic and aprotic solvents in combination. Majority of the aprotic solvents like chloroform, diethyl ether, dioxane, dichloromethane and n-hexane have been recently substituted with ethyl alcohol being much safer than the earlier solvents (Shakeri and Sahebkar 2016). Since ethyl alcohol leaves less residues and causes minimal damage, it may be a valuable solvent when the phospholipid complex yield is sufficiently high (Patel et al. 2009).

7.3.2.4 Stoichiometric Proportion of Active Constituents and Phospholipids

Usually, plant-phospholipid compounds are produced from interaction between natural/synthetic phospholipids and active herbal component in the molar ratios ranging from 0.5 to 2.0. But, according to literature and previous studies, stoichiometric proportion of 1:1 is considered as the most proficient proportion for formulating complexes of phospholipid (Bombardelli et al. 1991; Bombardelli 1994). This is evident from a study where quercetin-phospholipid complexes were formulated by mixing quercetin and Lipoid S100 at a molar proportion of 1:1 (Zhang et al. 2016). However, various other proportions of phospholipids and phytoconstituents have also been used.

7.3.3   Properties of Phytosomes

Basically, phytosome is an herbal drug-phospholipid complex formed by reacting stoichiometric amounts of phospholipid and drug in a suitable solvent. Data from spectroscopic studies reveals the key phospholipid-drug interaction owing to the development of hydrogen bonds between the hydrophilic heads of phospholipids (ammonium and phosphate groups) and polar groups of the drug. Phytosomes assumes micellar shape due to the presence of water, which make them look like liposomes constructions. Undoubtedly, phytosomes are better forms of natural products which are stable, easily absorbed and thus produce better results than the conventional herbal extracts (Kumar et al. 2017).

Phytosomes should not to be puzzled with liposomes. Liposomes are formed by mixing a water-soluble substance with phosphatidylcholine without forming any chemical bond, and there may be numerous phosphatidylcholine molecules surrounding the water-soluble compound, whereas in case of phytosomes, the PC and individual phytoconstituent form a 1:1 or a 2:1 complex involving a hydrogen bond formation in presence of suitable solvent (Suryawanshi 2011). The comparison between a liposome and a phytosome has been shown in Fig. 7.3 and Table 7.2 (Gandhi et al. 2012; Kumar et al. 2017).

Fig. 7.3 Comparison between phytosome and liposome representing phytosome consisting phospholipid-drug complex (where drug is linked to phosphatidylcholine through chemical bond in the ratio of 1:1 or 1:2) and liposome where several PC molecules surround the water soluble drug (Akki et al. 2019)

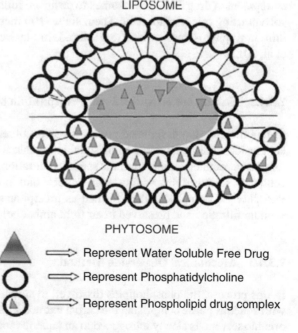

Table 7.2 Comparison of phytosome with liposome based on linkage through bonds, degree of absorption and bioavailability alongwith molecular arrangements

Characteristics	Phytosome	Liposome
Bond linkage	It is a complex of few molecules united through chemical bonds.	In liposomes, no chemical bond is formed.
Absorption and bioavailability	It affords better absorption and bioavailability.	The bioavailability and absorption parameters of liposomes are slighter than phytosomes.
Organisation of molecules	Phospholipid and every phytoconstituent are present in 1:1 or 2:1 proportion.	A large number of phospholipids border the water-soluble phytoconstituents.

7.3.4   Methods of Preparation

There are different techniques or methods employed in the preparation of phytosomes.

7.3.4.1   Solvent Evaporation Method

Here, both the phytoactive constituents and phospholipid are mixed in a container having organic solvent and is usually kept at 40 °C for 1 h to achieve maximum entrapment of drug in the phytosomes so designed followed by removing the organic solvent using rotary evaporator. Then, sieve (100 mesh size) is used for separating thin film phytosomes, which are stored in desiccators for overnight (Habbu et al. 2013).

7.3.4.2   Salting out or Anti-solvent Precipitation Method

In this method, the designated phytoactive constituent and phospholipid both are kept inside a flask containing some suitable organic solvent and then, the mixture is refluxed at defined temperature for specified duration. Later, the solution gets concentrated followed by adding anti-solvent like n-hexane (Saoji et al. 2016). Phospholipid complex is then formed as precipitates which are further subjected vacuum filtration and preserved in air tight amber coloured glass containers.

7.3.4.3   Mechanical Dispersion Method

In this process, the phospholipids dissolved in suitable solvent are kept in vicinity with aqueous phase comprising the herbal extract. Firstly, phospholipid dissolved in organic solvent, is slowly introduced to an aqueous solution containing the phytoactive constituents to be complexed with. Consecutive removal of the organic solvent under vacuum results in phyto-phospholipid complex development. Recent approaches for formulating such complex include super critical fluids, further encompassing compressed antisolvent process, gas antisolvent method and super-critical antisolvent methods (Li et al. 2008).

7.3.4.4   Lyophilization Technique

In this method, both phospholipid and phytoactive constituent are dissolved in respective solvents and further, the solution containing phytoconstituent is added to a solution having phospholipid followed by stirring till complex formation occurs which is separated by lyophilization (Mascarella 1993).

Various stages employed in formulation of phytosomes are depicted in Fig. 7.4.

7.3.5   Release of Drug from Phytosomes

Phytosomes are preferable while delivering lipophilic herbal drug components. Due to hydrolytic digestion, lipophilic components are likely to develop clusters in the intestine which would obstruct the controlled and constant release of the drug into systemic circulation. This problem is rectified when lipophilic drug are delivered through phytosomal complexes where the phospholipid develops a monolayer in the

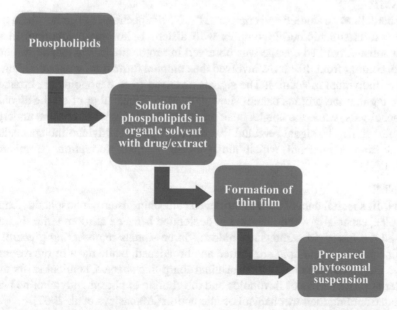

Fig. 7.4 General steps employed in preparation of a phytosome i.e. (i) Dissolving phospholipids in an organic solvent containing the drug, (ii) Removing the organic solvent by drying followed by filtration and (iii) Formation of thin film of phytosomes

digestive track, deters the cluster development, and boosts the dissemination of lipophilic drugs through the small intestine (Hamed et al. 2017). The entrapment efficiency of the phytosome reported previousely lies in the range of 86–98%, which could be due to the bond formation between the phytoconstituent and the polar head of the charged phospholipid boosting the association of phytoconstituents with the polar heads. The release of drug from phytosome complex is reported to be time-dependent and diffusion controlled. The maximum amount of drug release was estimated nearly 80–85% from the phytosomal complex. However, the rate of drug release from phytosomes is slower than the liposome is due to the linkage of the drug with the phosphatidyl head (Amit et al. 2016; Vu et al. 2018; Singh et al. 2018). The reason behind firmness of the phytosomal shell can be credited to the time-dependent drug release from the complex.

7.3.6   Characterization of Phytosomes

7.3.6.1   Spectroscopic Evaluation

Phytoactive constituents-phospholipid complexation and their molecular relationships are investigated using different spectral techniques like ^{1}H-NMR, ^{13}C-NMR, and IR (Dasgupta et al. 2015).

^{1}H-NMR
Bombardelli et al. studied and compared the NMR spectra of (+) cathechin, a flavanoid and its stoichiometric complex with disteroyl-phosphatidylcholine. In nonpolar solvent, marked change was observed in proton nuclear magnetic resonance signal coming from the atoms involved the complex formation, without adding signals of individual molecules. The signals observed from flavonoids are broadened depicting that the protons cannot show along with broadening of all the signals in phospholipids, while the singlet peak analogous to N-methyl of choline undergoes an uplift shift. The signal revealed that disteroyl-phosphatidylcholine conceals the signal from polyphenol which attributes in complex formation (Bombardelli et al. 1991).

^{13}C NMR
^{13}C-NMR spectral data of above mentioned cathechin-disteroylphosphatidylcholine complex, especially when recorded in deuterated benzene at room temp. Indicates that all the flavonoid carbons are hidden. Some signals representing glycerol and choline portion of the lipid are shifted and broadened, while most of the resonance of the fatty acid chain holds their original sharp line shape. Chemical shifts of ^{13}C nuclei in various rings of flavonoids and the choline of phosphatidylcholine helped in understanding their mechanism of interaction (Afanas'eva et al. 2007).

FTIR

The recorded IR spectral values are interpreted for different functional groups at their respective wave number (Udapurkar et al. 2018). It confirms the phytosomal complex formation by comparing the IR values of individual constituent and their physical mixture with those of complex. This technique is also used to determine and ensure the permanency of prepared complexes. The permanency is further ascertained by comparing the spectrum of phytosomal formulation (solid form) with that of micro-dispersion (in water) after lyophilisation at different time intervals.

7.3.6.2   Visualization

The Phytosomal formulations can be visualized either by TEM or SEM techniques. In SEM, the sample (dried) is placed on electron microscope brass stub, coated with gold in an ion sputter whereas in TEM, transmitted electrons are used to divulge the surface morphology with interior structure and crystallographic nature of the sample along with it is used to depict the size of phytosomal vesicles (Gnananath et al. 2017).

7.3.6.3   Morphological Evaluation

Morphological parameters like particle size of phytosome etc. are measured by dynamic light scattering particle size analyser (Alexander et al. 2016).

7.3.6.4   Entrapment Efficiency

Entrapment efficiency is can be estimated by centrifugation method (Kumari et al. 2011). Phytosomal formulations kept inside centrifuge tube are centrifuged at 14,000 rpm for 30 min. The collected supernatant (1 mL) is diluted with phosphate-buffered solution having pH 7.4 or distilled water. Then, the concentration of entrapped phytoactive constituent (drug) is determined by UV-visible spectrophotometer at a wavelength at which the maximum peak (λ max) is obtained for that constituent.

$$\% \text{Entrapment} = \text{Total drug} \quad \text{Diffused drug} / \text{Total drug} \times 100.$$

7.3.6.5 Vesicle Stability

The average size and structure of vesicles contributing towards vesicle stability are determined by Dynamic Light Scattering and Transmission Electron Microscopy, respectively (Mamta et al. 2015).

7.3.6.6 Crystallinity and Polymorphism

X-ray diffraction (XRD) and Differential scanning calorimetry (DSC) are widely used techniques for determining crystalinity and polymorphism. In DSC, interactions in phyto-phospholipid complex are characteristically identified with the appearance of new peaks, abolition of endothermic peaks, change in peak intensity, shape and its onset, relative peak area, peak temperature and melting point, or enthalpy etc. In XRD, phyto-phospholipid complexes are analysed either by reduction or complete absence of the intensity of large diffraction peaks corresponding to its crystalline drug (Chaturvedi et al. 2015).

7.3.6.7 *In-vitro* Drug Release

Determination of the permeation rate can be achieved by *in-vitro* drug release study. For calculating the drug release, phytosomal suspension is placed in Franz diffusion chamber where the samples are collected at different time intervals. Then, from the amount of drug entrapped at 0 time as the initial amount, the quantity of drug unconfined is measured indirectly (Kadu and Apte 2017).

7.3.7 Advantages of Phytosomes

Phytosomes have remarkable advantages over conventional dosage forms in terms of several parameters (Bombardelli et al. 1991; Kidd and Head 2005) some of the advantages are mentioned below:

- Potential enhancement of bioavailability of phytoconstituents
- Production of small size cells thereby securing the phytoactive constituents from harm by gastric enzymes/secretory chemicals/gastrointestinal bacteria.
- Delivers both hydrophilic and lipophilic drugs via mouth as well as skin displaying improved absorption, and enhanced therapeutic effects.
- Enhances drug loading efficiency to a level at which the drug itself in conjugation with lipids forms vesicles.
- Appreciable drug entrapment occurs.
- Enhanced drug absorption obviously requires lower amount of drug for the same therapeutic effect.

- Phytosomes are more stable formulations owing to chemical bonds formation between PC molecules and phytoactive constituents (Gabetta et al. 1989).
- Phosphatidylcholine employed in phytosome process, not only acts as a transporter but also provides nourishment to the skin.
- Since the cellular uptake of phytoconstituents is enhanced, consequently the required dose is reduced (Gizzi et al. 2016).
- Phytosomes show liposome-like structures inside water. But, they show advantages over liposomes, mainly including the following benefits: (a) the size of phytosomes is much smaller as compared to that of liposomes; (b) New H-bonds exists in phytosomes whereas no chemical bonds are observed in liposomes; (c) the molar ratio of PC and natural substances is 1:1 or 2:1 for phytosomes depending on the chemical bond-forming element; (d) In liposomes, the natural active ingredients are dissolved in the solvents and entrapped by the liposomal membrane, whereas in phytosomes, the phytoactive compound is associated to the polar portion of PC via chemical bond (Tripathy et al. 2013; Dhase and Saboo 2015).

The extracts of some plants including ginseng green tca, grape seed etc. have been successfully developed in to phytosomal formulations (Barzaghi et al. 1990) where their phytoconstituents are effectively associated to PC (Ajazuddin and Saraf 2010).

7.3.8 Limitations of Phytosomes

The main disadvantage associated with phytosomes is ascribed to the pH sensitivity of phospholipids due to their Zeta potential values, which must be considered while preparing the phytosomes (Ghanbarzadeh et al. 2016). Some other limitations are: (a) Both superficial and bulk drug-lipid interaction is mandatory; (b) Covalent type of linkage is needed to curb drug seepage; (c) They are prone to get merged or aggregated or hydrolysed by chemicals on preservation.

In recently developed drug delivery systems, phytosomes offer an innovative and upgraded technology for the delivery of phytoactive constituents at the targeted site, and at present, a number of phytosomal formulations are in clinical use (Babazadeh et al. 2018).

7.3.9 Recent Research Work on Phytosomes

1. **Soy phytosomebased thermogel as topical anti-obesity formulation:** Nano lipovesicles thermogel of Soybean, *Glycine max* (L.) Merrill, was prepared by solvent evaporation, co-solvency and salting out techniques and its optimized formulation was then selected for further analysis by FTIR and Zeta analyser.

Prepared phytosomes were then incorporated into a thermogel formulation which showed local anti-obesity effect in male albino rats (El-Menshawe et al. 2018).

2. **Development of antidiabetic phytosomes:** Antidiabetic phytosomes of methanolic fruit extract of three plants using a three-factor, three-level Box-behnken design (17 batches) were optimized and characterized. TEM data revealed their improved steadiness and a spherical shape. Optimized formulation gave maximum yield and showed highest entrapment efficiency. Their antidiabetic effect was equivalent to standard drug (metformin) at low dose level (Rathee and Kamboj 2017).

3. **Nano-phytosomes of Silibinin and Glycyrrhizic acid:** Ochi et al. (2016) developed nano-phytosoms of silibinin and glycyrrhizic acid to target liver HepG2 cell lines. Reports showed that such encapsulation of both drugs not only improved the therapeutic potential and permanency of silibinin, but also synergistically increased the therapeutic effect.

4. **L-carnosine phytosomes for ocular delivery:** Novel phytosomal formulation was prepared for ocular delivery of L-carnosine by blending hyaluronic acid hydrogel and phospholipid by solvent evaporation method and the results witnessed improvement in spreading capacity, sustained infusion, rheological and tolerability features for successful delivery of the drug (Abdelkader et al. 2016).

5. **Sinigrin phytosomes for *in-vitro* skin permeation:** Mazumder et al. developed phytosome complex of sinigrin with *in-vitro* skin permeation ability and showed desired release of drug from the complex and also proposed probability of exploiting this formulation for significant delivery of this drug to the skin (Mazumder et al. 2016).

6. **Phyto-liposomes of Silybin-phospholipid complex:** Silybin phospholipid complex by reverse-phase evaporation technique targeted human hepatoma cells and expressed three hundred times more potent biological effect (Angelico et al. 2014).

7. **Development of self-nano emulsifying drug delivery system (SNEDDS) of Ellagic acid:** Avachat and Patel (2015) prepared phytosome complex of ellagic acid-phospholipid by antisolvent process, followed by developing SNEDDS. Study outcome reported that this technique can be served as prominent method to develop formulations of phytoconstituents with limited bioavailability (Avachat and Patel 2015).

8. **Phytosomal formulation of Polyphenolic extracts of Persimmon (*Diospyros kaki* L.) fruit**: Phytosomes (less than 300 nm) of polyphenolic rich fruit extract were successfully prepared encapsulating 97.4% of total phenolics, and exhibited higher antioxidant activity (Direito, et al. 2019).

9. Enhanced therapeutic efficacy of phytosomal complexes for anticancer, antiinflammatory, antioxidant and hepatoprotective etc. activities have been established by formulating phospholipid complexes of potential drugs like daidzein, salvianolic acid B, rutin, 10-hydroxycamptothecin, luteolin, curcumin and silybin (Federico et al. 2006; Maiti et al. 2007; Xu et al. 2009; Peng et al. 2010; Wei et al. 2010; Zhang et al. 2011; Singh et al. 2012).

7.4 Conclusion

Phytoactive constituents are usually polyphenolics, and on complexation with phospholipids like PC form a newer delivery system known as "Phytosomes". The phytosome technology establishes a linkage between the traditional and novel drug delivery systems while dealing with active phytoactive constituents. Phytosomes may serve as a prospective technique in enhancing the absorption or bioavailability of drugs with extremely hydrophilic or lipophilic properties (Gnananath et al. 2017). The greater steadiness of phytosomes seems because of establishment of chemical bonds between phospholipid and phytoactive components. The bio-compatibility, biodegradability, ease of manufacturing, and availability of naturally occurring phospholipids are important advantages of phytosome technology. Moreover, the flexibility of phytosomal bilayer structure allows incorporation of compounds with different structures.

Phytosomes possesses better pharmacokinetic and biological parameters, which qualify them suitable for various other beneficial applications such as anti-diabetic, anti-inflammatory, anticancer, immune-modulator, cardiovascular etc. These formulations also depict their usefulness in cosmetics as anti-aging and for non-pathogenic skin problems. Finally, enhancement of systemic bioavailability of phytochemicals using phytosomal formulations enables dose reduction and thus helpful to improve the cost-effectiveness of treatment. Such systems (phytosomes) provides a stable carrier which ensures enhanced stability, improved hydrophilic drug loading and reduced drug leakage of the system. Undoubtedly, phytosomes are better than the conventional drug delivery systems in terms of the pharmacokinetic and pharmacodynamic and properties (Selvan 2019), but it is evident that this system can still be modified and used in plentiful ways to boost the overall bioavailability of drugs.

References

Abdelkader H, Longman M, Alany R, Pierscionek B (2016) Phytosome-hyaluronic acid systems for ocular delivery of L-carnosine. Int J Nanomedicine 11:2815–2827. https://doi.org/10.2147/IJN.S104774

Afanas'eva YG, Fakhretdinova E, Spirikhin L, Nasibullin R (2007) Mechanism of interaction of certain flavonoids with phosphatidylcholine of cellular membranes. Pharm Chem J 41(7):354–356. https://doi.org/10.1021/np9904509

Ajazuddin A, Saraf S (2010) Application of novel drug delivery system for herbal formulations. Fitoterapia 81(7):680–689. https://doi.org/10.1016/j.fitote.2010.05.001

Akki R, Sri KN, Govardhani KL, Ramya MG (2019) Phytosomes: a novel drug delivery for herbal extracts. Res J Life Sci Bioinforma Pharm Chem Sci 5(2):1069–1082. https://doi.org/10.26479/2019.0502.80

Alexander, A., Ajazuddin, Patel, R.J., Saraf, S., Saraf, S. (2016) Recent expansion of pharmaceutical nanotechnologies and targeting strategies in the field of phytopharmaceuticals for the deliv-

ery of herbal extracts and bioactives. J Control Release 10:110–124. https://doi.org/10.1016/j.
jconrel.2016.09.017

Amit P, Tanwar Y, Rakesh S, Poojan P (2013) Phytosome: Phytolipid drug delivery system for
improving bioavailability of herbal drug. J Pharm Biol Sci 3(1):51–57

Angelico R, Ceglie A, Sacco P, Colafemmina G, Ripoli M, Mangia A (2014) Phyto-liposomes as
nanoshuttles for water-insoluble silybin-phospholipid complex. Int J Pharm 471(1–2):173–181.
https://doi.org/10.1016/j.ijpharm.2014.05.026

Aqil F, Munagala R, Jeyabalan J, Vadhanam MV (2013) Bioavailability of phytochemicals and its
enhancement by drug-delivery systems. Cancer Lett 334(1):133–141. https://doi.org/10.1016/j.
canlet.2013.02.032

Avachat A, Patel V (2015) Self nanoemulsifying drug delivery system of stabilized ellagic
acid-phospholipid complex with improved dissolution and permeability. Saudi Pharm J
23(3):276–289. https://doi.org/10.1016/j.jsps.2014.11.001

Azeez N, Sivapriya V, Sudarshana D (2017) Phytosomes: emergent promising nano vesicular drug delivery system for targeted tumor therapy. Int Res J Pharm 8(34):1–4. https://doi.
org/10.22159/ajpcr.2017.v10i10.20424

Babazadeh A, Zeinali M, Hamishehkar H (2018) Nano-phytosome: a developing platform for
herbal anti-cancer agents in cancer therapy. Curr Drug Targets 19:170–180. https://doi.org/1
0.2174/1389450118666170508095250

Barzaghi N, Crema F, Gatti G, Pifferi G, Perucca E (1990) Pharmacokinetic studies on IdB
1016, a silybin-phosphatidylcholine complex, in healthy human subjects. Eur J Drug Metab
Pharmacokinet 15(4):333–338. https://doi.org/10.1007/BF03190223

Bhattacharya S (2009) Phytosomes: emerging strategy in delivery of herbal drugs and nutraceuticals. Pharma Times 41(3):8–12. https://doi.org/10.1155/2013/348186

Bhupinder K, Reena G, Sachin K, Monica G, Saranjit S (2018) Prodrugs, phospholipids and vesicular delivery – an effective triumvirate of pharmacosomes. Adv Colloid Interf Sci 5(2):233–235.
https://doi.org/10.1016/j.cis.2018.01.003

Bombardelli E (1991) Phytosome: new cosmetic delivery system. Boll Chim Farm 130(1):431–438

Bombardelli E (1994) Phytosome in functional cosmetics. Fitoterapia 65(5):387–401

Bombardelli E, Spelta M, Loggia D, Sosa S, Tubaro A (1991) Aging skin: protective effect of
Silymarin-phytosome. Fitoterapia 62(2):115–122

Bonifacio B, Bento P, Dos M, Ramos S, Bauab T, Chorilli M, Negri S (2014) Nanotechnology-
based drug delivery systems and herbal medicines: a review. Int J Nanomedicine 9(1):1–15.
https://doi.org/10.2147/IJN.S52634

Byeon JC, Ahn JB, Jang WS, Lee SE, Choi JS, Park JS (2019) Recent formulation approaches to
oral delivery of herbal medicines. J Pharm Investig 49(1):17–26

Chaturvedi M, Kumar M, Sinhal A, Saifi A (2015) Recent development in novel drug delivery systems of herbal drugs. Int J Green Pharm 5(2):87–94. https://doi.org/10.4103/0973-8258.85155

Chaurio R, Janko C, Munoz L, Frey B, Herrmann M, Gaipl U (2009) Phospholipids: key players
in apoptosis and immune regulation. Molecules 14(12):4892–4914. https://doi.org/10.3390/
molecules14124892

Cott J (1995) NCDEU update. Natural product formulation available in europe for psychotropic
indications. Psychopharmacol Bull 31(4):745–751

Dasgupta T, Mello P, Bhattacharya D (2015) Spectroscopic & chromatographic methods for quan-
titative analysis of phospholipid complexes of flavonoids-a comparative study. Pharm Anal
Acta 6(1):319–322. https://doi.org/10.4172/2153-2435.1000322

De Smet PAGM (1997) The role of plant derived drugs and herbal medicines in health care. Drugs
54(6):801–840

Dewan N, Dasgupta D, Pamdit S, Ahmed P (2016) Review on Herbosomes, a new arena for drug
delivery. J Pharmacogn Phytochem 5(4):104–108

Dhase AS, Saboo SS (2015) Preparation and evaluation of phytosomes containing methanolic
extract of leaves of *Aegle marmelos* (Bael). Int J PharmTech Res 8:231–240

Direito R, Reis C, Roque L et al (2019) Phytosomes with persimmon (*Diospyros kaki* L.) extract: preparation and preliminary demonstration of in vivo tolerability. Pharmaceutics 11(6):196. https://doi.org/10.3390/pharmaceutics11060296

El-Menshawe S, Ali A, Rabeh M (2018) Nanosized soy phytosome-based thermogel as topical anti-obesity formulation: an approach for acceptable level of evidence of an effective novel herbal weight loss product. Int J Nanomedicine 13:307–318. https://doi.org/10.2147/IJN.S153429

Federico A, Trappoliere M, Tuccillo C et al (2006) A new silybin vitamin-E phospholipid complex improves insulin resistance and liver damage in patients with non-alcoholic fatty liver disease: preliminary observations. Gut 55:901–902. https://doi.org/10.1136/gut.2006.091967

Gabetta B, Zini GF, Pifferi G (1989) Spectroscopic studies on IdB 1016, a new flavanolignan complex. Planta Med 55(7):615. https://doi.org/10.1055/s-2006-962168

Gandhi A, Dutta A, Pal A, Bakshi P (2012) Recent trends of phytosomes for delivering herbal extract with improved bioavailability. J Pharmacogn Phytochem 1(4):6–14

Ghanbarzadeh B, Babazadeh A, Hamishehkar H (2016) Nano-phytosome as a potential food-grade delivery system. Food Biosci 5:126–135. https://doi.org/10.1016/j.fbio.2016.07.006

Giriraj K (2011) Herbal drug delivery systems: an emerging area in herbal drug research. J Chronother Drug Deliv 2(3):113–119. https://doi.org/10.1016/j.fbio.2016.07.006

Gizzi C, Belcaro G, Gizzi G et al (2016) Bilberry extracts are not created equal: the role of non-anthocyanin fraction. Discovering the "dark side of the force" in a preliminary study. Eur Rev Med Pharmacol Sci 20:2418–2424

Gnananath K, Sri Nataraj K, Ganga Rao B (2017) Phospholipid complex technique for superior bioavailability of phytoconstituents. Adv Pharm Bull 7(1):35–42. https://doi.org/10.15171/apb.2017.005

Habbu P, Madagundi S, Kulkarni R, Jadav S, Vanakudri R, Kulkarni V (2013) Preparation and evaluation of Bacopa-phospholipid complex for antiamnesic activity in rodents. Drug Invent Today 5(1):13–21. https://doi.org/10.1016/j.dit.2013.02.004

Hamed M, Abolfazl S, Bahman R, Amin J, Zarrin B, Amirhossein S (2017) Phytosomal curcumin: a review of pharmacokinetic, experimental and clinical studies. Biomed Pharmacother 85:102–112. https://doi.org/10.1016/j.biopha.2016.11.098

Handa SS, Khanuja SPS, Longo KG, Rakesh DD (eds) (2008) Extraction technologies for medicinal and aromatic plants. International Centre for Science and High Technology, Italy

Kadu A, Apte M (2017) Phytosome : a novel approach to enhance the bioavailability of phytoconstituent. Asian J Pharm 5(2):453–461. https://doi.org/10.22377/ajp.v11i03.1445

Karatas A, Turhan F (2015) Phyto-phospholipid complexes as drug delivery system for herbal extracts. Turkish J Pharm Sci 12(1):93–102

Khan A, Alexander J, Ajazuddin A, Saraf A (2013) Recent advances and future prospects of phyto-phospholipid complexation technique for improving pharmacokinetic profile of plant actives. Pharm Rev 168(1):50–60. https://doi.org/10.1016/j.jconrel.2013.02.025

Kidd P, Head K (2005) A review of the bioavailability and clinical efficacy of milk thistle phytosome: a silybin phosphatidylcholine complex. Altern Med Rev 10(3):193–203

Kumar A, Kumar B, Singh S et al (2017) A review on phytosomes: novel approach for herbal phytochemicals. Asian J Pharm Clin Res 10(10):41–47. https://doi.org/10.22159/ajpcr.2017.v10i10.20424

Kumar S, Baldi A, Sharma DK (2019) Phytosomes: a modernistic approach for novel herbal drug delivery-enhancing bioavailability and revealing endless frontier of phytopharmaceuticals. J Dev Drugs 8(1):1–8. https://doi.org/10.4172/2329-6631.1000193

Kumari P, Singh N, Cheriyan B, Neelam J (2011) Phytosome: a noval approach for phytomedicine. Int J Inst Pharm Life Sci 1(2):89–100

Li Y, Yang D, Chen S, Chen S, Chan A (2008) Process parameters and morphology in puerarin, phospholipids and their complex microparticles generation by supercritical antisolvent precipitation. Int J Pharm 359(1–2):35–45. https://doi.org/10.1016/j.ijpharm.2008.03.022

Li J, Wang X, Zhang T, Wang C, Huang Z, Luo X (2015) A review on phospholipids and their main applications in drug delivery systems. Asian J Pharm Sci 10(2):81–98. https://doi.org/10.1016/j.ajps.2014.09.004

Lu M, Qiu Q, Luo X et al (2019) Phyto-phospholipid complexes (phytosomes): a novel strategy to improve the bioavailability of active constituents. Asian J Pharm Sci 14(3):265–269. https://doi.org/10.1016/j.ajps.2018.05.011

Maiti K, Mukherjee K, Gantait A, Saha BP, Mukherjee PK (2007) Curcumin–phospholipid complex: preparation, therapeutic evaluation and pharmacokinetic study in rats. Int J Pharm 330:155–163. https://doi.org/10.1016/j.ijpharm.2006.09.025

Mamta B, Anar J, Shah R (2015) An overview of Ethosome as advanced herbal drug delivery system. Int J Res Rev Pharm Appl Sci 2(1):1–14

Manach C, Scalbert A, Morand C (2004) Polyphenols: food source and bioavailability. Am J Clin Nutr 79:727–747

Mascarella S (1993) Therapeutic and anti-lipoperoxidant effects of silybin-phosphatidylcholine complex in chronic liver disease. Curr Ther Res 53(1):98–102

Mazumder A, Dwivedi A, Fox L, Brümmer A, Preez J, Gerber M (2016) In vitro skin permeation of sinigrin from its phytosome complex. J Pharm Pharmacol 68(12):1577–1583. https://doi.org/10.1111/jphp.12594

Musthaba S, Baboota S, Ahmed S, Ahuja A, Ali J (2009) Status of novel drug delivery technology for phytotherapeutics. Expert Opin Drug Deliv 6(6):625–637. https://doi.org/10.1517/17425240902980154

Obeid MA, Al Qaraghuli MM, Alsaadi M, Alzahrani AR, Niwasabutra K, Ferro VA (2017) Delivering natural products and biotherapeutics to improve drug efficacy. Ther Deliv 8(11):947–956. https://doi.org/10.4155/tde-2017-0060

Ochi M, Amoabediny G, Rezayat S, Akbarzadeh A, Ebrahimi B (2016) In vitro co-delivery evaluation of novel pegylated nano-liposomal herbal drugs of silibinin and glycyrrhizic acid (nano-phytosome) to hepatocellular carcinoma cells. Cell J 18(2):135–148. https://doi.org/10.22074/cellj.2016.4308

Olsson N, Salem N (1997) Molecular species analysis of phospholipids. J Chromatogr B Biomed Sci Appl 692(2):245–256. https://doi.org/10.1016/S0378-4347(96)00507-5

Patel J, Patel R, Khambholja K, Patel N (2009) An overview of phytosomes as an advanced herbal drug delivery system. Asian J Pharm Sci 4(6):363–371

Patel A, Tanwar Y, Rakesh S, Patel P (2013) Phytosome: Phytolipid drug delivery system for improving bioavailability of herbal drug. J Pharm Sci Biosci Res 3:51–57

Pawar HA, Bhangale BD (2015) Phytosome as a novel biomedicine: a microencapsulated drug delivery system. J Bioanal Biomed 7(1):006–012. https://doi.org/10.4172/1948-593X.1000116

Peng Q, Zhang ZR, Sun X et al (2010) Mechanisms of phospholipid complex loaded nanoparticles enhancing the oral bioavailability. Mol Pharm 7:565–575. https://doi.org/10.1021/mp900274u

Rathee S, Kamboj A (2017) Optimization and development of antidiabetic phytosomes by box-benhken design. J Liposome Res 3(2):1–27. https://doi.org/10.1080/08982104.2017

Saoji S, Belgamwar V, Dharashivkar S, Rode A, Mack C, Dave V (2016) The study of the influence of formulation and process variables on the functional attributes of simvastatin–phospholipid complex. J Pharm Innov 11(3):264–278. https://doi.org/10.1007/s12247-016-9256-7

Saraf A (2010) Applications of novel drug delivery system for herbal formulations. Fitoterapia 81:680–689. https://doi.org/10.1016/j.fitote.2010.05.001

Sarika D, Khar R, Chakraborthy G, Saurabh M (2016) Phytosomes: a brief overview. J Pharm Res 15(2):56–62. https://doi.org/10.18579/jpcrkc/2016/15/2/94471

Seimiya T, Ohki S (1973) Ionic structure of phospholipid membranes, and binding of calcium ions. Biochim Biophys Acta Biomembr 298(3):546–561. https://doi.org/10.1016/0005-2736(73)90073-4

Selvan ST (2019) Phytosomes: a newer trend in herbal drug delivery. Mintage J Pharm Med Sci 8(3):1–4

Semalty A, Semalty M, Rawat B, Singh D, Rawat M (2009) Pharmacosomes: the lipid-based novel drug delivery system. Expert Opin Drug Deliv 6(6):599–612. https://doi.org/10.1517/17425240902967607

Shakeri A, Sahebkar A (2016) Phytosome: a fatty solution for efficient formulation of phytopharmaceuticals. Recent Pat Drug Deliv Formul 10(1):7–10. https://doi.org/10.2174/1872211309666150813152305

Sharma S, Sikarwar M (2005) Phytosome: a review. Planta Indica 1:1–3

Singh D, Rawat MS, Semalty A, Semalty M (2012) Rutin-phospholipid complex: an innovative technique in novel drug delivery system- NDDS. Curr Drug Deliv 9:305–314. https://doi.org/10.2174/156720112800389070

Singh RP, Gangadharappa H, Mruthunjaya K (2017) Phospholipids: unique carriers for drug delivery systems. J Drug Deliv Sci Technol 39:166–179. https://doi.org/10.1016/j.jddst.2017.03.027

Singh R, Gangadharappa H, Mruthunjaya K (2018) Phytosome complexed with chitosan for gingerol delivery in the treatment of respiratory infection: in vitro and in vivo evaluation. Eur J Pharm Sci 122:214–229. https://doi.org/10.1016/j.ejps.2018.06.028

Suryawanshi JAS (2011) Phytosome: an emerging trend in herbal drug treatment. J Med Genet Genomics 3(6):109–114

Suriyakala P, Babu N, Rajan D, Prabakaran L (2014) Phospholipids as versatile polymer in drug delivery systems. Int J Pharm Sci 6(1):8–11

Talware N, Dias R, Gupta V (2018) Recent approaches in the development of Phytolipid complexes as novel drug delivery. Curr Drug Deliv 15:755–764. https://doi.org/10.2174/1567201815666180209114103

Tripathy S, Patel D, Baro L, Nair S (2013) A review on phytosomes, their characterization, advancement and potential for transdermal application. J Drug Deliv Ther 3:147–152. https://doi.org/10.22270/jddt.v3i3.508

Udapurkar PP, Dhusnure OG, Kamble SR (2018) Diosmin Phytosomes: development, optimization and physicochemical characterization. Indian J Pharm Educ Res 52(4):S29–S36

Venkatesan N, Babu B, Vyas S (2000) Protected particulate drug carriers for prolonged systemic circulation – a review. Indian J Pharm Sci 62:327–333

Vu H, Hook S, Siqueira S, Müllertz A, Rades T, McDowell A (2018) Are phytosomes a superior nanodelivery system for the antioxidant rutin? Int J Pharm 548(1):82–91. https://doi.org/10.1016/j.ijpharm.2018.06.042

Wei W, Shi SJ et al (2010) Lipid nanoparticles loaded with 10-hydroxycamptothecin-phospholipid complex developed for the treatment of hepatoma in clinical application. J Drug Target 18:557–566. https://doi.org/10.3109/10611861003599461

Xu K, Liu B, Ma Y et al (2009) Physicochemical properties and antioxidant activities of Luteolin-phospholipid complex. Molecules 14:3486–3493. https://doi.org/10.3390/molecules14093486

Zhang Z, Huang Y, Gao F et al (2011) Daidzein-phospholipid complex loaded lipid nanocarriers improved oral absorption: *in vitro* characteristics and *in vivo* behavior in rats. Nanoscale 3:780–1787. https://doi.org/10.1039/c0nr00879f

Zhang K, Zhang M, Liu Z (2016) Development of quercetin-phospholipid complex to improve the bioavailability and protection effects against carbon tetrachloride-induced hepatotoxicity in SD rats. Fitoterapia 113:102–109. https://doi.org/10.1016/j.fitote.2016.07.008

Chapter 8
Albumin as Natural Versatile Drug Carrier for Various Diseases Treatment

Hitesh Kumar Dewangan

Abstract Serum albumin is the most abundant protein in blood. It shows a variety of activities including antioxidant activities and binding activity toward lipophilic hormones, marker for glucose intolerance and protective effects. Albumin is used for the treatment of many diseases such as cancer, liver disease, renal disease, inflammatory disease, and cardiovascular infection. Over the last decade, various drug delivery systems have been created to improve the treatment of various disorders and diseases. Albumin has been generally studied as a protein carrier for the delivery of the drug. Especially, in nanotechnology, it is utilized for the delivery of various drug molecules for the sustained, control release and enhancement of bioavailability. So, the demand for albumin increased annually worldwide due to the different applications of albumin.

This chapter reviews recent progress in several key areas relevant to natural products in nano and micro delivery systems for biomedical applications and its production, purification, and distribution in the human body. Also the present chapter describes the special features of albumin as a drug carrier, and thereby provides scientists with knowledge of goal-driven design on albumin-based nanomedicine.

Keywords Albumin · Nanoparticles · Nanotechnology · Carrier · Drug-delivery · Protein · Application · Treatment

H. K. Dewangan (✉)
Institute of Pharmaceutical Research (IPR), GLA University, Mathura, Uttar Pradesh, India
e-mail: hitesh.dewangan@gla.ac.in

© The Editor(s) (if applicable) and The Author(s), under exclusive license to
Springer Nature Switzerland AG 2020
A. Saneja et al. (eds.), *Sustainable Agriculture Reviews 43*, Sustainable
Agriculture Reviews 43, https://doi.org/10.1007/978-3-030-41838-0_8

8.1 Introduction

The albumin is derived from the Latin word "Albumen" *i.e.* egg white; it is a family of a globular protein and most common serum albumin. It is soluble in water and moderately soluble in concentrated salt solution. Substance containing albumins, such (egg) white, are called albuminoids. Albumin is the abundant constituents of plasma protein and having the concentration is about 3.5–5.0 g/dl. The molecular weight of the albumin protein is about 66,500 Dalton and the elimination half-life of albumin is about 20 days. Albumin present in the plasma as well as in the extracellular space with a concentration of 40% and 60% respectively. Serum albumin is mainly present as the plasma protein and having a variety of functions such as antioxidant, buffering agent and efficacy for binding towards non-polar hormones and transported various molecules (Larsen et al. 2016; Nicholson et al. 2000). Albumin is also used as the macromolecular carrier which is biodegradable, non-toxic, non-irritant, biocompatible and non-immunogenic. Albumin nanoparticles based drug delivery systems are very precise and effective for active as well as passive targeting. In vivo drug delivery with the albumin requires the modification for better effectiveness and also for diagnosis (Sleep et al. 2013; Fanali et al. 2012).

8.2 Structure of Albumin

Chemically albumin containing three structures such as primary structure, secondary structure, and tertiary structure, and not having carbohydrate moiety. It has an elastic constitution and easily regains shape (Fig. 8.1).

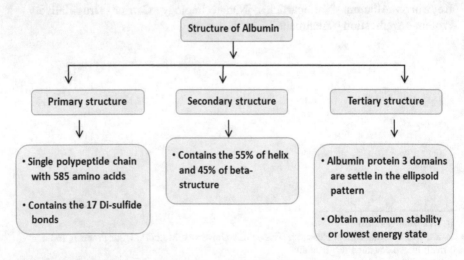

Fig. 8.1 Chemical structures of albumin: primary structure, secondary structure, and tertiary structure, and its alignment

Primary structure: The primary structure of albumin is made up of a single chain of a polypeptide that contains around 585 amino acids. The molecules of the peptide chain are bind with 17 disulfide bonds (Farrugia 2011).

Secondary structure: Albumin secondary structure having helix shape (55%) and beta-structure (45%).

Tertiary structure: In these three domains are situated in ellipsoid pattern and proteins molecules are blended and twisted in specific patterns. Comparable to primary and secondary structure, tertiary is more stable and lower energy state (Lee et al. 2014).

8.3 Types of Albumins

There are three types of naturally occurred albumin (Raoufinia et al. 2016a, b; Sokołowska et al. 2009):

  I. Ovalbumin (OVA)
 II. Bovine serum albumin (BSA)
III. Human serum albumin (HSA)

I. Ovalbumin Ovalbumin is the food protein and widely used in the designing of the food matrix. Ovalbumin chemically consisting of phosphor-glycoprotein in which 365 amino acid residues are found and each molecule bonded by the disulfide bonds. The molecular weight of ovalbumin is about 47,000 Dalton and having the isoelectric point at 4.8 (Elzoghby et al. 2012). Ovalbumin is the most frequently used carrier system for various drug deliveries due to its low cost as a comparison to other protein. Ovalbumin has the quality to form gels and also used as a stabilizer for the biphasic liquid dosages form such as emulsion and various pharmaceutical foam. Ovalbumin is pH and temperature-dependent, due to these properties it is also used in the controlled as well as targeted drug delivery (Wongsasulak et al. 2010).

II. Bovine Serum Albumin Bovine serum albumin is used in the transport of many drugs and used as a nano-carrier because it has low cost and acceptable in the pharmaceutical industry due to highly biocompatible. The purification of bovine serum albumin protein is very easy compared to the other one. The molecular weight of bovine serum albumin is about 69,323 Dalton and isoelectric point is 4.7 in the water at 25 °C (Tantra et al. 2010; Hu et al. 2006).

III. Human Serum Albumin Serum albumin is a very common plasma protein and having the elimination half-life is about 19 days which is very soluble, consisting of 585 amino acids (Hirose et al. 2010; Kratz, 20,008). Human serum albumin has a molecular weight of 66,500 Dalton and stable at the pH range of a 4–9 and temperature at 60 °C for 10 h. Human serum albumin is the best carrier for the transport of drug molecules due to having the ability to biodegradability and reduction in the toxicity (Kratz, 20,008; Fasano et al. 2005).

8.4   Albumin Synthesis, Distribution, and Degradation

Albumin is synthesized in the liver and daily 12 g of albumin is produced, which represents 25% of total hepatic protein synthesis. After synthesis of albumin in hepatocytes cells, it does not store in a liver, it circulates in portal. About 12–25 g of albumin is synthesis per day in normal human and its natural half-life is 19 days. The normal concentration of albumin is 2.9–5.5 g/dl in youngsters and 3.5–5 g/dl in adults. 35% of the total albumin is store and found in the intravascular compartment. The albumin digestion and synthesis all are regulated by a specific gene. For example, in case of injury, diabetes, hepatic ailments, sepsis, low corticosteroids and reduce growth hormone, the process of gene transcription is reduced likewise the albumin synthesis is also decreased and disaggregation of the ribosome (Bernardi et al. 2012).

Initially, albumin is synthesized by pre-protein. Liver hepatocyte cells firstly synthesized one type of protein called as pre-pro-albumin. In the presence of enzyme signal peptidase pre-pro-albumin is converted to the pro-albumin and further pro-albumin converts to albumin in presence to furin (Fig. 8.2). This synthesized albumin protein circulates in intravascular and extravascular space followed by the lymphatic system. Around 40% albumin presented in intravascular space and 60% found in extravascular space. After long circulation albumin is depredated by muscle, kidney, liver, and GIT (Bernardi et al. 2012; Brzoska et al. 2004).

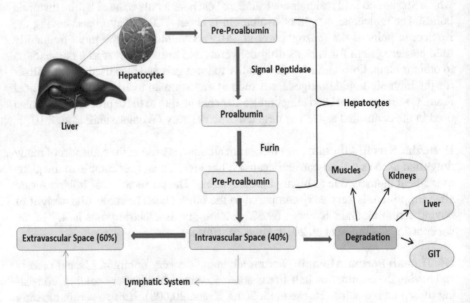

Fig. 8.2 Synthesis, distribution, and degradation of albumin. Albumin is synthesized in the hepatocytes cells and circulates. Albumin is found in the intravascular space (40%) and extravascular space (60%) and degraded in organs like muscle, kidney, and liver

Serum albumin is produced on a particular ribosome-messenger ribonucleic acid (mRNA) complex. Which are further bound to the ER (endoplasmic reticulum) in the hepatocyte and only 30 min is required for the gathering of molecules of the albumin and then it is discharged into the systemic circulation. However, the time required to take in this step is unidentified. The most significant perspective in the guideline of albumin creation is nitrogen intake. Short term fasting quickly exhausts the measure of albumin synthesis. Chronic ailing health likewise is related to a low level of albumin synthesis, even though this issue by infection or due to altered hormone level and as often as possible by raised globulin levels (Gradishar 2006).

Decrease synthesis limit should include a changed in the amount of polysome wherein also reduction of mRNA concentration, or maybe diminished the production of an inhibitor to ribonuclease which ensuing the damage of RNA and disaggregation of the ribosome from mRNA. Since albumin is accepted to be made and discharged on an mRNA-ribosome film formation. Fasting may be relied upon to change this piece of the intracellular mechanism. Despite that, fasting does not decrease explicitly the amount of endoplasmic polysome, even as all RNA has decreased the segment between free (non-membrane bound) RNA. However, the free or unbound polysome and the film bound polysome are disaggregated (mRNA with fewer connected ribosomes) into littler subunits which are prudently inactive in synthesized protein (Bernardi et al. 2012).

A protein inhibitor to ribonuclease is available in the cell sap. The lacking protein production inside the cell could allow more prominent ribonuclease activity by diminishing the synthesis of this inhibitor. To re-stimulate and balance out this framework to a sustained state it isn't important to utilize a reasonable eating routine. They demonstrated that the re-aggregation of ribosomes with mRNA happens inside hours following the organization of abundance tryptophan and isoleucine. Thus, both of the amino acids result in expanded of albumin production. The stimulatory system of activity for two amino acids isn't recognized. An abnormal state after the administration of these amino acids, it is conceivable that a portion of the mRNA coded for albumin is enduring and accessible for re-aggregation. Qualities for the half-existence of the albumin mRNA is broadly dissimilar (Varshney et al. 2010).

8.4.1 Albumin Degradation

Albumin degradation is likewise fairly consistent in man and several minor creature species. There are two viewpoints of degradation which must be isolated simply. The partial degradative rate for the piece of plasma albumin or the whole interchangeable albumin mass which is metabolized day by day and unqualified quantity. Experimentally the partial rate may remain. Generally consistent despite huge changes in the supreme rate and albumin debasement might be viewed as unaltered. In hypoalbuminemia, the fractional rate is frequently ordinary, especially when the hypoalbuminemia is related to hyper gammaglobulinemia.

Scarcely few examinations have been performed utilizing patients in hyper albuminemic states since this is a clinical condition once in a while observed except maybe in intense lack of hydration. In any case, after albumin infusion, an increment in absolute derivative rate happens and is related to an expansion in the partial rate of debasement. These progressions were adequate to represent most of the overabundance colloid injected. No adjustment in albumin synthesis happened. It shows that albumin synthesis and degradation are not interdependent. While albumin synthesis is influenced intensely by a reduction in sustenance admission, albumin degradation falls simply after the albumin pool has diminished (Bernardi et al. 2012; Varshney et al. 2010).

8.4.2 Albumin Distribution

Albumin is circulated between the plasma & extra plasma pools in around a 40:60 proportion in man. The majority of the extravascular albumin is situated in the skin in the human and the grouping of albumin in the extravascular extracellular skin water approaches in the plasma. These abnormal states play a role in controlling the osmotic balance of the skin during water exchange. In the liver the measure of albumin situated in hepatic interstitial water is little and, accepting an even circulation of albumin in hepatic interstitial space, the concentration would be short of one-tenth of that in plasma or lymph. In any case, the hepatic interstitial space is comprised of a protein-polysaccharide grid which may exclude albumin from its area. Albumin has comparative unequal dissemination in the pulmonary interstitial volume and different tissues likely will demonstrate varieties of equal degree. Albumin has been disconnected in tears, sweat, salivation, gastric juice, ascites, and edema and assuredly is available in each body liquid and the concentration will be varied (Bernardi et al. 2012; Varshney et al. 2010).

8.5 Sources of Albumin

Albumin protein is natural polymers obtain from a variety of sources and used in various drug delivery systems. Albumin is either obtained from animal or plant sources. Animal sources like a human serum, eggs, chicken, fish, yogurts, meats substitute, and nutritional drinks, etc. Plant sources such as peanut cheese, potatoes, tofu, burgers, castor bean, etc.

8.6   Methods of Purification of Albumin Protein

8.6.1   Plasma Fractionation Method with Ethyl Alcohol

Purification method was first developed by the Cohn and colleagues at 60 years ago by using the fractionation method. This method is based on the difference in solubility of albumin by comparing with other proteins present in the plasma and albumin protein will precipitate at low pH and high concentration. The separation of protein is obtained by increasing the concentration of ethyl alcohol in every fraction and albumin was found in the last fraction. The process will continue or done in a bacteriostatic condition which is used as an active molecule. This method applies to large scale production in the industry (Buchacher and Iberer 2006).

8.6.2   Combination of Cohn Method and Liquid Chromatography

This method is worldwide applicable and the technique was demonstrated given by Tanaka et al in this higher volume of plasma and low price is fractionated in the Cohn technique, therefore, Victoria and their colleagues increases the quality as well as purity by incorporating the liquid chromatography. This method is said to be an advanced method of plasma protein purification. The main importance of this method is that low cost and a high degree of purification can increase productivity and the technique can increase the yield at least 99% compared to Cohn methods. Purified albumin protein present as by-product and other proteins such as factor VIII and immunoglobulin and others are also isolated (Tanaka et al. 1998; Ramin et al. 2016).

8.6.3   Purification from Placenta

The strategy that has been created for the generation of albumin is cleaning from the human placenta produced by the Raoufinia and his associates. In this strategy technique utilizing ethanol-based solvent precipitation and ion-exchange chromatography methods. Since placenta accumulation is accessible, filtration from the placenta is a standout amongst the most proficient methods for the decontamination of albumin (Raoufinia et al. 2016a, b).

8.6.4   Affinity Precipitation

Cao and Ding suggest an affinity precipitation process for purification of albumin. In this method utilizing a soluble-insoluble polymer joined with an affinity ligand that relies upon the polymer highlights, for example- pH, temperature and light reaction. With utilizing butyl acrylate, N-methylol acrylamide and N-isopropyl acrylamide as monomers, a thermo-reaction copolymer was blended which was called PNBN (Ramin et al. 2016).

8.6.5   Heat Shock Method

Albumin protein shows the stability of increasing the temperature in comparison with the other protein found in the plasma so it can be purified by the heat shock method. On increasing the temperature in the range of 60 °C albumin shows the resistant and the microorganism may be inhibited or inactivated at this temperature, 0.04 M caprylic acid is used to stabilize at the pH range of 5 and also stabilize an increase in temperature up to 60 °C. In this situation known as an increase in temperature causes denaturation and precipitation in the solution of other proteins found in the plasma but not albumin protein, so albumin protein was concentrated by using the ultrafiltration and purity is about 98% was obtained (Ramin et al. 2016).

8.6.6   Precipitation of Ammonium Sulfate Conjugates with Liquid Chromatography

This method of purification of albumin protein Precipitation of ammonium sulfate conjugate with liquid chromatography can yield 90% of purified albumin. Applying these method for separation of albumin from the immunoglobulin. When treated with ice-cold acetone around ammonium sulfate (50%) and lipid contents are removed. Then, perform separation of transferrin in the next steps. Especially two chromatography was used such as ion exchange and size exclusion chromatography. Purified albumin was free from the DNAse and immunoglobulin proved by assays such as DNAse and immunoglobulin detection assay and albumin was evaluated by using the blotting and chemiluminescence method (Odunuga and Shazhko 2013).

8.6.7 Chromatography Method for the Purification

Mostly in industry ethyl alcohol fractionation method is utilized for the fractionation and identification of plasma proteins. This method is accurate and given the purity of compounds. Various chromatographic methods are available for the purification and identification of albumin or other proteins. According to the need and desirability method is selected. Various chromatographic methods such as ion-exchange chromatography (IEC), simulated moving bed chromatography (SMBC), steric exclusion chromatography (SXC), expanded bed adsorption chromatography and affinity chromatography have used the identification of albumin (Varshney et al. 2010).

8.7 Albumin Properties

Albumin, a multifunctional protein has ligand-restricting capacities with regards to cancer prevention agent properties. It also has free radical-catching properties in egg whites. Impairments in egg whites are cancer prevention agents. Other properties such as limits oxidation, high availability, biodegradability, no any poisonous quality, medical significance, most abundant and easy of decontamination (Sleep 2015; Taverna et al. 2013).

8.7.1 Enzymatic Properties

Serum plasma albumin has fascinating enzymatic properties such as enolase action, esterase action, impacts on eicosanoids, aryl acyl amidase action, stereospecificity, buildup responses, official and initiation of medication conjugates. Likewise, a few enzymatic properties of egg whites ligand such as heme and hemin–human albumin (Kragh-Hansen 2013).

8.8 Basic Functions of Albumin

Albumin works basically as a transporter protein for steroids, unsaturated fats, and thyroid hormones in the blood. A main important role in settling extracellular liquid volume by contributing to colloid osmotic pressure of plasma because littler creatures (like rodents) work at a lower circulatory strain, they need less oncotic strain to adjust this, and hence need less albumin to keep up legitimate liquid distribution. The function of albumin is presented in Fig. 8.3. According to the importance of the albumin there following function are performed:

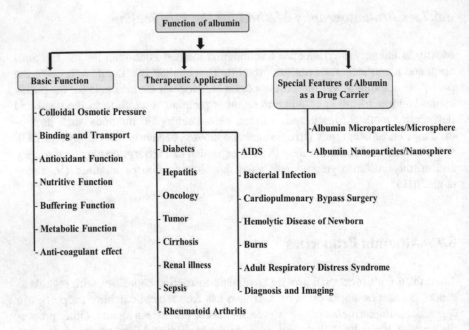

Fig. 8.3 Various functions of albumin: basic function, therapeutic function and special function as a drug carrier

8.8.1 Colloidal Osmotic Pressure

Albumins significantly help in maintaining the osmotic pressure inside the blood capillaries. Change in the concentration of albumin can cause a change in the intra-vascular colloidal osmotic pressure, albumin is found in extra-vascular site and highest concentration or amount found in the interstitial. Many drugs have the good binding capability to albumin and this will show the better delivery of a drug to a tissue site, metabolism as well as the elimination of therapeutic molecules. Naturally found colloids are fresh frozen plasma as well as frozen plasma which are infused in animal medicines (Evans 2002; Tazhbayev et al. 2019).

8.8.2 Binding and Transport Functions

Albumins have many binding sites for binding of therapeutic molecules such as non-polar anions and long chain of fatty acids. A low lipophilic therapeutic molecule does not bind with albumin more frequently and having low affinities such as ascorbate and tryptophan. But a higher lipophilic therapeutic molecule such as L-tryptophan has more binding towards the albumin. The monovalent molecules are not bound to the albumin bur divalent ions have a more binding affinity. Albumin proteins consisting of negative ions so acidic therapeutic molecules are not bound to albumin whereas basic drugs have better binding capability towards albumin. Albumin also used as the primary and secondary carrier for some therapeutic molecules, which is steroidal like vitamin D and thyroxin (Varshney et al. 2010; Curry et al. 1999). Steroidal drugs have low binding affinity towards albumin but albumin present in the highest concentration in blood circulation so the required amount of drug binds to albumin. The delivery the drugs at their specific site this has clinically significant applicable. (Varshney et al. 2010). Albumin ties with bilirubin and gets a breakdown, which resulted in the formation of heme. It also acts as binding as well as a vehicle to carry metals, especially in the blood. It particular bind with copper (II) and nickel (II) and generally nonspecific ties with calcium (II) and zinc (II). It also ties water, cations, (Ca^{2+}, Na^+, and K^+), unsaturated fats, hormones, bilirubin, thyroxine (T4). Its principle capacity is to control the colloidal osmotic pressure of blood (Mendez et al. 2005; Bertucci and Domenici 2002). Human serum albumin has multiple functions with its binding properties such as albumin work as the solubilizing agent for unsaturated long-chain fatty acid and it is the most basic requirement for digestion of lipids.

8.8.3   Anti-oxidant Function

In normal physiological conditions, plasma serum albumin has played a remarkable role such as antioxidant potential inside the cell. It is engaged with searching for oxygen free radicals, which is involved in the pathogenesis of provocative elements. Physiological arrangements of human serum egg whites have been appeared to repress the generation of oxygen free radicals by polymorphonuclear leukocytes (Roche et al. 2008). These are a significant scavenger of oxidizing specialists, for example, hypochlorous acid (HOCl) shaped from the chemical myeloperoxidase, which is discharged by initiated neutrophils. Albumin protein has a more binding affinity towards the copper whereas less binding towards the iron, free radicals such as hydrochloric acid and peroxynitrite. They may cause a reduction of pro-oxidant which undergoes oxidization for prevention to other molecules from being oxidized. The human serum albumin has ligand binding as well as trapping of free radicals (Faure et al. 2008; Van Belle et al. 2010; Garcia-Martinez et al. 2015).

8.8.4   Nutritive Function

Albumin provides nutrition to the body because it is protein and proteins are said to be building blocks of the body. So a decrease in the level of albumin causes albumin. The main causes of albumin are rapid degradation or elimination of albumin protein from the body, improper function of the liver, Congestive heart failure. Others reason like tuberculosis, sarcoma or amyloidosis.

8.8.5   Buffering Function

Albumin proteins also act as a buffering agent that maintains the fluid balance in the outer and inner side of the cell by concentrations gradient. Cells are like balloons that expel air or fluid and require filling again and the cells are burst when cells are more filled so the albumin is required acting as a buffering agent. Albumin has a negative charge, the nearness of albumin particles many charged has been buildups and the overall wealth of albumin in plasma implies that it can go about as a powerful plasma buffer found at the physiological pH.

8.8.6   Metabolic Function

Albumin protein has binding as well as transport function but additionally, it also causes the inactivation of small molecules such as disulfiram which is antibiotics bind irreversibly with the albumin protein and metabolized by the acetylation. Penicillin is also an antibiotic which has binding to the albumin protein and forms the complex and causes the allergy known as bis-albuminemia. Albumin protein also causes inactivation or biotransformation of endogenous substances lipids and eicosanoids and these molecules are bind to albumin. Albumin proteins also cause the stabilization of some eicosanoids like prostaglandin 12 and thromboxane A2.

8.8.7   Anticoagulant Effects

Plasma serum albumin has consequences for blood coagulation. It appears to apply a heparin-like activity, which may be identified with comparability in the structures of the two atoms. Heparin has negatively charged sulfate groups that bind to negatively charged gatherings on antithrombin III, in this way applying an anticoagulant effect. It has numerous negatively charged groups and showed a negative relationship between albumin concentrations. The heparin, which is a necessity in patients have experiencing hemodialysis. These examinations have demonstrated a heparin-like activity of albumin, through an upgrade of the balance of factor Xa by antithrombin III (Elena et al. 2012; Kim et al. 1999).

8.9   Therapeutic Applications

The use of albumin as active molecules carrier such as drug-albumin complex, drug-albumin ligands, drug-albumin nano or microparticles, antibody conjugates, and drug-albumin conjugate, etc. Various active drug molecules, polypeptide or prodrug is covalently or non-covalently bind with plasma albumin protein. When targeting ligands moiety is attached with the active constituents, the shape of the system is additionally changed for example prodrugs combined with protein surface. Other examples like specific antibodies and nanobodies. But in the case of micelle, nanoparticles, microbubbles with lipophilic medicine and diagnostic agents are firstly solubilized in the appropriate solvent system and then this formulation is administered in system circulation by various routes.

8.9.1   Diabetes

Insulin plays a vital role in the treatment of diabetes. The mechanism involved diabetes to control the insulin production in the body and this insulin reduces blood glucose level in the body for over 24 h. Nowadays in advance technology insulin subsequently binds with 5–7 fatty acids. In this their binding site is present in the albumin for the enhancement of bioavailability (Loureiro et al. 2016; Julio et al. 2009). Levemir® (insulin detemir) has developed by Novo Nordisk which is the analog of insulin, applicable for the treatment of diabetes type 1 & 2 and approved in 2004. In this, a saturated fatty acid is present in the chemical form of an amino acid at the B29 position. It is administered by sub coetaneous (SC) route as aqueous vehicles. The pharmacokinetic profile and bioavailability are highly appropriate and its onset of action is 26 h. Introduce another new product Liraglutide (Victoza®) is an albumin-binding by-product of GLP-1. The product is protected from enzymatic metabolic degradation due to plasma albumin protein binding. It is also administered by the SC route and half-life is 11–15 h (Fiume et al. 1997; Wang et al. 2004; Elsadek and Kratz 2012; Kratz 2014).

8.9.2   Hepatitis

Plasma serum albumin is also involved in the treatment of hepatitis. Human Genome Science has built up a new albumin protein-based substances. It has a therapeutic active protein or peptide which will be hereditarily intertwined to recombinant human albumin. The lead compound of this innovation has Albuferon®. Albuferon® is an albumin conjugate for liver focusing on. 'Albinterferon-α-2b', a combination macromolecule of albumin and interferon-α-2b (INFα-2b) that is utilized for the treatment of hepatitis C contamination. INFα-2b contains a mass of ~19 kDa. It restricted to use thusly and is expected to infuse as often as possible (every day or thrice week by week). By the hereditary combination of recombinant INFα-2b with recombinant serum albumin, a combination macromolecule with a mass of 85.7 kDa was created (Fiume et al. 1997).

8.9.3 Oncology

Albumin proteins are most effectively used as carrier in cancer because it will deposit on the site of tumor and uptake effectively due to enhancement in the metabolic activity and vascular permeability. Albumin and transferrin conjugate are pH-sensitive protein, so the anticancer category drugs like alkylating and anthracyclines are produced as conjugate which may deliver the drug to cancerous cells because cancerous cells are acidic. In-vivo, doxorubicin albumin conjugate was manufactured with glutaraldehyde has efficacy to treat cancer (Kratz et al. 1998). One of the kind properties of albumin responsible for the gathering in solid tumors is given:

(a) Improved retention time and permeability of albumin in passive tumor targeting (EPR effect).
(b) Two albumin binding proteins accessible on the tumor endothelium gp60 receptor and furthermore the SPARC (Secreted Protein Acidic and Rich in Cysteine), that might be a discharged glycoprotein with the high binding ability to albumin inside the tumor and, improve its retention time.
(c) Serum albumin levels got decreased in a cancer patient understanding given the extra demand of amino acids by the multiplying tumor mass that is accomplished by digest albumin. The state of a diminished level of albumin is comprehended as hypoalbuminemia (Hauser et al. 2006; Cho et al. 2008).

INNO-206 for Sarcoma and Gastric Cancer (in Clinical Trials)
Felix demonstrated that a maleimide bearing prodrug of doxorubicin was rapidly and by choice bound to the cysteine-34 position of endogenous albumin. Due to those examinations, 6-male imidocaproyl hydrazone side-effect of doxorubicin (DOXOEMCH, renamed INNO-206) is showing up as a clinical competitor because of high plasma stability in its albumin-bound structure. INNO-206 has unrivaled adequacy against malignant growth and gastric disease with decreased cardiotoxicity (Loureiro et al. 2016; Ajaj et al. 2009; Kratz 2014).

8.9.4 Tumor

Albumin collects in dangerous and inflamed tissue because of a leaky capillary joined with a missing or faulty lymphatic waste framework. Tumor take-up in pre-clinical models can be effectively pictured by infusing the dye Evans Blue, that ties quickly and tightly to circulating albumin and makes subcutaneously developing tumors turn blue inside in few hour post-infusion. As a choice to drug targeting, conjugating restorative peptides or cytokines with albumin is an alluring methodology of improving their pharmacokinetic profile because of the long-half-life of albumin in the body clinically. Others like methotrexate-albumin conjugate, an albumin-binding prodrug of doxorubicin (Gupta and Lis 2010).

Abraxane® for Treating Solid Tumors
Abraxane® is an albumin paclitaxel nanoparticle framework and furthermore the most progressive medication result of grab innovation created by American bioscience. In grab innovation, human albumin and lipophilic medication have experienced a stream diminish high pressure to make paclitaxel nanoparticles with a mean molecule estimate range 130 nm. Resulting paclitaxel nanoparticle, Abraxane®, is fluid. This Abraxane® nanoparticle is stable in the galenic formulation. And this nanoparticle breaks down quickly when intravenous implantation promoting soluble albumin tie paclitaxel buildings having a size like that of endogenous albumin. These albumin-paclitaxel complex aggregate are expanded the permeation and potential effect inside the solid tumors. Extra albumin transport pathway oversees by the glycoprotein. However, specific receptors are presented on the endothelial cell surface (Gradishar 2006; Gradishar et al. 2009).

8.9.5 Cirrhosis

In cirrhotic patients suffering from paracentesis, albumin implantations prevent speedy ascetic liquid accumulation and bring down the danger of post-paracentesis related circulatory dysfunction (Bernardi et al. 2012; Guevara et al. 2012).

8.9.6 Renal Illness

Post-translational alterations, similar to the discontinuity of albumin, are found in nephrotic and diabetic patients. Proteolytic fracture of serum albumin is because of the oxidative pressure. Albumin discontinuity may apply a pathophysiological activity in uremic disorder (Fouque et al. 2008).

8.9.7 Sepsis

Hypoalbuminemia normal in the emergency unit be brought about by declined hepatic albumin creation as well as improved proteolytic debasement and clearance of albumin. Human serum albumin may advantage for hypoalbuminemia sick patients and improve organ work (Dellinger et al. 2013).

8.9.8 Rheumatoid Arthritis

The inflammation in the joints known as rheumatoid arthritis and it affected millions of people throughout the world. Methotrexate is a therapeutic molecule that is used in both rheumatoid arthritis and cancer. By forming the methotrexate albumin conjugate which is done by combing methotrexate to lysine residue of human serum albumin that increases the specificity and half-life at the site of inflamed tissue. This can decrease the degradation of cartilage and the attack of synovial fibroblasts (Alderson et al. 2004).

A novel antibody-based innovation has been coming by the Belgium pharma organization Ablynx. It incorporates work of albumin binding nanobodies for the assembling of remarkable formulation and reaches to phase II clinical preliminary examinations for the treatment of rheumatoid joint pain. The trivalent antibody contains two antitumor death factor (TNF-α) nanobodies (TR2) and one albumin restricting nanobodies (ARI). For example, ATN-103 (presently Ozoralizumab) with an MW 45 kD effectively created preclinically. TNF-α is one among the key mediators of the inflammatory response and has recently been used because the molecular target for the development of three approved immunoglobulins namely Enebrel® (etanercept), Remicade® (infliximab), and Humira® (adalimumab), which is used for the treatment of rheumatoid arthritis wherever they are used alone or together with immunosuppressant agent (Kim et al. 1999; Kratz 2014).

8.9.9 AIDS

The susceptibility of currently available therapeutic peptides to degradation by peptidase, lack of bioavailability and non-targeted distribution are some reasons that restrict it as a therapeutic agent. Therefore develop an albumin-based carrier moiety which prevents the drawback of old treatment. Mainly focused on rising their pharmacokinetic profile, creating use of the albumin-binding ways represented previously *i.e.* attachment of a maleimide moiety for covalently binding to albumin or myristic acid for physical binding to the fatty acid-binding sites on albumin. This type of technology has been used for antiviral applications that show their polarity of some peptide influences the effectiveness of albumin-binding and also for the development of albumin-binding peptides enhances their therapeutic action (Kratz 2008).

PC-1505: An Albumin-Conjugate of an Antiviral C34 Peptide That Inhibits HIV-1 Entry

Roche and his applied science group have been developed an albumin-conjugated antiviral C34 peptide for preventing the drawback of currently available HIV infection. It is a specific type of antiviral drug which is now available in the market. It is brand name is Fuzeon® and approved by the Food and Drug Administration in March 2003. It is made up of artificial peptide that targets the specific gp41. In this subunit, the glycoprotein is still non-covalently bound to gp120 and in the second step enters in HIV T-cells, which helps enhance antiviral activity. Fuzeon® is only one of the marketed compounds which are directly targeted to specific cells and have conformational rearrangements of gp41 and thus prevent the entry of HIV into uninfected cells (Elsadek and Kratz 2012).

8.9.10   Bacterial Infections

Staphylococcus aureus (S. Aureus) is nosocomial infections (hospital-acquired infections) causes microorganism which is one of the bacteria which generate faster multidrug-resistant. Currently, available inhibitory marketed antibacterial drug is FARKGALRQ, which is the cationic peptide nonapeptides. But their application is very limited due to their extremely low bioavailability and susceptibility to degradation. Scientist division at Diamond light Ltd. has been working on this field to prevent this drawback of nonapeptides. The scientist group derivatized the myristic acid to create new N-myristoylated nonapeptide RH01which is called as Myr-FARKGALRQ. And the technology used for the preparation is called SAFETY™. The benefit of these newly developed nano peptides is having quality to bind with plasma serum albumin which increases the bioavailability as well as activity against the S. aureus (Elsadek and Kratz 2012; Poca et al. 2012).

8.9.11   Cardiopulmonary Bypass Surgery

In the present scenario albumin also given in Cardiopulmonary bypass surgery and will be effective to maintain the colloidal pressure at normal (Russell et al. 2004).

8.9.12   Hemolytic Disease of Newborn

Hemolytic disease in the newborn occurs due to the blood group are not compatible, disturbance in the red blood cells which contain the specific antigen and this important antigen is not found in the mother so a disturbance in the antibody production. Then albumin was given that binds and removes the bilirubins which are unconjugated and responsible to cause the disease (Larsen et al. 2016).

8.9.13 Burns

Albumin proteins are widely used in case of burns because due to the burning there is loss of protein, electrolytes as well as fluids so for immediate treatment albumin is given (Blunt et al. 1998).

8.9.14 Adult Respiratory Distress Syndrome

This situation is known as hypoproteinemic which causes interstitial respiratory edema, so 25% albumin solution was given with diuretics to treat the pulmonary burden that occurs by the deficiency of albumin protein (Larsen et al. 2016; Elsadek and Kratz 2012).

8.9.15 Diagnosis and Imaging

The safety of albumin *in-vivo* can be stabilized by the application and significance in clinical nano-therapeutics such as Abraxane and other diagnosis molecules. The manufacturing of albumin-based nanoparticles and bioimaging has better biocompatibility and such material has the clinical benefit for approval. The therapeutic molecules for cancer therapy, enzyme inhibitor or inhibitors are loaded, encapsulated and entrapped in albumin protein as nanocarrier used for the MRI, PTT, PDT, combination therapy and theranostics (Arasteh et al. 2014).

Albumin has been utilized as nuclear medicine for Injectable radiopharmaceutical 99mTc aggregate at nearly 30 years. It has gamma radiation emitted radionuclide imaging agent which is consists of human plasma serum albumin particles labeled sterile liquid suspension of 99mTc. It's accessible in many tools from several manufacturers given different trade names such as 99mTc-Nanocoll®, 99mTc-Albures®, Tc-99 m-Microalbumin®, 99mTc Human blood serum Albumin®, and Technetium-99 m albumin Colloid®. These developed kits are employed for diagnosis of inflammation scanning, bone marrow scanning, lookout node detection in breast cancer, identification of different solid tumors mass, rheumatoid arthritis and leg swelling, etc. These kits are also used for imaging of respiratory organs to assess and check the existence of respiratory organ emboli and diagnosing for various internal organs and lymphoscintigraphy. Another is an albumin-based AFL-HAS promising fluorescence loaded marker which is mostly used for the recognition and detection of brain tumors during surgery. In these human serum albumin has been linked with 5-([4, 6- dichlorotriazin-2 yl] amino) fluoresceine. This AFL-HAS fluorescein-labeled albumin conjugate has already existed in the market made by Orpegen pharmaceutical company Ltd. (Heidelberg, Germany). A further Vasovist® has been developed which is principally used as a contrast agent for identification and detection of malignant lesions in humans and preclinical tumor models estimation. Vasovist® from Beyer Schering, produced by randy Lauffer and B-22956/1 from Bracco Imaging, are serum albumin bonded gadolinium (III) which is worked on magnetic resonance imaging contrast agents exploited to accumulate high-resolution imaging pictures as well as an enhanced vessel structure delineation and vascular diseases (Elsadek and Kratz 2012).

8.10 Albumin as Drug Delivery for Natural Products

Albumin has a various characteristic which is used in the treatment of various diseases like diabetes, cancer, and rheumatoid arthritis, etc. To treat the diabetes human insulin which is peptide in nature are manufactured and approved. For the treatment of cancer albumin nanoparticle of Abraxane@ was also approved, which is used in breast cancer (Green et al. 2006). Albumin is assuming an important role as a medication carrier in the clinical setting. Principally, three delivery technologies can be recognized: coupling of low-sub-atomic weight medications to exogenous or endogenous albumin, conjugation with bioactive proteins, and encapsulation of medications into albumin nanoparticles. Delivery of the Abraxane@ as an albumin nanocarrier provides a better response as compared to the taxol in the treatment of breast cancer. The formulation of the Abraxane@ founds in obtained a single size of albumin particles. Therefore, after administration rapidly dissociate due to high permeability and retention. For the targeting of the tumor the legends targeting well be modified on the surface of albumin protein as nanocarrier which can increase the efficacy as well as safety in drug molecules. The albumin nanoparticles are classified into five categories they are given below-

- Template
- Nanocarrier
- Scaffold
- Stabilizer
- Albumin polymeric conjugate

Various types of polymers are also used in the delivery of various drug molecules either it may be hydrophilic or lipophilic. Albumin based micro and nano-particulate delivery systems are most widely used for the delivery of therapeutic molecules to the targeted sites. Delivery of therapeutic molecules with the albumin particulate system protects the therapeutic molecules from the degradation and increases the absorption. Also, alters the pharmacokinetic properties and improves the intracellular penetration and distribution in the body (Tao et al. 2019; Suri et al. 2007; Roco 2003). Albumin based nano-particle drug delivery systems are more frequently used. Most of the therapeutic molecules are incorporated or in-capsulated into albumin and it will attractive because different types of binding sites are present in the albumin molecules (Patil 2003). The albumin primary structure contains the numbers of charged amino acids so the albumin nanoparticles provide the electrostatic absorption of both positive molecules as well as negatively molecules (Irache et al. 2005; Weber et al. 2000). Human serum albumin nanoparticles are used in the genes therapy because it avoids the interaction with the serum (Brzoska et al. 2004; Simoes et al. 2004). It is also are used for the delivery of polar and non-polar molecules because albumin has various numbers of binding sites. Albumin nano-particle or nano-carrier is most frequently used in delivery due to its defined size and surface modification is done according to the delivery and sites. Albumins have both high stability as well as high solubility in the water and also biodegradable. Albumins nano-carrier systems are most efficiently used in cancer treatment and it was confirmed (Elzoghby et al. 2012; Fernandez-Urrusuno et al. 1999). Some of the albumin-based nano and micro-carrier delivery system is listed in Table 8.1.

Albumin is popular for its substance binding capability. It has a flexible macromolecule with immunomodulatory, inhibitor, detoxifying properties and may act as the best potent drug carrier. Plasma serum albumin simple may be used to associate endogenous or exogenous protein for the treatment of varied diseases such as autoimmune disease, cancer, polygenic disease, and infectious disease. These delivery systems contain nanoparticles, protein, prodrugs and amide derivatives which covalently bind to albumin and also physically bind to the protein fragments and therapeutic active peptides. In the last few years ago, the utilization of albumin has been a supply of talk specializing in whether albumin will act as a more robust drug carrier for the delivery of macromolecules or not. Albumin can bind various vary of molecules. Following are its features and its main function which tells as a drug carrier (Neumann et al. 2010). There are some following special futures are found in albumin, therefore used for the application of particulate system:

- Albumin bind active molecules provide a new kind of compound that offered for enhancement of their bioavailability in plasma. The phenomena of the formation of albumin bind active molecules compound are easy due to the presence of negative charge in serum albumin. In this, bind with ligands and drug compounds with serum albumin by electrostatic binding force.
- The other main important function that is performed by albumin is a transport function. Mostly drug molecules are bind with the tertiary structure of albumin. The material transport by the albumin includes the micro and nano-sized amount of medicine, hormones, bilirubin, bile, metals, long-chain fatty acids, anions, L-thyroxine, endotoxins, nitric oxide and different microorganism products like protein G-like albumin-binding molecule.
- The binding capacity of serum albumin with active drug molecules is different; it only depends upon the capacity of the active molecules. According to this various bond is formed like electrostatic interactions, hydrophobic bonding or covalent bonding. Many binding sites are presented on plasma serum albumin. For example, in fatty acid seven binding sites are represented. The foremost eminent websites for drug binding are the site I and site II settled on the sub-domains IIA and IIIA, severally.

Table 8.1 List of some albumin-based nano-carrier drug delivery system and its outcome

Drug	Carriers	Outcome	References
Thymoquinone	Nanoparticle	Enhances the thermal stability of the active molecule and decreases its toxicity	Kazan et al. (2019)
Epigallocatechin-3-gallate	Nanoparticle	Provided slow and sustained release of drug with improved pharmacokinetics and bioavailability of drug	Ramesh and Mandal (2019)
Curcumin	Solid lipid nanoparticle	Protective effects against scopolamine-induced passive avoidance memory retrieval deficit	Lari et al. (2019)
Curcumin	Nanoparticle	NPs provide targeted delivery of drug and promising candidate for the treatment of human epithelial growth factor receptor 2 positive cancer cells	Saleh et al. (2019)
Doxorubicin	Nanoparticles	Albumin nanoparticles provide an alternative, more specific way to overcome transporter-mediated resistance	Onafuye et al. (2019)
Vinblastine	Nanoparticle	Improving the precision and timing of drug release, easy operation, and higher compliance for pharmaceutical applications	Huang et al. (2019)
Irinotecan	Nanoparticle	Efficiently delivering the drug to the tumor site without causing any toxicity	Taneja and Singh (2018)
Doxorubicin	Nanoparticle	The developed formulation has novel and high potential nano-carrier for management of non-resistance and multidrug-resistant for cancer cells	Kayani et al. (2018)
Camptothecin	Nano colloidal particles	Increase the solubility of drugs, reduce the toxicity, improve the stability of the drug	Lian et al. (2017)
Vincristine	Nanoparticle	Developed drug delivery system have high drug loading, high cancer cell accumulation, and cancer cell targeting	Taghdisi et al. (2016)
Oridonin	Nanoparticle	Developed for a liver targeting carrier which seemed to be a stable delivery system for poorly soluble drug	Li et al. (2013)
Camptothecin	Nanoparticle	Enhance the stable of developed formulation and have the potential for targeted delivery of anticancer drugs	Li et al. (2011)
Noscapine	Nanoparticle	Developed formulation have efficacy on breast cancer cells	Sebak et al. (2010)

- High-affinity binding and complex formation are primarily reversible, other hand covalent bound albumin with active molecules could happen during a reversible or irreversible manner.
- Medication restricting strongly influences the conveyance of bound medication to tissue location and the removal of the drug. Highly bound medications have just a little level of serum albumin in free form. Different variables that are

significant in medication egg whites collaborations and might be in charge of the wide inter-individual variety seen age, pH, temperature and ionic quality, which can influence the quantity of binding site in vitro, and Displacement of medications from their coupling site by different medications or by any endogenous substances happens and may change their distribution, pharmacological activity, digestion and excretion.

For examples, manufacturing and production of albumin microsphere of NSAIDs category drugs will enhance the drug release of these category drugs as well decreases the chances of side effect (Tuncay 2000). Some albumin micro and nanoparticulate formulation are also excised in the market, represented in Table 8.2 (Elzoghby et al. 2012).

The various mechanisms involved in the absorption and distribution of albumin particles. When any albumin particle when enters in the body, firstly internalized by endocytosis process. The process of endocytosis mainly depends on particle size, shape, and surface charge. After endocytosis, the internalized materials finally depredated and transferred by endosomes to lysosomes. Hence, degraded exogenous antigens further followed the MHC class II pathway and this pathway is involved in antigen antibody-mediated immune responses. While existing antigens by MHC class I pathway generated cytotoxic T-lymphocyte (CTL) response (O'Hagan 1998; Storni et al. 2005; Shen et al. 2006). The prevention of active molecules internalization by endosomes is a significant issue in the control of processing pathways. For the successful release of micro and nanoparticles from endosomes, membrane disruptive agents are required to release internalized micro and nanoparticles into the cytoplasm. Some agents which help in membrane-penetration such as pathogen-derived pore-forming proteins, some peptide, lipid or polymer which disrupts the endosomal membrane (endosome escaping) and reduced the pH inside the cells compartment. Therefore, in recent year various pH-sensitive particles have been developed for enhancement of cytoplasmic delivery (Plank et al. 1998; Shai 1995; Yessine and Leroux 2004).

The osmotic colloidal mechanism also plays a significant role in the degradation of particles. It quickly degrades into several smaller fragments. An increase in osmotic pressure of endosomes causes immediate entry of water through the

Table 8.2 Albumin micro and nanoformulation available in market

Drug	Dosage form	Application
Cisplatin	Albumin microsphere	Lung cancer
Paclitoxel	Albumin microsphere	Lung cancer
5-Flurouracil	Albumin magnetic microsphere	Breast cancer
Cyclophosphamide	Albumin magnetic microsphere	Breast cancer
Abraxen	Albumin nanoparticles	Breast cancer
Tamoxifen	Albumin nanoparticles	Breast cancer
Albendazole	Albumin nanoparticles	Ovarian cancer
Levemir	Albumin solution	Diabetes

membrane and disrupts the membrane. Some research articles proved that the acid-sensitive nanoparticles induced antigen-specific response and produced showed antitumor activity (Murthy et al. 2003; Chen et al. 2005). Polycations are also responsible for the acidification of endosomes that break these vesicles. The mechanism involved in the absorption of protons by acid organelles and generation of osmotic pressure, which results in swelling and rapture of endosomes, and release of internalized materials into the cell cytoplasm (Standley et al. 2004). Mostly substance contains hydrophobic alkyl and carboxyl groups and present protonated ions at internal endosomal pH. At low pH, hydrophobicity gets increased and polymers can easily enter into the endosomal membranes and cause disruption (Boussif et al. 1995). When drug molecules are encapsulated in albumin polymeric shell attach by covalent bond, the covalently bonded micro or nanoparticles offers the following advantages:

- Albumin micro or nanoparticles are easy to prepare, biodegradable and reproducible.
- It is easily bound with various types of drug substance due to high binding affinity and available binding sites to form the matrix of albumin particles.
- A low dose of drug molecules is sufficient for better responses and easily stimulates the presenting cells for further processing. It shows good stability when kept in a long time (Gengoux and Leclerc 1995).
- Active molecules encapsulated nano/microparticles show higher efficiency and bioavailability properties (Kersten and Hirshberg 2004).

8.11 Contraindication and Side Effects

Plasma albumin is used in different clinical conditions. Restoration of blood volume, crisis treatment of stun, and different circumstances related to hypovolemia are a portion of the clinical utilization of egg whites. Knowing this, Serum albumin is a respectability biomarker for liver function synthesis and a few sicknesses such as incendiary issue, mind tumors, rheumatoid joint pain, myocardial ischemia, malignant growth, blood cerebrum boundary (BBB) harm, kidney illness, cerebrovascular infection, cardiovascular risk disease. In clinically, albumin is 4–5% involved in the emergency treatment of cardiopulmonary bypass, hypovolemic stun, sequestration of protein-rich fluids, intense liver failure, etc. Other clinical signs of albumin 20–25% arrangement may incorporate hypoproteinemia, hypovolemia, grown-up respiratory trouble disorder (ARDS), intense nephrosis, intense liver disappointment, hemolytic infection of the infant (HDN), renal dialysis, burn treatment, cardiopulmonary bypass, hypovolemic stun, sequestration of protein-rich liquids, and erythrocyte resuspension. Along with the advantageous effects of serum plasma albumin, several side effects and contraindication is observed when introduced intravenously injection. Some of the contraindicated side effects are listed in Table 8.3.

Table 8.3 Side effects of albumin when administered intravenously

Organ	Side effects
Skin irritation	Urticarial, pruritus, edema, skin rash, erythema
Gastrointestinal tract	Increased salivation, nausea and vomiting
Nervous system	Headache and febrile reactions

It is well-known plasma albumin is a highly valuable therapeutic product in controlling a broad range of various disorders. But in some situations, it may be dangerous when administered intramuscularly. To control the problem of albumin follow the albumin prescription and checked, the level of more than 2.5 g/dl (hyperalbuminemia) which seems to cause contraindication. In some disease conditions like acute or chronic pancreatitis, severe anemia or specific pulmonary edema, cardiac and renal failures avoided the administration of albumin because of chances of acute circulatory overload, additionally also avoided in patients who are suffering from nervous system disorders (Zhou et al. 2013).

8.12 Conclusion

The above chapter introduces to extending the area of clinical and preclinical drug applications/advancements that, utilize albumin as a protein transporter to improve the pharmacokinetic profile of the active molecules. Plasma albumin is turning out as a standout amongst the most significant medication transporter for therapeutic molecules, and antibodies. It is utilized particularly for the diagnosis and treatment of viral infections, cancer, and arthritis. It is a perfect endogenous serum protein for enhancement of bioavailability, stability, high flexibility, targeting, and half-life. Serum albumin having the quality to bind with the FcRn receptors which control the half-life of formulated albumin-based active therapeutic molecules or diagnostic agents. Therefore, considering the profitable achievement of a product that utilizes albumin as a continuous clinical trial and it attracts the attention of scientists. All things considered, the pharmaceutical, biotechnological, clinical utilization of albumin will be completely investigated in the coming decade and the field will be extended to advance signs compare to others.

References

Ajaj AK, Graeser R, Fichtner I, Kratz F (2009) In vitro and in vivo study of an albumin-binding prodrug of doxorubicin that is cleaved by cathepsin B. Cancer Chemother Pharmacol 64:413–418. https://doi.org/10.1007/s00280-009-0942-8

Alderson P, Bunn F, Li Wan Po A, Li L, Pearson M, Roberts I, Schierhout G (2004) Human albumin solution for resuscitation and volume expansion in critically ill patients. Cochrane Database Syst Rev 4:310–319. https://doi.org/10.3748/wjg.v17.i30.3479

Arasteh A, Farahi S, Habibi-Rezaei M, Moosavi-Movahedi AA (2014) Glycated albumin: an overview of the in vitro models of an in vivo potential disease marker. J Diabetes Metab Disord 13:49–58. https://doi.org/10.1186/2251-6581-13-49

Bernardi M, Maggioli C, Zaccherini G (2012) Human albumin in the management of complications of liver cirrhosis. Crit Care 16:211–220. https://doi.org/10.1186/cc11218

Bertucci C, Domenici E (2002) Reversible and covalent binding of drugs to human serum albumin: methodological approaches and physiological relevance. Curr Med Chem 9:1463–1481. https://doi.org/10.2174/0929867023369673

Blunt MC, Nicholson JP, Park JR (1998) Serum albumin and colloid osmotic pressure in survivors and non-survivors of prolonged critical illness. Anesthesia 53:755–761. https://doi.org/10.1046/j.1365-2044.1998.00488.x

Boussif O, Lezoualch O, Zanta MA, Mergny MD, Scherman D, Demeneix B, Behr JP (1995) A versatile vector for gene and oligonucleotide transfer into cells in culture and in vivo. Proc Natl Acad Sci U S A 92:7297–7301. https://doi.org/10.1073/pnas.92.16.7297

Brzoska M, Langer K, Coester C, Loitsch S, Wagner TO, Mallinckrodt C (2004) Incorporation of biodegradable nanoparticles into human airway epithelium cells in vitro study of the suitability as a vehicle for drug or gene delivery in pulmonary diseases. Biochem Biophys Res Commun 318:562–570. https://doi.org/10.1016/j.bbrc.2004.04.067

Buchacher A, Iberer G (2006) Purification of intravenous immunoglobulin g from human plasma aspects of yield and virus safety. Biotechnol J 1(2):148–163. https://doi.org/10.1002/biot.200500037

Chen R, Yue Z, Eccleston ME, Williams S, Slater NK (2005) Modulation of cell membrane disruption by pH-responsive pseudo-peptides through grafting with hydrophilic side chains. J Control Release 108:63–72. https://doi.org/10.1016/j.biomaterials.2008.12.036

Cho K, Wang X, Nie S, Chen ZG, Shin DM (2008) Therapeutic nanoparticles for drug delivery in cancer. Clin Cancer Res 14:1310–1316. https://doi.org/10.1158/1078-0432.CCR-07-1441

Curry S, Brick P, Franks NP (1999) Fatty acid binding to human serum albumin: new insights from crystallographic studies. Biochim Biophys Acta 1441(2–3):131–140. https://doi.org/10.1016/s1388-1981(99)00148-1

Dellinger RP, Levy MM, Rhodes A, Annane D, Gerlach H, Opal SM, Sevransky JE, Sprung CL, Douglas IS, Jaeschke R, Osborn TM, Nunnally ME, Townsend SR, Reinhart K, Kleinpell RM, Angus DC, Deutschman CS, Machado FR, Rubenfeld GD, Webb SA, Beale RJ, Vincent JL, Moreno R (2013) Surviving sepsis campaign: international guidelines for management of severe sepsis and septic shock, 2012. Intensive Care Med 41(2):580–637. https://doi.org/10.1097/CCM.0b013e31827e83af

Elena S, Claude N, Robert K, Andreas T, Ingrid PF, Christine V, Iris J, Massimo M (2012) Safety and pharmacokinetics of a novel recombinant fusion protein linking coagulation factor IX with albumin (rIX-FP) in hemophilia B patients. Blood 120(12):2405–2411. https://doi.org/10.1182/blood-2012-05-429688

Elsadek B, Kratz F (2012) Impact of albumin on drug delivery – new applications on the horizon. J Control Release 157:4–28. https://doi.org/10.1016/j.jconrel.2011.09.069

Elzoghby AO, Samy WM, Elgindy NA (2012) Albumin-based nanoparticles as potential controlled release drug delivery systems. J Control Release 157(2):168–182. https://doi.org/10.1016/j.jconrel.2011.07.031

Evans TW (2002) Review article: albumin as a drug–biological effects of albumin unrelated to oncotic pressure. Aliment Pharmacol Ther 16:6–11. https://doi.org/10.1046/j.1365-2036.16.s5.2.x

Fanali G, di Masi A, Trezza V, Marino M, Fasano M, Ascenzi P (2012) Human serum albumin: from bench to bedside. Mol Asp Med 33:209–290. https://doi.org/10.1016/j.mam.2011.12.002

Farrugia A (2011) Falsification or paradigm shift? Toward a revision of the common sense of transfusion. Transfusion 51(1):216–224. https://doi.org/10.1111/j.1537-2995.2010.02817.x

Fasano M, Curry S, Terreno E, Galliano M, Fanali G, Narciso P, Notari S, Ascenzi P (2005) The extraordinary ligand binding properties of human serum albumin. IUBMB Life 57:787–796. https://doi.org/10.1080/15216540500404093

Faure P, Wiernsperger N, Polge C, Favier A, Halimi S (2008) Impairment of the antioxidant properties of serum albumin in patients with diabetes: protective effects of metformin. Clin Sci (Lond) 114:251–256. https://doi.org/10.1042/CS20070276

Fernandez-Urrusuno R, Calvo P, Remunan-Lopez C, Vila-Jato JL, Alonso MJ (1999) Enhancement of nasal absorption of insulin using chitosan nanoparticles. Pharm Res 16:1576–1581. https://doi.org/10.1023/a:1018908705446

Fiume L, Di Stefano G, Busi C, Mattioli A, Bonino F, Torrani-Cerenzia M, Verme G, Rapicetta M, Bertini M, Gervasi GB (1997) Liver targeting of antiviral nucleoside analogues through the asialoglycoprotein receptor. J Viral Hepat 4:363–370. https://doi.org/10.1046/j.1365-2893.1997.00067.x

Fouque D, Kalantar-Zadeh K, Kopple J, Cano N, Chauveau P, Cuppari L, Franch H, Guarnieri G, Ikizler TA, Kaysen G, Lindholm B, Massy Z, Mitch W, Pineda E, Stenvinkel P, Treviño-Becerra A, Wanner C (2008) A proposed nomenclature and diagnostic criteria for protein-energy wasting in acute and chronic kidney disease. Kidney Int 73(4):391–398. https://doi.org/10.1038/sj.ki.5002585

Garcia-Martinez R, Andreola F, Mehta G, Poulton K, Oria M, Jover M, Soeda J, Macnaughtan J, De Chiara F, Habtesion A, Mookerjee RP, Davies N, Jalan R (2015) The immunomodulatory and antioxidant function of albumin stabilizes the endothelium and improves survival in a rodent model of chronic liver failure. J Hepatol 62(4):799–806. https://doi.org/10.1016/j.jhep.2014.10.031

Gengoux C, Leclerc C (1995) In vivo induction of CD4 + T cell responses by antigen covalently linked to synthetic micro-spheres does not require adjuvant. Int Immunol 7:45–53. https://doi.org/10.1093/intimm/7.1.45

Gradishar WJ (2006) Albumin-bound paclitaxel: a next-generation taxane. Expert Opin Pharmacother 7:1041–1053. https://doi.org/10.1517/14656566.7.8.1041

Gradishar WJ, Krasnojon D, Cheporov S, Makhson AN, Manikhas GM, Clawson A, Bhar P (2009) Significantly longer progression-free survival with nab-paclitaxel compared with docetaxel as first-line therapy for metastatic breast cancer. J Clin Oncol 27:3611–3619. https://doi.org/10.1200/JCO.2008.18.5397

Green MR, Manikhas GM, Orlov S, Afanasyev B, Makhson AM, Bhar P, Hawkins MJ (2006) Abraxane, a novel Cremophor-free, albumin-bound particle form of paclitaxel for the treatment of advanced non-small-cell lung cancer. Ann Oncol 17(8):1263–1268. https://doi.org/10.1093/annonc/mdl104

Guevara M, Terra C, Nazar A, Solà E (2012) Albumin for bacterial infections other than spontaneous bacterial peritonitis in cirrhosis. A randomized, controlled study. J Hepatol 57:759–765. https://doi.org/10.1016/j.jhep.2012.06.013

Gupta D, Lis CG (2010) Pretreatment serum albumin as a predictor of cancer survival: a systematic review of the epidemiological literature. Nutr J 9:69. https://doi.org/10.1186/1475-2891-9-69

Hauser CA, Stockler MR, Tattersall MH (2006) Prognostic factors in patients with recently diagnosed incurable cancer: a systematic review. Support Care Cancer 14:999–1011. https://doi.org/10.1007/s00520-006-0079-9

Hirose M, Tachibana A, Tanabe T (2010) Recombinant human serum albumin hydrogel as a novel drug delivery vehicle. Mater Sci Eng C 30:664–669. https://doi.org/10.1016/j.msec.2010.02.020

Hu YJ, Liu Y, Sun TQ, Bai AM, Lü JQ, Pi ZB (2006) Binding of anti-inflammatory drug cromolyn sodium to bovine serum albumin. Int J Biol Macromol 39:280–285. https://doi.org/10.1016/j.ijbiomac.2006.04.004

Huang KS, Yang CH, Wang YC, Wang WT, Lu YY (2019) Microfluidic synthesis of vinblastine-loaded multifunctional particles for magnetically responsive controlled drug release. Pharmaceutics 11:1–14. https://doi.org/10.3390/pharmaceutics11050212

Irache JM, Merodio M, Arnedo A, Campanero MA, Mirshahi M, Espuelas S (2005) Albumin nanoparticles for the intravitreal delivery of anti-cytomegaloviral drugs. Mini-Rev Med Chem 5:293–305. https://doi.org/10.2174/1389557053175335

Julio R, Jane R, Mark B, Fred Y, Murray S (2009) Potential of albiglutide, a long-acting GLP-1 receptor agonist, in type 2 diabetes: a randomized controlled trial exploring weekly, biweekly, and monthly dosing. Diabetes Care 32:1880–1886. https://doi.org/10.2337/dc09-0366

Kayani Z, Firuzi O, Bordbar AK (2018) Doughnut-shaped bovine serum albumin nanoparticles loaded with doxorubicin for overcoming multidrug-resistant in cancer cells. Int J Biol Macromol 107:1835–1843. https://doi.org/10.1016/j.ijbiomac.2017.10.041

Kazan A, Celiktas OY, Zhang YS (2019) Fabrication of thymoquinone loaded albumin nanoparticles by microfluidic particle synthesis and their effect on planarian regeneration. Macromol Biosci 14:182–192. https://doi.org/10.1002/mabi.201900182

Kersten G, Hirshberg H (2004) Antigen delivery systems. Expert Rev Vaccines 3:453–462. https://doi.org/10.1586/14760584.3.4.453

Kim SB, Chi HS, Park JS, Hong CD, Yang WS (1999) Effect of increasing serum albumin on plasma D-dimer, von Willebrand factor, and platelet aggregation in CAPD patients. Am J Kidney Dis 33:312–317. https://doi.org/10.1016/s0272-6386(99)70306-9

Kragh-Hansen U (2013) Molecular and practical aspects of the enzymatic properties of human serum albumin and albumin-ligand complexes. Biochim Biophys Acta 1830(12):5535–5544. https://doi.org/10.1016/j.bbagen.2013.03.015

Kratz F (2008) Albumin as a drug carrier: design of prodrugs, drug conjugates, and nanoparticles. J Control Release 132:171–183. https://doi.org/10.1016/j.jconrel.2008.05.010

Kratz F (2014) A clinical update of using albumin as a drug vehicle a commentary. J Control Release 190:331–336. https://doi.org/10.1016/j.jconrel.2014.03.013

Kratz F, Beyer U, Roth T, Schu MT, Unold A, Fiebig HH, Unger C (1998) Albumin conjugates of the anticancer drug chlorambucil: synthesis, characterization, and in vitro efficacy. Arch Pharm Pharm Med Chem 331:47–53. https://doi.org/10.1002/(sici)1521-4184(199802)331:2<47::aid-ardp47>3.0.co;2-r

Lari RS, Moezi L, Pirsalami F, Abkar M, Moosavi M (2019) Curcumin-loaded BSA nanoparticles protect more efficiently than natural curcumin against scopolamine-induced memory retrieval deficit. Basic Clin Neurosci 10(2):157–164. https://doi.org/10.32598/bcn.9.10.255

Larsen MT, Kuhlmann M, Hvam ML, Howard KA (2016) Albumin-based drug delivery: harnessing nature to cure disease. Mol Cell Ther 4(1):1–13. https://doi.org/10.1186/s40591-016-0048-8

Lee E, Eom JE, Jeon KH, Kim TH, Kim E, Jhon GJ, Kwon Y (2014) Evaluation of albumin structural modifications through cobalt-albumin binding (CAB) assay. J Pharm Biomed Anal 91:17–23. https://doi.org/10.1016/j.jpba.2013.12.003

Li Q, Liu C, Zhao X, Zu Y, Wang Y, Zhang B, Zhao D, Zhao Q, Su L, Gao Y, Sun B (2011) Preparation, characterization and targeting of micronized 10-hydro-oxycamptothecin-loaded folate-conjugated human serum albumin nanoparticles to cancer cells. Int J Nanomedicine 6:397–405. https://doi.org/10.2147/IJN.S16144

Li C, Zhang D, Guo H, Hao L, Zheng D, Liu G, Shen J, Tian X, Zhang Q (2013) Preparation and characterization of galactosylated bovine serum albumin nanoparticles for liver-targeted delivery of oridonin. Int J Pharm 448(1):79–86. https://doi.org/10.1016/j.ijpharm.2013.03.019

Lian B, Li Y, Zhao X, Zu Y, Wang Y, Zhang Y, Li Y (2017) Preparation and optimization of 10-hydro-oxy-camptothecin nano colloidal particles using the antisolvent method combined with high-pressure homogenization. J Chem 41:1–10. https://doi.org/10.1155/2017/5752090

Loureiro A, Azoia NG, Gomes AC, Cavaco-Paulo A (2016) Albumin-based nanodevices as drug carriers. Curr Pharm Des 22(10):1371–1390. https://doi.org/10.2174/13816128226661601251149000

Mendez CM, Mcclain CJ, Marsano LS (2005) Albumin therapy in clinical practice. Nutr Clin Pract 20:314–320. https://doi.org/10.1177/0115426505020003314

Murthy N, Xu M, Schuck S, Kunisawa J, Shastri N, Frechet MJ (2003) A macromolecular delivery vehicle for protein-based vaccines: acid-degradable protein-loaded microgels. Proc Natl Acad Sci U S A 29:4995–5000. https://doi.org/10.1073/pnas.0930644100

Neumann E, Frei E, Funk D, Becker MD, Schrenk HH, Müller-Ladner U, Fiehn C (2010) Native albumin for targeted drug delivery. Expert Opin Drug Deliv 7:915–925. https://doi.org/10.151 7/17425247.2010.498474

Nicholson JP, Wolmarans MR, Park GR (2000) The role of albumin in critical illness. Br J Anaesth 85(4):599–610. https://doi.org/10.1093/bja/85.4.599

O'Hagan DT (1998) Recent advances in immunological adjuvants: the development of particulate antigen delivery systems. Expert Opin Investig Drugs 7:349–359. https://doi.org/10.1517/13543784.7.3.349

Odunuga OO, Shazhko A (2013) Ammonium sulfate precipitation combined with liquid chromatography is sufficient for purification of bovine serum albumin that is suitable for most routine laboratory applications. Biochem Compounds 1(1):03–09. https://doi.org/10.7243/2052-9341-1-3

Onafuye H, Pieper S, Mulac D, Jr JC, Wass MN, Langer K, Michaelis M (2019) Doxorubicin-loaded human serum albumin nanoparticles overcome transporter-mediated drug resistance in drug-adapted cancer cells. Beilstein J Nanotechnol 10:1707–1715. https://doi.org/10.3762/bjnano.10.166

Patil GV (2003) Biopolymer albumin for diagnosis and in drug delivery. Drug Dev Res 58:219–247. https://doi.org/10.1002/ddr.10157

Plank C, Zauner W, Wagner E (1998) Application of membrane-active peptides for drug and gene delivery across cellular membranes. Adv Drug Deliv Rev 34:21–35. https://doi.org/10.1016/s0169-409x(98)00005-2

Poca M, Concepción M, Casas M, Alvarez-Urturi C, Gordillo J, Hernández-Gea V, Román E, Guarner-Argente C, Gich I, Soriano G, Guarner C (2012) Role of albumin treatment in patients with spontaneous bacterial peritonitis. Clin Gastroenterol Hepatol 10:309–315. https://doi.org/10.1016/j.cgh.2011.11.012

Ramesh N, Mandal AK (2019) Encapsulation of epigallocatechin-3-gallate into albumin nanoparticles improves pharmacokinetic and bioavailability in rat model. Biotech 9(6):238–246. https://doi.org/10.1007/s13205-019-1772-y

Ramin R, Ali M, Neda K, Fatemeh S, Sara S, Jalal A (2016) Overview of albumin and its purification methods. Adv Pharm Bull 6(4):495–507. https://doi.org/10.15171/apb.2016.063

Raoufinia R, Mota A, Keyhanvar N, Safari F, Shamekhi S, Abdolalizadeh J (2016a) Overview of albumin and its purification methods. Adv Pharm Bull 6(4):495–507. https://doi.org/10.15171/apb.2016.063

Raoufinia R, Mota A, Nozari S, Aghebati Maleki L, Balkani S, Abdolalizadeh J (2016b) A methodological approach for purification and characterization of human serum albumin. J Immunoass Immunochem 37(6):623–635. https://doi.org/10.1080/15321819.2016.1184163

Roche M, Rondeau P, Singh NR, Tarnus E, Bourdon E (2008) The antioxidant properties of serum albumin. FEBS Lett 582(13):1783–1787. https://doi.org/10.1016/j.febslet.2008.04.057

Roco MC (2003) Nanotechnology: convergence with modern biology and medicine. Curr Opin Biotechnol 14:337–346. https://doi.org/10.1016/s0958-1669(03)00068-5

Russell JA, Navickis RJ, Wilkes MM (2004) Albumin versus crystal- crystalloid for pump-priming in cardiac surgery: a meta-analysis of controlled trials. J Cardiothorac Vasc Anesth 18:429–437. https://doi.org/10.1053/j.jvca.2004.05.019

Saleh T, Saudi T, Shojaosadati SA (2019) Aptamer functionalized curcumin-loaded human serum albumin (HSA) nanoparticles for targeted delivery to HER-2 positive breast cancer cells. Int J Biol Macromol 1(130):109–116. https://doi.org/10.1016/j.ijbiomac.2019.02.129

Sebak S, Mirzaei M, Malhotra M, Kulamarva A, Prakash S (2010) Human serum albumin nanoparticles as an efficient noscapine drug delivery system for potential use in breast cancer: preparation and in vitro analysis. Int J Nanomedicine 5:525–532. https://doi.org/10.2147/ijn.s10443

Shai Y (1995) Mechanism of the binding, insertion, and destabilization of phospholipid bilayer membranes by a-helical antimicrobial and cell non-selective membrane-lytic peptides. Biochim Biophys Acta 1462:55–70. https://doi.org/10.1016/s0005-2736(99)00200-x

Shen H, Ackerman AL, Cody V, Giodini A, Hinson ER, Cresswell P, Edelson RL, Saltzman WM, Hanlon DJ (2006) Enhanced and prolonged cross-presentation following the endosomal escape of exogenous antigens encapsulated in biodegradable nanoparticles. Immunology 117:78–88. https://doi.org/10.1111/j.1365-2567.2005.02268.x

Simoes S, Slepushkin V, Pires P, Gaspar R, de Lima RMC, Duzgunes N (2004) Human serum albumin enhances DNA transfection by lipoplexes and confers resistance to inhibition by serum. Biochim Biophys Acta 1463:459–469. https://doi.org/10.1016/s0005-2736(99)00238-2

Sleep D (2015) Albumin and its application in drug delivery. Expert Opin Drug Deliv 12(5):793–812. https://doi.org/10.1517/17425247.2015.993313

Sleep D, Cameron J, Evans LR (2013) Albumin is a versatile platform for drug half-life extension. Biochim Biophys Acta 1830:5526–5534. https://doi.org/10.1016/j.bbagen.2013.04.023

Sokołowska M, Wszelaka-Rylik M, Poznański J, Bal W (2009) Spectroscopic and thermodynamic determination of three distinct binding sites for Co(II) ions in human serum albumin. J Inorg Biochem 103:1005–1013. https://doi.org/10.1016/j.jinorgbio.2009.04.011

Standley SM, Kwon TJ, Murthy N, Kunisawa J, Shastri N, Guillaudeu SJ, Lau L, Fréchet JMJ (2004) Acid-degradable particles for protein-based vaccines: enhanced survival rate for tumor-challenged mice using ovalbumin model. Bioconjug Chem 15:1281–1288. https://doi.org/10.1021/mp800068x

Storni T, Kundig TM, Senti G, Johansen P (2005) Immunity in response to particulate antigen delivery systems. Adv Drug Deliv Rev 57:333–355. https://doi.org/10.1016/j.addr.2004.09.008

Suri SS, Fenniri H, Singh B (2007) Nanotechnology-based drug delivery systems. J Occup Med Toxicol 1:02–16. https://doi.org/10.1186/1745-6673-2-16

Taghdisi SM, Danesh NM, Ramezani M, Abnous K (2016) Targeted delivery of vincristine to T-cell acute lymphoblastic leukemia cells using an aptamer-modified albumin conjugate. RSC Adv 6(52):46366–46371. https://doi.org/10.1039/C6RA08481H

Tanaka K, Shigueoka EM, Sawatani E, Dias GA, Arashiro F, Campos TC, Nakao HC (1998) Purification of human albumin by the combination of the method of Cohn with liquid chromatography. Braz J Med Biol Res 31(11):1383–1388. https://doi.org/10.1590/s0100-879x1998001100003

Taneja N, Singh KK (2018) Rational design of polysorbate 80 stabilized human serum albumin nanoparticles tailored for high drug loading and entrapment of irinotecan. Int J Pharm 536(1):82–94. https://doi.org/10.1016/j.ijpharm.2017.11.024

Tantra R, Tompkins J, Quincey P (2010) Characterisation of the de-agglomeration effects of bovine serum albumin on nanoparticles in aqueous suspension. Colloids Surf B: Biointerfaces 75:275–281. https://doi.org/10.1016/j.colsurfb.2009.08.049

Tao C, Chuah YJ, Xu CJ, Wang DA (2019) Albumin conjugates and assemblies as versatile biofunctional additives and carriers for biomedical applications. J Mater Chem B 7:357–367. https://doi.org/10.1039/C8TB02477D

Taverna M, Marie AL, Mira JP, Guidet B (2013) Specific antioxidant properties of human serum albumin. Ann Intensive Care 3(1):04–11. https://doi.org/10.1186/2110-5820-3-4

Tazhbayev Y, Mukashev M, Burkeev M, Kreuter J (2019) Hydroxyurea-loaded albumin nanoparticles: preparation, characterization, and in vitro studies. Pharmaceutics 11(8):410–422. https://doi.org/10.3390/pharmaceutics11080410

Tuncay M (2000) In vitro and in vivo evaluation of diclofenac sodium loaded albumin microspheres. J Microencapsul 17(2):145–155. https://doi.org/10.1080/026520400288382

Van Belle E, Dallongeville J, Vicaut E, Degrandsart A, Baulac C, Montalescot G (2010) Ischemia-modified albumin levels predict long-term outcome in patients with acute myocardial infarction. The French Nationwide OPERA study. Am Heart J 159:570–576. https://doi.org/10.1016/j.ahj.2009.12.026

Varshney A, Sen P, Ahmad E, Rehan M, Subbarao N, Khan RH (2010) Ligand binding strategies of human serum albumin: how can the cargo be utilized? Chirality 22(1):77–87. https://doi.org/10.1002/chir.20709

Wang W, Ou Y, Shi Y (2004) AlbuBNP, a recombinant B-type natriuretic peptide and human serum albumin fusion hormone, as a long-term therapy of congestive heart failure. Pharm Res 21:2105–2111. https://doi.org/10.1023/b:pham.0000048203.30568.81

Weber C, Coester C, Kreuter J, Langer K (2000) Desolvation process and surface characteristics of protein nanoparticles. Int J Pharm 194:91–102. https://doi.org/10.1016/s0378-5173(99)00370-1

Wongsasulak S, Patapeejumruswong M, Weiss J, Supaphol P, Yoovidhya T (2010) Electrospinning of food-grade nanofibers from cellulose acetate and egg albumin blends. J Food Eng 98:370–376. https://doi.org/10.1016/j.jfoodeng.2010.01.014

Yessine MA, Leroux JC (2004) Membrane-destabilizing polyanions: interaction with lipid bilayers and endosomal escape of bio-macromolecules. Adv Drug Deliv Rev 56:999–1021. https://doi.org/10.1016/j.addr.2003.10.039

Zhou T, Lu S, Liu X, Zhang Y, Xu F (2013) Review of the rational use and adverse reactions to human serum albumin in the people's Republic of China. Patient Prefer Adherence 7:1207–1212. https://doi.org/10.2147/PPA.S53484

Printed in the United States
by Baker & Taylor Publisher Services